Lecture Notes in Computer Science 13771

The series Lecture Notes in Computer Science (LNCS), including its subseries Lecture Notes in Artificial Intelligence (LNAI) and Lecture Notes in Bioinformatics (LNBI), has established itself as a medium for the publication of new developments in computer science and information technology research, teaching, and education.

LNCS enjoys close cooperation with the computer science R & D community, the series counts many renowned academics among its volume editors and paper authors, and collaborates with prestigious societies. Its mission is to serve this international community by providing an invaluable service, mainly focused on the publication of conference and workshop proceedings and postproceedings. LNCS commenced publication in 1973.

David W. Silva · Eckhard Hitzer · Dietmar Hildenbrand
Editors

Advanced Computational Applications of Geometric Algebra

First International Conference, ICACGA 2022 Denver, CO, USA, October 2–5, 2022 Proceedings

Springer

Editors
David W. Silva
Algemetric
Colorado Springs, CO, USA

Eckhard Hitzer
International Christian University
Tokyo, Japan

Dietmar Hildenbrand
TU Darmstadt
Darmstadt, Germany

ISSN 0302-9743 ISSN 1611-3349 (electronic)
Lecture Notes in Computer Science
ISBN 978-3-031-34030-7 ISBN 978-3-031-34031-4 (eBook)
https://doi.org/10.1007/978-3-031-34031-4

This Springer imprint is published by the registered company Springer Nature Switzerland AG
The registered company address is: Gewerbestrasse 11, 6330 Cham, Switzerland

Paper in this product is recyclable.

Preface

As General Chair of the International Conference of Advanced Computational Applications of Geometric Algebra, I am extremely proud of all the outstanding contributions in this proceedings. It has been a long and arduous process to bring this conference to life, and I am honored to have been part of it.

It is my utmost conviction that all manuscripts in this publication are of great relevance and importance, as they genuinely contribute to advancing the field of Geometric Algebra and related fields. I want to thank all the authors, editors, and reviewers who contributed their time and effort to make this conference a success. I would also like to recognize all sponsors, volunteers, and other supporters who directly and indirectly contributed to the conference and its proceedings.

I have great expectations for the contributions presented in this work, and I believe they will have a lasting impact on the field of Geometric Algebra. I am confident that the ideas and perspectives presented here will significantly interest anyone who reads it.

Thank you all for your hard work and dedication to making this conference a reality. I am honored to have been part of this incredible journey.

Part I on geometric applications begins with Yao, Mann and Li, using the double conformal GA (DCGA) model for ray tracing. In particular, the intersection of a line with a cyclide is explored, and how four points can be extracted from the resulting colinear point quadruple. Conversely, a colinear point quadruple can be constructed from four points, and the line containing these points can be found. Next, Derevianko and Vašík construct an algorithm for similarity search between 2D images using Geometric Algebra for Conics (GAC) with intrinsic transformations of rotation, translations, dilations and isotropic scaling. Image objects are represented by ellipses fitted into the contour points and the is search for transformations between GAC objects. The result is a low-cost similarity test applied to real object images. Then, Havel reviews in CGA the representation of line-bound vectors, plane-bound bivectors, tetrahedra and of some elementary geometric concepts, and discusses their connections to a recently discovered extension of Heron's triangle area formula to the volume of a tetrahedron. Furthermore, Sobczyk defines universal geometric algebras of compatible null vectors following rules of addition and multiplication of real and complex square matrices. Then he defines the concepts of Grassmann algebra, dual Grassmann algebra, associated real and complex geometric algebras, and isomorphic real or complex coordinate matrix algebras. Finally, he discusses the horosphere, as well as affine and conformal transformations. Part I concludes with Hitzer's study of the inner product of oriented points in conformal geometric algebra and its geometric meaning. The inner product of two general oriented points is computed, and analyzed (including symmetry) in terms of distance, and angles between the distance vector and the local point orientations. Examples illustrate the result, followed by a generalization to n dimensions.

Part II is about Computer Science applications based on GA. Pepe et al. demonstrate GA as a powerful tool to model and predict the structure of proteins which is important for their properties. This avoids expensive and time-consuming experiments. Neto et al. show that Clifford Convolutional Neural Networks are able to outperform real valued networks of equivalent size in their application: the diagnosis of acute lymphoblastic leukemia (ALL) which is a type of cancer identified by malformed lymphocytes, known as lymphoblasts in the bloodstream. Riter et al. present a new approach to the problem of recognizing a Euclidean distance matrix based on GA. This can be used to model problems for instance in sensor network localization, molecular geometry and GPS modelling. Neumann et al. present their algorithm called GAAlign for point cloud registration using GA. They are able to show that their sampling-based algorithm outperforms the conventional methods in terms of robustness. Hrdina et al. present a new mathematical system for quantum computing based on GA. This Quantum Register Algebra QRA is able to directly use the Dirac notation and its computations can be executed by the GA algorithms optimizer GAALOP.

Part III deals with technological applications based on GA. Montoya and Eid present a geometric procedure for computing differential characteristics of multi-phase electrical signals using GA. It is shown how the concept of instantaneous frequency in electrical networks can be intimately linked to the so-called Darboux bivector. Byrtus and Frolik use conformal transformations in the kinematic chain of a specific planar mechanism in order to describe the forward kinematics of generalized robotic snakes. They present two possible ways to describe the configuration of a three-link generalised snake robot using Compass Ruler Algebra, which allows us to point out the nature of the generalized snake as an extension of the classic snake robot. Zamora et al. use the IMU sensor of their camera for hand-eye calibration based on QGA. In the proposed method, the linear acceleration and angular velocity of the sensor attached to the camera is employed with the kinematic model of the robot articulation to find the relative camera position and orientation.

In Part IV, on applications to physics and mathematics, Hamilton shows how fermions and all forces of nature fit elegantly into a supergeometric algebra in 11+1 spacetime dimensions, proposed as a new language of physics. Next, Xambó-Descamps reanalyzes the fundamental Stern-Gerlach (SG) experiment by elucidating the Hermitian structure of the algebra of geometric quaternions (the even algebra of the Geometric Algebra of the Euclidean 3D space) which allows us to regard it as the Hilbert space of a q-bit, and checks that the computed probabilities obey the statistics of the SG experiments. Then, Filimoshina and Shirokov study Lie groups in degenerate geometric algebras. These Lie groups preserve the even and odd subspaces under the adjoint and twisted adjoint representations. The results are important for the study of generalized spin groups in degenerate cases. Furthermore, Wieser and Lasenby apply computing with the universal properties of Clifford algebras and even subalgebras. They observe that for operations defined in terms of coordinates on a multivector basis it can be difficult to rigorously show coordinate-independence. They apply the "universal property" to ensure "coordinate-free" operations by construction. Parallels are seen to the process of writing recursive programs. They derive a universal property of the even subalgebra, and apply it to explicitly construct equivalences between Clifford algebras. These ideas can be formalized in a theorem prover. Finally in Part IV, Mcleod Price and Staples explore binary linear codes via Zeon and sym-Clifford algebras. Zeon ("nil-Clifford") and "sym-Clifford" methods are used to reformulate essential concepts of binary linear coding theory, i.e., to generate linear codes and to illustrate Clifford-algebraic formulations of encoding, decoding and error-correction.

A unique feature of ICACGA was a round table on Geometric Algebra applied to Power System Engineering by Prado (Spain), Eid (Egypt), Arsenovic (USA), Montoya (Spain), and Mira (Spain). Understanding power systems and power theory or circuit analysis is crucial for reliable and effective operation. Generation, transmission and distribution power systems have a multidimensional nature, and GA excellently tackles such problems, better than matrix algebra and complex numbers in traditional electrical engineering, which has difficulties in the proper interpretation of the physical phenomena involved. GA unifies the tools of tensors, quaternions or differential forms and interprets power flows in power systems from an intuitive, physical point of view. Major applications developed to date were presented and discussed by experts.

A total of 24 papers were sent for review in a double-blind process, of these 18 were selected for presentation at the conference, and inclusion in these proceedings.

October 2022

Dietmar Hildenbrand
Eckhard Hitzer
David W. Silva

Organization

Program Chairs

Dietmar Hildenbrand	Technische Universität Darmstadt, Germany
Eckhard Hitzer	International Christian University, Tokyo, Japan
David W. Silva	Algemetric, Colorado, USA

General Chair

David W. Silva	Algemetric, Colorado, USA

Scientific Committee

Derek Abbott	University of Adelaide, Australia
Rafael Alves	Federal University of ABC, São Bernardo do Campo, Brazil
Sven Buchholz	TH Brandenburg, Germany
Leo Dorst	University of Amsterdam, The Netherlands
Andrew J. S. Hamilton	University of Colorado, Boulder, USA
Jaroslav Hrdina	Brno University of Technology, Brno, Czech Republic
Jonathan Katz	University of Maryland, USA
Anthony Lasenby	University of Cambridge, UK
Joan Lasenby	University of Cambridge, UK
Carlile Lavor	University of Campinas, Campinas, Brazil
Eleftheria Makri	KU Leuven, Belgium
Vincent Nozick	Gustave Eiffel University, Champs sur Marne, France
Alyn Rockwood	Boulder Graphics, USA
Vaclav Skala	University of West Bohemia, Pilsen, Czech Republic
G. Stacey Staples	Southern Illinois University, Edwardsville, USA
Petr Vašík	Brno University of Technology, Czech Republic
Sebastià Xambó	Technical University of Catalonia, Barcelona, Spain
Shouhuai Xu	University of Colorado, Colorado Springs, USA
Zhaoyuan Yu	Nanjing Normal University, Nanjing, China

Advisory Committee

Daniel Apon	MITRE and University of Maryland, College Park, MD, USA
Chris Doran	University of Cambridge, UK
Sahan Gamage	ARM, UK
Wilder Lopes	GraphStax.ai, USA
Timothy Mattson	Intel, USA
Marcelo Xavier	Ford Motor Company, USA
Julio Zamora	Intel, Mexico

Invited Talks

Modeling and Computing with Geometric Algebra (GA)

David Hestenes

Abstract. Concepts of "Circle" and "Sphere" have been central to geometric thinking since ancient times. Today we proclaim that their conceptual power can be vastly enhanced by formulation with Geometric Algebra and implementation with computers.

The Block sphere is basic to Quantum Computing while the Poincaré sphere is basic to optics. But it is not commonly understood that these spheres are more than mere analogues, they are expressions of basic structure in the electron itself. Geometric Algebra makes that manifest as we aim to explain today.

The primacy of GA in understanding the electron is ample reason for advocating it as the lingua franca across the entire physics and engineering curriculum. Moreover, GA curriculum materials are already well developed for (a) Rigid body dynamics, (b) Electrodynamics, (c) Quantum mechanics, (d) Gauge gravity, (e) physics pedagogy, and even (f) neuroscience.

The task remains to develop GA software that cultivates geometric and algebraic intuition for students at all grade levels. Tools are available at: bivector.net.

A discussion of how to use and improve them will be a prime subject for this conference.

Toward the Open Metaverse

Michael Kass

Abstract. In 1992 science fiction writer Neal Stephenson coined the term "metaverse" for a large-scale shared virtual world where people can interact with one another in a simulation of reality. The notion is certainly compelling. Why connect to each other in the 2D world of the current internet when we have the technology to interact in 3D? Today, as virtual worlds pervade the world of entertainment and as a variety of businesses start to deploy "Digital Twins" to simulate and control real buildings, factories and infrastructure, it's clear that the metaverse no longer belongs in the realm of science fiction. It has entered the realm of the possible. But how will the metaverse come into existence, and how can it end up being based on open standards? Can our current web evolve into the metaverse, or will it need to be something fundamentally different? At the core of any true 3D web or the open Metaverse, we need an open, powerful, flexible and efficient way to describe a shared virtual world. In this talk we will explore how NVIDIA Omniverse makes use of such a description to take some key steps toward turning the notion of the open metaverse into a genuinely useful reality.

Interpolation and Design: Lessons and Opportunities from CAGD for Geometric Algebra

Alyn Rockwood

Abstract. Computer-aided geometric design (CAGD) is primarily concerned with the design of curves and surfaces in CAD/CAM/CAE, animation, analysis, styling, simulation, etc. In the last half-century it has developed into a mature, widely used, and indispensable technology. Initially, it was thought that CAGD was a simple application of interpolation and approximation theory. Although CAGD relies heavily on such concepts, many surprising subtleties, questions, and ramifications arose that had not been anticipated, nor easily handled within the traditional theory. As GA develops its own version of interpolation theory it may be well to understand the lessons from CAGD, as well as the opportunities it provides. We give a review of some of those issues, which have direct implications for GA.

Illustrating Geometric Algebra and Differential Geometry in 5D Color Space

Werner Benger

Abstract. Geometric Algebra is popular for its immediate geometric interpretations of its algebraic objects and operations based on Clifford Algebra on vector spaces. For instance, in Euclidean 3D space quaternions are known to be numerically superior to rotation matrices. Geometric Algebra allows for an intuitive interpretation in terms of planes of rotations and extends this concept to arbitrary dimensions. The space of colors forms a vector space, too, though one of non-geometrical nature, and spun by the primary colors red, green, blue. The formalism of Geometric Algebra can be applied here as well, amalgamating surprisingly with the notion of vectors and co-vectors known from differential geometry: tangential vectors on a manifold correspond to additive colors red, green, blue, whereas co-vectors from the co-tangential space correspond to subtractive primary colors magenta, yellow, cyan.

Geometric Algebra in turn considers vectors, bi-vectors and anti-vectors as part of its general multi-vector scheme. In 3D space vectors, anti-vectors, bi-vectors and co-vectors are all three-dimensional objects that can be identified with each other, so their distinction is not obvious. Higher dimensional spaces exhibit the differences more clearly, but our intuition of higher dimensional geometry limits our understanding. However, using color spaces we can intuitively go further by considering "transparency" as an independent, four-dimensional property of a color vector. We can thereby explore four-dimensional Geometric (Clifford) algebra independent of spacetime or special/general relativity which is usually used for 4D GA. Though even in 4D space, ambiguities remain between vectors, co-vectors, bi-vectors and bi-co-vectors. Bi-vectors and bi-co-vectors – both six-dimensional objects – are visually equivalent. They become uniquely different only in five or higher dimensions.

Envisioning five-dimensional geometry is challenging to the human mind, but in color space we can add another property, "texture," to constitute a five-dimensional vector space. The properties of a bi-vector and a bi-co-vector become evident visually by inspecting the possible combinations of colors/transparency/texture. This higher-dimensional yet intuitive approach demonstrates the need to distinguish among different kinds of vectors before identifying them in special situations, such as 3D Euclidean space.

Closing the GA^{2}P: Making Geometric Algebra a Regular Expression of Aerospace Engineering

Todd Ell

Abstract. Herein is a discussion of aerospace industries' key initiatives and the subsequent use of digital engineering through model-based design and analysis. This points to the 'goodness of fit' of Geometric Algebra models and techniques to describe various cyber-physical systems prevalent in upcoming aerospace systems. A Geometric Algebra Adaptation Program (G[A]^{2}P) is being stood up within Collins Aerospace to address these evolving needs. This requires giving attention to the set of tools and training material needed to enable that adaption. The talk with will briefly touch on the several early applications within Collins Aerospace and Geometric Algebra tools being crafted to meet those application needs. Early observations about the adaptation program will be shared to inform the strengths & weaknesses in regard to in academic/technical material currently available. Finally, future technical challenges, the solutions that would be of benefit to aerospace in general, will be described.

The Initial NIST Post-Quantum Cryptography Standards, and What's Next?

Daniel Apon

Abstract. In recent years, there has been a substantial amount of research on quantum computers – machines that exploit quantum mechanical phenomena to solve mathematical problems that are difficult or intractable for conventional computers. If large-scale quantum computers are ever built, they will be able to break many of the public-key cryptosystems currently in use. This would seriously compromise the confidentiality and integrity of digital communications on the Internet and elsewhere. The goal of post-quantum cryptography (also called quantum-resistant cryptography) is to develop cryptographic systems that are secure against both quantum and classical computers, and can interoperate with existing communications protocols and networks.

The question of when a large-scale quantum computer will be built is a complicated one. While in the past it was less clear that large quantum computers are a physical possibility, many scientists now believe it to be merely a significant engineering challenge. Some engineers even predict that within the next twenty or so years, sufficiently large quantum computers will be built to break essentially all public key schemes currently in use. Historically, it has taken almost two decades to deploy our modern public key cryptography infrastructure. Therefore, regardless of whether we can estimate the exact time of the arrival of the quantum computing era, we must begin now to prepare our information security systems to be able to resist quantum computing.

In this talk, we briefly survey the state-of-the-art in post-quantum cryptography, at the point that the survivors of this half-decade, international process have just recently emerged from the NIST PQC competition to become the initial post-quantum cryptographic standards. We discuss the new standards, their design, their security, as well as what else may be coming in post-quantum cryptography in the near future.

Introduction to Quantum Computing

Tristan Müller

Abstract. Quantum computing is a rapidly growing field that has drawn considerable interest in the recent years. While development of quantum computers was mostly driven by fundamental and academic research not too long ago, quantum computing is now on the edge of widespread industry adoption – as shown by an increasing number of vendors offering access to quantum computers as well as an ever-increasing number of users from different industry areas.

In my talk I will present why quantum computing is fundamentally different to classical computing and how this could provide drastic speedups for specific problems. To that end, I will first review some fundamental concepts from quantum mechanics. Afterwards, I will provide a brief overview of quantum computing hardware and common qubit architectures, as well as projected future developments. In addition, I will present IBM's software stack, including the IBM Quantum Tools.

Computational Image Processing Workflows Using Quaternions

Nek Valous

Abstract. Image pixels can be encoded by a linear combination of the three basis vectors in a hypercomplex algebra framework; this encoding provides the opportunity to process color images in a geometric way. The proposed approach is based on a rapid and versatile method, using quaternions, that can enhance computational image processing workflows applied to natural and biomedical images. This pixel-based approach is computationally efficient, thus taking advantage of parallel architectures in modern computing systems, and has applications either as a standalone tool or integrated in image processing pipelines. Essentially, the method demonstrates that feature-rich mathematical frameworks can provide efficient solutions for computational image processing.

Geometric Algebra Computing for Computer Graphics Using GAALOP

Dietmar Hildenbrand

Abstract. Two courses at the SIGGRAPH conferences in 2019 and 2022 demonstrate the increasing interest of the computer graphics community in Geometric Algebra and especially in Projective Geometric Algebra PGA. GAALOP (Geometric Algebra Algorithms Optimizer) is a tool used to visualize Geometric Algebra as well as to generate optimized code for many programming languages such as C/C++, OpenCL, CUDA, Python, Matlab, Mathematica, and Rust. Based on GAALOP and its online version GAALOPWeb, we present how Geometric Algebra can advantageously be used for geometric operations such as intersections, reflections, and projections, and for transformations with geometric objects such as lines, planes, circles, spheres, conics, and quadrics. Leo Dorst in his presentations shows how PGA unifies the rotational and translational aspects of Euclidean motions resulting in easily integrable equations of motion. We present GAALOP examples dealing with these results.

Quantum Register Algebra: Quantum Circuits

Jaroslav Hrdina, Rafael Alves, Ales Navrat, Petr Vasik, Dietmar Hildenbrand,
Ivan Eryganov, Carlile Lavor, and Christian Steinmetz

Abstract. We introduce serial and parallel quantum gates in the Quantum Register Algebra (QRA) framework, which is an efficient tool for quantum computing. We present a GAALOP (Geometric Algebra Algorithms Optimizer) implementation of our approach. We illustrate these principles by presenting one example of serial gates and two examples of parallel gates.

Complementary Orientations in Geometric Algebras

Leo Dorst

Abstract. Oriented elements are part of geometry, and they come in two complementary types: intrinsic and extrinsic. Those different orientation types manifest themselves by behaving differently under reflection. Dualization in geometric algebras can be used to encode them; or, alternatively, orientation types inform the interpretation of dualization. We employ the Hodge dual, to include important algebras with null elements like PGA. Oriented elements can be combined using the meet operation, and the dual join (which is here introduced for that purpose). We suggest a visualization of all oriented elements of 3D DGA (the algebra of normal directions) and 3D PGA (the plane-based algebra of Euclidean motions). Using the proposed framework, software written to process one orientation type can be employed to process the complementary type consistently.

A Lightweight Implementation of GA in Julia

Chris Doran

Abstract. Julia is a modern programming language that is rapidly gaining popularity in the scientific community. It offers the ability to code in a high-level functional style while still achieving the performance of C/C++. There is a strong ecosystem in place, and tight integration with VSCode. With care, Julia is even capable of compiling the same code for both CPU and GPU implementation. In this talk I discuss a simple, lightweight implementation of various geometric algebras in Julia. Some of the ideas are drawn from earlier experience with Haskell, and the difficulties encountered with achieving high performance in that language. The talk concludes with a discussion of what hardware changes could yield even greater performance for low-dimensional geometric algebras.

Contents

Geometry Applications

Line–Cyclide Intersection and Colinear Point Quadruples in the Double Conformal Model

Huijing Yao[1], Stephen Mann[2(✉)], and Qinchuan Li[3]

[1] School of Artificial Intelligence, Anhui Polytechnic University, Wuhu 241000, Anhui, People's Republic of China

[2] Cheriton School of Computer Science, University of Waterloo, 200 University Avenue W., Waterloo N2L 3G1, Canada
smann@uwaterloo.ca

[3] Mechatronic Institute, Zhejiang Sci-Tech University, Hangzhou 310018, Zhejiang, People's Republic of China
lqchuan@zstu.edu.cn

Abstract. In this paper, we look at using the double conformal model for ray tracing. In particular, we explore the intersection of a line with a cyclide in the double conformal model, and how to extract the four points from the resulting colinear point quadruple. Further, we show how to directly construct a colinear point quadruple from four points, and we show how to find the line containing the points of a colinear point quadruple. We also briefly touch on barycentric coordinates in DCGA.

Keywords: Cyclides · double conformal model · colinear point quadruples

1 Introduction

The double conformal geometric algebra [3] (DCGA) allows for the representation of tori as well as planes, quadrics, and cyclides. Further, we can intersect these cyclides with lines, giving a form of a point quadruple. This suggests that the model can be used for ray tracing tori and more general cyclides. However, to be of use in ray tracing, we need to extract the points of intersection for the point quadruples, and perform a lighting calculation at the point closest to the origin of the ray.

In this paper, we explore how to extract these points of intersection, extracting a quartic equation from the point quadruple. Additionally, we look more closely at point quadruples, showing how to construct a point quadruple directly from four points. Further, we investigate barycentric coordinates in DCGA, for which we perform the computations by converting to the conformal geometric algebra and perform the computation there; we briefly present this conversion in this paper as barycentric coordinates are required in ray tracing. We verify the correctness of our ideas using a simple ray tracer.

Our focus in this paper is on the ideas required to implement this DCGA ray tracer, and we defer many of the proofs to a future publication.

Supported by NSERC and NSFC.

D. W. Silva et al. (Eds.): ICACGA 2022, LNCS 13771, pp. 3–14, 2024.
https://doi.org/10.1007/978-3-031-34031-4_1

2 Background

The *Double Conformal Geometric Algebra* (DCGA) contains two subspaces of the *Conformal Geometric Algebra* (CGA) [2]. Using the notation of Easter and Hitzer [3], the first CGA subspace (CGA1) is generated by the vectors

$$e_i \cdot e_j = \begin{cases} 1 \; i = j, \; 1 \leq i \leq 4 \\ -1 \; i = j = 5 \\ 0 \; i \neq j \end{cases} \tag{1}$$

while the second subspace (CGA2) is generated by the vectors

$$e_i \cdot e_j = \begin{cases} 1 \; i = j, \; 6 \leq i \leq 9 \\ -1 \; i = j = 10 \\ 0 \; i \neq j \end{cases} . \tag{2}$$

In both subspaces, there is a point at the origin,

$$\boldsymbol{e}_{o1} = (-\boldsymbol{e}_4 + \boldsymbol{e}_5)/2, \quad \boldsymbol{e}_{o2} = (-\boldsymbol{e}_9 + \boldsymbol{e}_{10})/2,$$

and a point at infinity

$$\boldsymbol{e}_{\infty 1} = (\boldsymbol{e}_4 + \boldsymbol{e}_5), \quad \boldsymbol{e}_{\infty 2} = (\boldsymbol{e}_9 + \boldsymbol{e}_{10}),$$

all four of which are null vectors,

$$\boldsymbol{e}_{o1} \cdot \boldsymbol{e}_{o1} = \boldsymbol{e}_{o2} \cdot \boldsymbol{e}_{o2} = \boldsymbol{e}_{\infty 1} \cdot \boldsymbol{e}_{\infty 1} = \boldsymbol{e}_{\infty 2} \cdot \boldsymbol{e}_{\infty 2} = 0.$$

Further, $\boldsymbol{e}_{o1} \cdot \boldsymbol{e}_{\infty 1} = \boldsymbol{e}_{o2} \cdot \boldsymbol{e}_{\infty 2} = -1$.

A point in CGA1 with an embedding vector $\boldsymbol{p}_{c^1} = x\boldsymbol{e}_1 + y\boldsymbol{e}_2 + z\boldsymbol{e}_3$ is represented as

$$\boldsymbol{P}_{C^1} = \boldsymbol{p}_{c^1} + \tfrac{1}{2}\boldsymbol{p}_{c^1}^2 \boldsymbol{e}_{\infty 1} + \boldsymbol{e}_{o1},$$

with a point in CGA2 defined analogously for a vector $\boldsymbol{p}_{c^2} = x\boldsymbol{e}_6 + y\boldsymbol{e}_7 + z\boldsymbol{e}_8$:

$$\boldsymbol{P}_{C^2} = \boldsymbol{p}_{c^2} + \tfrac{1}{2}\boldsymbol{p}_{c^1}^2 \boldsymbol{e}_{\infty 2} + \boldsymbol{e}_{o2}.$$

A sphere in CGA1 with center $\boldsymbol{C}_{c^1}$ and radius r is defined as

$$\boldsymbol{S}_{C^1} = \boldsymbol{C}_{c^1} - \tfrac{1}{2}r^2 \boldsymbol{e}_{\infty 1}.$$

Similarly, a sphere in CGA2 with center $\boldsymbol{C}_{c^2}$ and radius r is defined as

$$\boldsymbol{S}_{C^2} = \boldsymbol{C}_{c^2} - \tfrac{1}{2}r^2 \boldsymbol{e}_{\infty 2}.$$

In DCGA, the point at the origin is defined as

$$\boldsymbol{e}_o = \boldsymbol{e}_{o1} \wedge \boldsymbol{e}_{o2}$$

and the point at infinity is defined as

$$\boldsymbol{e}_\infty = \boldsymbol{e}_{\infty 1} \wedge \boldsymbol{e}_{\infty 2}.$$

Easter and Hitzer showed that

$$\boldsymbol{e}_o \cdot \boldsymbol{e}_o = 0, \quad \boldsymbol{e}_\infty \cdot \boldsymbol{e}_\infty = 0, \quad \text{and } \boldsymbol{e}_o \cdot \boldsymbol{e}_\infty = -1.$$

Given $\boldsymbol{t}_{c^1} = x\boldsymbol{e}_1 + y\boldsymbol{e}_2 + z\boldsymbol{e}_3$ and $\boldsymbol{t}_{c^2} = x\boldsymbol{e}_6 + y\boldsymbol{e}_7 + z\boldsymbol{e}_8$ (with the same x, y, z values for both), Easter and Hitzer define a point $\boldsymbol{P}_\mathcal{D} = \mathcal{D}(\boldsymbol{t}_c) = \mathcal{D}(x, y, z)$ in DCGA as

$$\mathcal{D}(x, y, x) = \boldsymbol{P}_\mathcal{D} = \boldsymbol{P}_{C^1} \wedge \boldsymbol{P}_{C^2}. \tag{3}$$

Expanding the DCGA point $\boldsymbol{P}_D$ gives

$$\begin{aligned}
\boldsymbol{P}_D &= (\boldsymbol{t}_{\varepsilon^1} + \tfrac{1}{2}\boldsymbol{t}^2\boldsymbol{e}_{\infty 1} + \boldsymbol{e}_{o1}) \wedge (\boldsymbol{t}_{\varepsilon^2} + \tfrac{1}{2}\boldsymbol{t}^2\boldsymbol{e}_{\infty 2} + \boldsymbol{e}_{o2}) \\
&= \boldsymbol{t}_{\varepsilon^1} \wedge \boldsymbol{t}_{\varepsilon^2} + \boldsymbol{t}_{\varepsilon^1} \wedge \boldsymbol{e}_{o2} + \boldsymbol{e}_{o1} \wedge \boldsymbol{t}_{\varepsilon^2} + \tfrac{1}{2}\boldsymbol{t}^2\boldsymbol{e}_{\infty 1} \wedge (\boldsymbol{t}_{\varepsilon^2} + \boldsymbol{e}_{o2}) \\
&\quad + \tfrac{1}{2}\boldsymbol{t}^2(\boldsymbol{t}_{\varepsilon^1} + \boldsymbol{e}_{o1}) \wedge \boldsymbol{e}_{\infty 2} + \frac{1}{4}\boldsymbol{e}_\infty + \boldsymbol{e}_o \\
&= \frac{x}{2}(\boldsymbol{t}^2 - 1)\boldsymbol{e}_{19} + \frac{x}{2}(\boldsymbol{t}^2 + 1)\boldsymbol{e}_{1,10} + \frac{x}{2}(\boldsymbol{t}^2 - 1)\boldsymbol{e}_{46} \\
&\quad + \frac{x}{2}(\boldsymbol{t}^2 + 1)\boldsymbol{e}_{56} + \frac{y}{2}(\boldsymbol{t}^2 - 1)\boldsymbol{e}_{29} + \frac{y}{2}(\boldsymbol{t}^2 + 1)\boldsymbol{e}_{2,10} \\
&\quad + \frac{y}{2}(\boldsymbol{t}^2 + 1)\boldsymbol{e}_{47} + \frac{y}{2}(\boldsymbol{t}^2 + 1)\boldsymbol{e}_{57} + \frac{z}{2}(\boldsymbol{t}^2 - 1)\boldsymbol{e}_{39} \\
&\quad + \frac{z}{2}(\boldsymbol{t}^2 + 1)\boldsymbol{e}_{3,10} + \frac{z}{2}(\boldsymbol{t}^2 - 1)\boldsymbol{e}_{48} + \frac{z}{2}(\boldsymbol{t}^2 + 1)\boldsymbol{e}_{58} \\
&\quad + xy\boldsymbol{e}_{17} + xy\boldsymbol{e}_{26} + yz\boldsymbol{e}_{28} + yz\boldsymbol{e}_{37} + xz\boldsymbol{e}_{18} + xz\boldsymbol{e}_{36} \\
&\quad + x^2\boldsymbol{e}_{16} + y^2\boldsymbol{e}_{27} + z^2\boldsymbol{e}_{38} + \frac{1}{4}(\boldsymbol{t}^4 - 1)\boldsymbol{e}_{4,10} \\
&\quad + \frac{1}{4}(\boldsymbol{t}^4 - 1)\boldsymbol{e}_{59} + \frac{1}{4}(\boldsymbol{t}^4 - 2\boldsymbol{t}^2 + 1)\boldsymbol{e}_{49} + \frac{z}{2}(\boldsymbol{t}^4 + \boldsymbol{t}^2 + 1)\boldsymbol{e}_{5,10}
\end{aligned}$$

where

$$\begin{aligned}
\boldsymbol{t} = \boldsymbol{t}_{\varepsilon^1} &= x\boldsymbol{e}_1 + y\boldsymbol{e}_2 + z\boldsymbol{e}_3, \\
\boldsymbol{t}_{\varepsilon^2} &= x\boldsymbol{e}_6 + y\boldsymbol{e}_7 + z\boldsymbol{e}_8, \boldsymbol{t}^2 = x^2 + y^2 + z^2, \\
\boldsymbol{t}^4 &= x^4 + y^4 + z^4 + 2x^2y^2 + 2y^2z^2 + 2z^2x^2.
\end{aligned}$$

Using the extraction operator $\boldsymbol{T}_a$ in Table 1, the scalar component a can be extracted from a point $\boldsymbol{P}_D$ as $a = \boldsymbol{T}_a \cdot \boldsymbol{P}_D$.

Table 1. The DCGA bivector extraction operators

$\boldsymbol{T}_x = \frac{1}{2}(\boldsymbol{e}_1\boldsymbol{e}_{\infty 2} + \boldsymbol{e}_{\infty 1}\boldsymbol{e}_6)$	$\boldsymbol{T}_y = \frac{1}{2}(\boldsymbol{e}_2\boldsymbol{e}_{\infty 2} + \boldsymbol{e}_{\infty 1}\boldsymbol{e}_7)$	$\boldsymbol{T}_z = \frac{1}{2}(\boldsymbol{e}_3\boldsymbol{e}_{\infty 2} + \boldsymbol{e}_{\infty 1}\boldsymbol{e}_8)$
$\boldsymbol{T}_{x^2} = \boldsymbol{e}_6\boldsymbol{e}_1$	$\boldsymbol{T}_{y^2} = \boldsymbol{e}_7\boldsymbol{e}_2$	$\boldsymbol{T}_{z^2} = \boldsymbol{e}_8\boldsymbol{e}_3$
$\boldsymbol{T}_{xy} = \frac{1}{2}(\boldsymbol{e}_7\boldsymbol{e}_1 + \boldsymbol{e}_6\boldsymbol{e}_2)$	$\boldsymbol{T}_{yz} = \frac{1}{2}(\boldsymbol{e}_7\boldsymbol{e}_3 + \boldsymbol{e}_8\boldsymbol{e}_2)$	$\boldsymbol{T}_{zx} = \frac{1}{2}(\boldsymbol{e}_8\boldsymbol{e}_1 + \boldsymbol{e}_6\boldsymbol{e}_3)$
$\boldsymbol{T}_{xt^2} = \boldsymbol{e}_1\boldsymbol{e}_{o2} + \boldsymbol{e}_{o1}\boldsymbol{e}_6$	$\boldsymbol{T}_{yt^2} = \boldsymbol{e}_2\boldsymbol{e}_{o2} + \boldsymbol{e}_{o1}\boldsymbol{e}_7$	$\boldsymbol{T}_{zt^2} = \boldsymbol{e}_3\boldsymbol{e}_{o2} + \boldsymbol{e}_{o1}\boldsymbol{e}_8$
$\boldsymbol{T}_1 = -\boldsymbol{e}_\infty$	$\boldsymbol{T}_{t^2} = \boldsymbol{e}_{o2}\boldsymbol{e}_{\infty 1} + \boldsymbol{e}_{\infty 2}\boldsymbol{e}_{o1}$	$\boldsymbol{T}_{t^4} = -4\boldsymbol{e}_o$

2.1 Objects in DCGA

In this paper, our primary interest is in lines, tori, and cyclides in general. Tori and cyclides will be objects that we ray trace (as well as quadrics and polygons), while lines will be used to represent rays when performing intersections with the objects in our scene. The following is a review of Easter and Hitzer's constructions for these objects.

For any implicit quadric surface

$$a_{x^2}x^2 + a_{y^2}y^2 + a_{z^2}z^2 + a_{xy}xy + a_{yz}yz + a_{zx}zx + a_x x + a_y y + a_z z + a_1 = 0$$

where a_i are scalar coefficients, Easter and Hitzer define the quadric surface as a bivector in DCGA:

$$\begin{aligned} \boldsymbol{Q}_d = \; & a_{x^2}\boldsymbol{T}_{x^2} + a_{y^2}\boldsymbol{T}_{y^2} + a_{z^2}\boldsymbol{T}_{z^2} + a_{xy}\boldsymbol{T}_{xy} + a_{yz}\boldsymbol{T}_{yz} + a_{xz}\boldsymbol{T}_{xz} \\ & + a_x\boldsymbol{T}_x + a_y\boldsymbol{T}_y + a_z\boldsymbol{T}_z + a_1\boldsymbol{T}_1, \end{aligned} \tag{4}$$

Easter and Hitzer also define quartic surfaces like torus and cyclides in DCGA. The implicit equation for a quartic surface is

$$\begin{aligned} & a_{t^4}\boldsymbol{t}^4 + a_{t^2}\boldsymbol{t}^2 + a_{xt^2}x\boldsymbol{t}^2 + a_{yt^2}y\boldsymbol{t}^2 + a_{zt^2}z\boldsymbol{t}^2 + a_{x^2}x^2 + a_{y^2}y^2 + a_{z^2}z^2 \\ & + a_{xy}xy + a_{yz}yz + a_{zx}zx + a_x x + a_y y + a_z z + a_1 = 0 \end{aligned}$$

where $\boldsymbol{t} = x\boldsymbol{e}_1 + y\boldsymbol{e}_2 + z\boldsymbol{e}_3$ is a test point and the a_i are scalar coefficients. Easter and Hitzer define the 2-vector quartic surface entity (more specifically, a Darboux cyclide quartic surface) $\boldsymbol{Q}_t$ as

$$\begin{aligned} \boldsymbol{Q}_t = \; & a_{t^4}\boldsymbol{T}_{t^4} + a_{t^2}\boldsymbol{T}_{t^2} + a_{xt^2}\boldsymbol{T}_{xt^2} + a_{yt^2}\boldsymbol{T}_{yt^2} + a_{zt^2}\boldsymbol{T}_{zt^2} \\ & + a_{x^2}\boldsymbol{T}_{x^2} + a_{y^2}\boldsymbol{T}_{y^2} + a_{z^2}\boldsymbol{T}_{z^2} + a_{xy}\boldsymbol{T}_{xy} \\ & + a_{yz}\boldsymbol{T}_{yz} + a_{zx}\boldsymbol{T}_{zx} + a_x\boldsymbol{T}_x + a_y\boldsymbol{T}_y + a_z\boldsymbol{T}_z + a_1\boldsymbol{T}_1. \end{aligned} \tag{5}$$

The implicit quartic equation for a circular torus positioned at the origin that surrounds the z-axis is

$$\boldsymbol{t}^4 + 2\boldsymbol{t}^2(R^2 - r^2) + (R^2 - r^2)^2 - 4R^2(x^2 + y^2) = 0 \tag{6}$$

where $\boldsymbol{t} = x\boldsymbol{e}_1 + y\boldsymbol{e}_2 + z\boldsymbol{e}_3$, R is the major radius, and r is the minor radius. Easter and Hitzer defined a corresponding the DCGA GIPNS 2-vector torus surface entity $\boldsymbol{O}$ as

$$\boldsymbol{O} = \boldsymbol{T}_{t^4} + 2(R^2 - r^2)\boldsymbol{T}_{t^2} + (R^2 - r^2)^2\boldsymbol{T}_1 - 4R^2(\boldsymbol{T}_{x^2} + \boldsymbol{T}_{y^2}). \tag{7}$$

An arbitrary torus can be obtained by rotating and translating $\boldsymbol{O}$.

A point $\boldsymbol{P}_D$ is on a surface $\boldsymbol{Q}$ if and only if

$$\boldsymbol{P}_D \cdot \boldsymbol{Q} = 0.$$

The DCGA GIPNS 4-vector line 1D surface entity $\boldsymbol{L}$ is defined as

$$\boldsymbol{L} = \boldsymbol{L}_{c^1} \wedge \boldsymbol{L}_{c^2} \tag{8}$$

where

$$L_{C^1} = D_{\mathcal{E}^1} - (p_{\mathcal{E}^1} \cdot D_{\mathcal{E}^1})e_{\infty 1} \tag{9}$$

$$L_{C^2} = D_{\mathcal{E}^2} - (p_{\mathcal{E}^2} \cdot D_{\mathcal{E}^2})e_{\infty 2}. \tag{10}$$

Thus, the DCGA line L is the wedge of the line as represented in CGA1 with the same line as represented in CGA2. L is also called a *bi-CGA GIPNS line entity*. The D are unit bivectors perpendicular to the line, and p is any point on the line. The dual unit vector $d = D/I_3$, or $d_{\mathcal{E}^1} = D_{\mathcal{E}^1}/I_{\mathcal{E}^1} = dxe_1 + dye_2 + dze_3$ and $d_{\mathcal{E}^2} = D_{\mathcal{E}^2}/I_{\mathcal{E}^2} = dxe_6 + dye_7 + dze_8$, is in the direction of the line, where $I_{\mathcal{E}^1} = e_1 \wedge e_2 \wedge e_3$, $I_{\mathcal{E}^2} = e_6 \wedge e_7 \wedge e_8$ are the Euclidean 3D unit pseudoscalars.

The DCGA dual of an object T is defined as T^*:

$$T^* = T/I_D = TI_D^{-1} = T \cdot I_D^{-1} = -T \cdot I_D \tag{11}$$

where $I_D = I_{C^1} \wedge I_{C^2}$, $I_{C^1} = e_1 \wedge e_2 \wedge e_3 \wedge e_4 \wedge e_5$, $I_{C^2} = e_6 \wedge e_7 \wedge e_8 \wedge e_9 \wedge e_{10}$.

In our derivations, we will use the following properties of the inner product and of the outer product [7]:

$$(a_1 \wedge \cdots \wedge a_r) \cdot (b_1 \wedge \cdots \wedge b_s) = \begin{cases} ((a_1 \wedge \cdots \wedge a_r) \cdot b_1) \cdot (b_2 \wedge \cdots \wedge b_s) & r \geq s \\ (a_1 \wedge \cdots \wedge a_{r-1}) \cdot (a_r \cdot (b_1 \wedge \cdots \wedge b_s)) & r < s \end{cases} \tag{12}$$

and

$$(a_1 \wedge \cdots \wedge a_r) \cdot b = \sum_{n=1}^{N} (-1)^{r-i} a_1 \wedge \cdots a_{i-1} \wedge (a_i \cdot b) \wedge a_{i+1} \wedge \cdots \wedge a_r \tag{13}$$

$$a \cdot (b_1 \wedge \cdots \wedge b_s) = \sum_{n=1}^{N} (-1)^{i-1} b_1 \wedge \cdots b_{i-1} \wedge (a \cdot b_i) \wedge b_{i+1} \wedge \cdots \wedge b_s \tag{14}$$

2.2 Transformations

While our ray tracer uses transformations of canonical representations of spheres and tori to model our scene, we otherwise do not use transformations in this paper. We refer the reader to Easter and Hitzer's paper for the relevant details of transformations in DCGA [3].

2.3 Surface Normals

For the lighting calculation in ray tracing, we need the surface normal at the point of intersection. We use the differential operators and normal formula of Breuils et al. [1] (see also [3] for an earlier form),

$$D_x = (e_1 \wedge e_{\infty 1} + e_6 \wedge e_{o2})$$
$$D_y = (e_2 \wedge e_{\infty 1} + e_7 \wedge e_{o2})$$
$$D_z = (e_3 \wedge e_{\infty 1} + e_8 \wedge e_{o2}),$$

with Breuils et al.'s CGA1 normal formula to the DCGA surface T at the DCGA point P_D being

$$n = ((D_x \times T) \cdot P_D)e_1 + ((D_y \times T) \cdot P_D)e_2 + ((D_z \times T) \cdot P_D)e_3,$$

where $\times$ is the commutator product.

3 Line–Cyclide Intersection

In ray tracing, we need to compute the intersection of rays from the eye with all the objects in our scene, and in the case of multiple intersections, we need to determine the closest point of intersection to the eye. At times, we can use the DCGA line containing the ray to compute the intersections, but at other times, we need a parametric representation of the ray, $\boldsymbol{r}(t) = \boldsymbol{P} + t\boldsymbol{v},\ t > 0$, where $\boldsymbol{P}$ is a point and $\boldsymbol{v}$ is a Euclidean vector. DCGA does not have a strong concept of a Euclidean vector, so in our ray tracer, we construct our ray $\boldsymbol{r}(t)$ in CGA, requiring us at times to map back and forth from DCGA and CGA1.

To compute the intersection of the ray with the torus (or cyclide in general), we use the method of the Easter and Hitzer [3] to compute the intersection of the DCGA line containing $\boldsymbol{r}(t)$ with the cyclide. However, while Easter and Hitzer show that the outer product of a line with a cyclide gives the intersection between the two objects, they do not show how to extract the points of intersection of the resulting object. In this section, we show how to extract these points of intersection.

The intersection of an arbitrary cyclide $\boldsymbol{O}$ and a line $\boldsymbol{L}$ is

$$\boldsymbol{OL} = \boldsymbol{O} \wedge \boldsymbol{L}. \tag{15}$$

Let $\boldsymbol{v} = (v_x, v_y, v_z)$ be the CGA1 direction of $\boldsymbol{L}$ (where $\boldsymbol{v} = \boldsymbol{L} \cdot (\boldsymbol{e}_\infty \wedge \boldsymbol{e}_o \wedge (\boldsymbol{e}_6 + \boldsymbol{e}_7 + \boldsymbol{e}_8))$, and let $\boldsymbol{\mathcal{D}}(x_0, y_0, z_0)$ be a point on $\boldsymbol{L}$. A point on a line through $\boldsymbol{\mathcal{D}}(x_0, y_0, z_0)$ in direction v is (expanding (3))

$$\begin{aligned}
&\boldsymbol{\mathcal{D}}(x_0 + tv_x, y_0 + tv_y, z_0 + tv_z) = \\
&\quad \Big(((x_0 + tv_x)\boldsymbol{e}_1 + (y_0 + tv_y)\boldsymbol{e}_2 + (z_0 + tv_z)\boldsymbol{e}_3) + \\
&\quad \tfrac{1}{2}((x_0 + tv_x)\boldsymbol{e}_1 + (y_0 + tv_y)\boldsymbol{e}_2 + (z_0 + tv_z)\boldsymbol{e}_3)^2\boldsymbol{e}_{\infty 1} + \boldsymbol{e}_{o1}\Big) \wedge \\
&\quad \Big(((x_0 + tv_x)\boldsymbol{e}_6 + (y_0 + tv_y)\boldsymbol{e}_7 + (z_0 + tv_z)\boldsymbol{e}_8) + \\
&\quad \tfrac{1}{2}((x_0 + tv_x)\boldsymbol{e}_6 + (y_0 + tv_y)\boldsymbol{e}_7 + (z_0 + tv_z)\boldsymbol{e}_8)^2\boldsymbol{e}_{\infty 2} + \boldsymbol{e}_{o2}\Big) \\
&= \Big([x_0\boldsymbol{e}_1 + y_0\boldsymbol{e}_2 + z_0\boldsymbol{e}_3 + \tfrac{1}{2}(x_0^2 + y_0^2 + z_0^2)\boldsymbol{e}_{\infty 1} + \boldsymbol{e}_{o1}] + \\
&\quad [v_x\boldsymbol{e}_1 + v_y\boldsymbol{e}_2 + v_z\boldsymbol{e}_3 + (x_0v_x + y_0v_y + z_0v_z)\boldsymbol{e}_{\infty 1}]t + \\
&\quad [\tfrac{1}{2}(v_x^2 + v_y^2 + v_z^2)\boldsymbol{e}_{\infty 1}]t^2\Big) \wedge \\
&\quad \Big([x_0\boldsymbol{e}_6 + y_0\boldsymbol{e}_7 + z_0\boldsymbol{e}_8 + \tfrac{1}{2}(x_0^2 + y_0^2 + z_0^2)\boldsymbol{e}_{\infty 2} + \boldsymbol{e}_{o2}] + \\
&\quad [v_x\boldsymbol{e}_6 + v_y\boldsymbol{e}_7 + v_z\boldsymbol{e}_8 + (x_0v_x + y_0v_y + z_0v_z)\boldsymbol{e}_{\infty 2}]t + \\
&\quad [\tfrac{1}{2}(v_x^2 + v_y^2 + v_z^2)\boldsymbol{e}_{\infty 2}]t^2\Big)
\end{aligned} \tag{16}$$

The point $\boldsymbol{\mathcal{D}}(x_0 + tv_x, y_0 + tv_y, z_0 + tv_z)$ is on $\boldsymbol{OL} = \boldsymbol{O} \wedge \boldsymbol{L}$ if

$$\boldsymbol{\mathcal{D}}(x_0 + tv_x, y_0 + tv_y, z_0 + tv_z) \cdot \boldsymbol{OL} = 0. \tag{17}$$

Expanding Eq. 17 gives 100 terms. While merging duplicate terms reduces this to 89 terms, this number of terms would make implementation challenging. We can ease the implementation computing the intersection by assigning $\boldsymbol{K}, \boldsymbol{M}, \boldsymbol{N}, \boldsymbol{G}, \boldsymbol{H}, \boldsymbol{R}$ to each row of (16); i.e., let

$$\begin{aligned}
\boldsymbol{K} &= x_0\mathbf{e}_1 + y_0\mathbf{e}_2 + z_0\mathbf{e}_3 + \tfrac{1}{2}(x_0^2 + y_0^2 + z_0^2)\mathbf{e}_{\infty 1} + \mathbf{e}_{o1} \\
\boldsymbol{M} &= v_x\mathbf{e}_1 + v_y\mathbf{e}_2 + v_z\mathbf{e}_3 + (x_0v_x + y_0v_y + z_0v_z)\mathbf{e}_{\infty 1} \\
\boldsymbol{N} &= \tfrac{1}{2}(v_x^2 + v_y^2 + v_z^2)\mathbf{e}_{\infty 1} \\
\boldsymbol{G} &= x_0\mathbf{e}_6 + y_0\mathbf{e}_7 + z_0\mathbf{e}_8 + \tfrac{1}{2}(x_0^2 + y_0^2 + z_0^2)\mathbf{e}_{\infty 2} + \mathbf{e}_{o2} \\
\boldsymbol{H} &= v_x\mathbf{e}_6 + v_y\mathbf{e}_7 + v_z\mathbf{e}_8 + (x_0v_x + y_0v_y + z_0v_z)\mathbf{e}_{\infty 2} \\
\boldsymbol{R} &= \tfrac{1}{2}(v_x^2 + v_y^2 + v_z^2)\mathbf{e}_{\infty 2}
\end{aligned} \tag{18}$$

so that

$$\boldsymbol{\mathcal{D}}(x_0 + tv_x, y_0 + tv_y, z_0 + tv_z) = (\boldsymbol{K} + \boldsymbol{M}t + \boldsymbol{N}t^2) \wedge (\boldsymbol{G} + \boldsymbol{H}t + \boldsymbol{R}t^2).$$

Then expanding $\boldsymbol{\mathcal{D}}(x, y, z) \cdot \boldsymbol{OL}$ gives

$$\begin{aligned}
\boldsymbol{\mathcal{D}}(x, y, z) \cdot \boldsymbol{OL} &= ((\boldsymbol{K} + \boldsymbol{M}t + \boldsymbol{N}t^2) \wedge (\boldsymbol{G} + \boldsymbol{H}t + \boldsymbol{R}t^2)) \cdot \boldsymbol{OL} \\
&= t^0(\boldsymbol{K} \wedge \boldsymbol{G}) \cdot \boldsymbol{OL} + \\
&\quad t^1(\boldsymbol{K} \wedge \boldsymbol{H} + \boldsymbol{M} \wedge \boldsymbol{G}) \cdot \boldsymbol{OL} + \\
&\quad t^2(\boldsymbol{K} \wedge \boldsymbol{R} + \boldsymbol{M} \wedge \boldsymbol{H} + \boldsymbol{N} \wedge \boldsymbol{G}) \cdot \boldsymbol{OL} + \\
&\quad t^3(\boldsymbol{M} \wedge \boldsymbol{R} + \boldsymbol{N} \wedge \boldsymbol{H}) \cdot \boldsymbol{OL} + \\
&\quad t^4(\boldsymbol{N} \wedge \boldsymbol{R}) \cdot \boldsymbol{OL}.
\end{aligned} \tag{19}$$

Solving for the roots of Eq. 19 and substituting these roots into $\boldsymbol{\mathcal{D}}(x_0 + tv_0, y_y + ty_0, z_0 + tz_0)$ gives us the four points of intersection between the line and the cyclide. In a ray tracer, we are interested in the first intersection along the ray, which is given by the smallest, positive real root. If none of the roots are real numbers (or if all the real roots are negative), then our ray did not intersect the surface.

Remark 31. *Note that* (16) *is the parametric equation for a line in DCGA. Further note that* (16) *is a fourth degree equation. While it may seem odd that the parametric equation for a line in DCGA is a fourth degree equation, the high degree results from the line being constructed as the outer product of two CGA lines, each of which are degree two equations. If we apply the extraction operators* $\boldsymbol{T}_x, \boldsymbol{T}_y, \boldsymbol{T}_z$ *to* (16), *we find the linear equations we expect for a line.*

Remark 32. *While the solutions of* (19) *will likely work for the intersection of a line with an arbitrary quartic, in ray tracing we are primarily interested in finite volumes having a well-defined inside and outside; thus, in our ray tracer (Sect. 6), we only tested our method on cyclides.*

3.1 DCGA and Quadrics

In this section, we study the polynomial equation resulting from the intersection of a line with a quadratic in DCGA. While the resulting polynomial appears to be quartic, we note that the degree three and four terms of the latter are zero, and that the polynomial is only quadratic as expected.

The intersection of a quadric $\boldsymbol{Q}$ and a line $\boldsymbol{L}$ is

$$\boldsymbol{QL} = \boldsymbol{Q} \wedge \boldsymbol{L}. \tag{20}$$

Let $\boldsymbol{\mathcal{D}}(x_0, y_0, z_0)$ be a point on $\boldsymbol{L}$ and let $\boldsymbol{v} = (v_x, v_y, v_z)$ be the direction of $\boldsymbol{L}$. Then an arbitrary point $\boldsymbol{\mathcal{D}}(x, y, z)$ on $\boldsymbol{L}$ is

$$\begin{aligned} &\boldsymbol{\mathcal{D}}(x, y, z) = \boldsymbol{\mathcal{D}}(x_0 + tv_x, y_0 + tv_y, z_0 + tv_z) = \\ &(\boldsymbol{K} + \boldsymbol{M}t + \boldsymbol{N}t^2) \wedge (\boldsymbol{G} + \boldsymbol{H}t + \boldsymbol{R}t^2) \end{aligned}$$

with $\boldsymbol{K}, \boldsymbol{M}, \boldsymbol{N}, \boldsymbol{G}, \boldsymbol{H}, \boldsymbol{R}$ from (18). Then expanding $\boldsymbol{\mathcal{D}}(x, y, z) \cdot \boldsymbol{QL}$ gives

$$\begin{aligned} \boldsymbol{\mathcal{D}}(x, y, z) \cdot \boldsymbol{QL} = \quad & ((\boldsymbol{K} + \boldsymbol{M}t + \boldsymbol{N}t^2) \wedge (\boldsymbol{G} + \boldsymbol{H}t + \boldsymbol{R}t^2)) \cdot \boldsymbol{QL} \\ = \quad & t^0(\boldsymbol{K} \wedge \boldsymbol{G}) \cdot \boldsymbol{QL} + \\ & t^1(\boldsymbol{K} \wedge \boldsymbol{H} + \boldsymbol{M} \wedge \boldsymbol{G}) \cdot \boldsymbol{QL} + \\ & t^2(\boldsymbol{K} \wedge \boldsymbol{R} + \boldsymbol{M} \wedge \boldsymbol{H} + \boldsymbol{N} \wedge \boldsymbol{G}) \cdot \boldsymbol{QL} + \\ & t^3(\boldsymbol{M} \wedge \boldsymbol{R} + \boldsymbol{N} \wedge \boldsymbol{H}) \cdot \boldsymbol{QL} + \\ & t^4(\boldsymbol{N} \wedge \boldsymbol{R}) \cdot \boldsymbol{QL}. \end{aligned} \tag{21}$$

Expanding the t^3 and t^4 terms of Equation (21), we find that both are identical 0. Thus, Equation (21) is a quadratic polynomial. Since the degree three and degree four terms of (21) will be zero for a quadratic, if it is known that $\boldsymbol{QL}$ is a quadric (as is typically the case in a ray tracer), we can reduce the computational costs by not evaluating the inner product coefficients for these two terms.

4 Colinear Point Quadruples (CPQ)

The intersection of a line with a torus gives an object whose inner product with a point is zero at the four points of intersection between the torus and the line, and non-zero at any other point. We refer to this object as a *collinear point quadruple* (CPQ). In this section, we show how to create a collinear point quadruple directly from four points.

Theorem 1. *Given four collinear but distinct DCGA points,* $\boldsymbol{P}_{D1} = \boldsymbol{P}_{C_1^1} \wedge \boldsymbol{P}_{C_1^2} = \boldsymbol{\mathcal{D}}(x_1, y_1, z_1), \boldsymbol{P}_{D2} = \boldsymbol{P}_{C_2^1} \wedge \boldsymbol{P}_{C_2^2} = \boldsymbol{\mathcal{D}}(x_2, y_2, z_2), \boldsymbol{P}_{D3} = \boldsymbol{P}_{C_3^1} \wedge \boldsymbol{P}_{C_3^2} = \boldsymbol{\mathcal{D}}(x_3, y_3, z_3), \boldsymbol{P}_{D4} = \boldsymbol{P}_{C_4^1} \wedge \boldsymbol{P}_{C_4^2} = \boldsymbol{\mathcal{D}}(x_4, y_4, z_4)$, *where*

$$x_1^2 + y_1^2 + z_1^2 \neq x_2^2 + y_2^2 + z_2^2 \tag{22}$$

$$x_3^2 + y_3^2 + z_3^2 \neq x_4^2 + y_4^2 + z_4^2. \tag{23}$$

Let

$$\boldsymbol{CPQ} = ((\boldsymbol{Pp}_1 \cdot (\boldsymbol{e}_{o1} \wedge \boldsymbol{e}_{\infty 1})) \wedge (\boldsymbol{Pp}_2 \cdot (\boldsymbol{e}_{o2} \wedge \boldsymbol{e}_{\infty 2}))) + ((\boldsymbol{Pp}_1 \cdot (\boldsymbol{e}_{o2} \wedge \boldsymbol{e}_{\infty 2})) \wedge (\boldsymbol{Pp}_2 \cdot (\boldsymbol{e}_{o1} \wedge \boldsymbol{e}_{\infty 1}))) \tag{24}$$

where $\boldsymbol{Pp}_1 = \boldsymbol{P}_{D1} \wedge \boldsymbol{P}_{D2}$ *and* $\boldsymbol{Pp}_2 = \boldsymbol{P}_{D3} \wedge \boldsymbol{P}_{D4}$*. Then for an arbitrary DCGA point* $\boldsymbol{P}_D$*,*

$$\boldsymbol{P}_D \wedge (\boldsymbol{CPQ}/\boldsymbol{I}_D) = \boldsymbol{P}_D \cdot \boldsymbol{CPQ}^* = \begin{cases} 0 & \text{if } \boldsymbol{P}_D \in \{\boldsymbol{P}_{D1}, \boldsymbol{P}_{D2}, \boldsymbol{P}_{D3}, \boldsymbol{P}_{D4}\} \\ \text{non-0 otherwise} \end{cases} \tag{25}$$

and

$$\boldsymbol{CPQ} = (k_2^1 - k_1^1)(k_4^2 - k_3^2)(\boldsymbol{P}_{C_1^2} \wedge \boldsymbol{P}_{C_2^2} \wedge \boldsymbol{P}_{C_3^1} \wedge \boldsymbol{P}_{C_4^1} + \boldsymbol{P}_{C_1^1} \wedge \boldsymbol{P}_{C_2^1} \wedge \boldsymbol{P}_{C_3^2} \wedge \boldsymbol{P}_{C_4^2}).$$

where $k_i^j = \boldsymbol{e}_{oj} \cdot \boldsymbol{P}_{C_i^j} = -\frac{1}{2}(x_i^2 + y_i^2 + z_i^2)$*,* $\boldsymbol{e}_{\infty j} \cdot \boldsymbol{P}_{C_i^j} = -1$ $(i = 1 \cdots 4, j = 1, 2)$*.*

Remark 41. *We note that the representation of a collinear point quadruple is not unique. In particular, the* $\boldsymbol{CPQ}$ *constructed using point pairs* $\boldsymbol{Pp}_{12} = \boldsymbol{P}_{D1} \wedge \boldsymbol{P}_{D2}$ *and* $\boldsymbol{Pp}_{34} = \boldsymbol{P}_{D3} \wedge \boldsymbol{P}_{D4}$ *in Eq. 24 is not equal to the* $\boldsymbol{CPQ}$ *constructed using point pairs* $\boldsymbol{Pp}_{13} = \boldsymbol{P}_{D1} \wedge \boldsymbol{P}_{D3}$ *and* $\boldsymbol{Pp}_{24} = \boldsymbol{P}_{D2} \wedge \boldsymbol{P}_{D4}$ *in Eq. 24, although both* $\boldsymbol{CPQ}$*'s have the same zero set.*

Remark 42. *While the representation of a* $\boldsymbol{CPQ}$ *in Theorem 1 gives a fairly concise representation for a* $\boldsymbol{CPQ}$*, a* $\boldsymbol{CPQ}$ *created in this manner is coordinate system dependent. In particular, conditions* (22) *and* (23) *result in a* $\boldsymbol{CPQ}$ *that is identically zero in one coordinate system but valid in another coordinate system. This coordinate system dependence complicated the proof of Theorem 1.*

Remark 43. *If the set of four points fails the condition in Eq. 22 or Eq. 23, then reordering the points by swapping either of* $\boldsymbol{P}_{D1}, \boldsymbol{P}_{D2}$ *with either of* $\boldsymbol{P}_{D3}, \boldsymbol{P}_{D4}$ *will yield of set of four points that meets these conditions and can be used to construct a non-zero* $\boldsymbol{CPQ}$*.*

The method in Sect. 3 allows us to compute the four points of a $\boldsymbol{CPQ}$ given the line $\boldsymbol{L}$ containing the four points, a point on $\boldsymbol{L}$, and the direction of $\boldsymbol{L}$. However, we might want to find the four points on a $\boldsymbol{CPQ}$ without knowing the line $\boldsymbol{L}$ containing the $\boldsymbol{CPQ}$. While we do not have general results for an arbitrary CPQ, we can show that for CPQ's constructed either as in Theorem 1 or as the intersection of a line with a quartic, the CGA1 direction $\boldsymbol{d}_1$ of the line containing $\boldsymbol{P}_1, \boldsymbol{P}_2, \boldsymbol{P}_3, \boldsymbol{P}_4$ is

$$\boldsymbol{d}_1 + \alpha \boldsymbol{e}_{\infty 1} = \boldsymbol{CPQ} \cdot (\boldsymbol{e}_\infty \wedge \boldsymbol{e}_{o2}) \tag{26}$$

and

$$\boldsymbol{S} = \boldsymbol{CPQ} \cdot (\boldsymbol{e}_o \wedge \boldsymbol{e}_{\infty 2}) / \boldsymbol{CPQ} \cdot (\boldsymbol{e}_o \wedge \boldsymbol{e}_\infty) \tag{27}$$

is a CGA1 sphere through the origin whose center is a point on the line containing $\boldsymbol{P}_1, \boldsymbol{P}_2, \boldsymbol{P}_3, \boldsymbol{P}_4$.

5 Barycentric Coordinates in DCGA

In the homogeneous model ($e_i^2 = 1, e_o^2 = 0$) a 3D point with coordinates (x, y, z) is represented as $\boldsymbol{P} = x\boldsymbol{e}_1 + y\boldsymbol{e}_2 + z\boldsymbol{e}_3 + \boldsymbol{e}_o$. If a point $\boldsymbol{P}$ lies in a plane $\Delta P_0P_1P_2$, then $\boldsymbol{P}$ can be represented as a linear combination $\boldsymbol{P} = a_0\boldsymbol{P}_0 + a_1\boldsymbol{P}_1 + a_2\boldsymbol{P}_2$. a_0, a_1, a_2 are called *barycentric coordinates.* Barycentric coordinates are used for a variety of purposes in computer graphics; in particular, they are used when performing lighting calculations on triangles.

We can find the barycentric coordinates of a point $\boldsymbol{P}$ relative to a triangle $\Delta P_0P_1P_2$ by computing [2]:

$$a_0 = \frac{(\boldsymbol{P} - \boldsymbol{P}_2) \wedge (\boldsymbol{P}_1 - \boldsymbol{P}_2)}{(\boldsymbol{P}_0 - \boldsymbol{P}_2) \wedge (\boldsymbol{P}_1 - \boldsymbol{P}_2)},$$
$$a_1 = \frac{(\boldsymbol{P} - \boldsymbol{P}_2) \wedge (\boldsymbol{P}_0 - \boldsymbol{P}_2)}{(\boldsymbol{P}_1 - \boldsymbol{P}_2) \wedge (\boldsymbol{P}_0 - \boldsymbol{P}_2)},$$
$$a_2 = 1 - a_0 - a_1.$$

We note that the homogeneous model is a subspace of CGA, and we can map a point $\boldsymbol{P}$ in CGA to a point $\boldsymbol{P}'$ in homogeneous by

$$\boldsymbol{P}' = -(\boldsymbol{P} \wedge \boldsymbol{e}_\infty) \cdot \boldsymbol{e}_o = \boldsymbol{P} - \tfrac{1}{2}(x^2 + y^2 + z^2)\boldsymbol{e}_\infty.$$

Note that if $\boldsymbol{P} = \boldsymbol{e}_\infty$ or $\boldsymbol{P} = \boldsymbol{e}_0$, then $(\boldsymbol{P} \wedge \boldsymbol{e}_\infty) \cdot \boldsymbol{e}_o = 0$. For ray tracing, we are only interested in finite closed surfaces, so the case of $\boldsymbol{P} = \boldsymbol{e}_\infty$ is of no concern; when $\boldsymbol{P} = \boldsymbol{e}_0$, note that $\boldsymbol{P}' = \boldsymbol{P}$, and no computation needs to be done, but if an expression that works for all $\boldsymbol{P}$ is desired, see [3]. Regardless, we can find barycentric coordinates of a point $\boldsymbol{P}$ in CGA relative to a triangle $\Delta P_0P_1P_2$ in CGA by mapping all points to the homogeneous model and using the above method.

Likewise, in DCGA, we can map a point $\boldsymbol{P_D}$ in DCGA to a point $\boldsymbol{P}'$ in homogeneous1 by $(\boldsymbol{P_D} \wedge \boldsymbol{e}_{\infty 1}) \cdot \boldsymbol{e}_o$, and again find the barycentric coordinates of a point $\boldsymbol{P_D}$ in DCGA relative to a triangle $\Delta P_0P_1P_2$ in DCGA by applying the above method.

6 Ray Tracer Verification

We implemented the methods described in this paper, and integrated them in a simple ray tracer [5] in gaigen [4]. Our focus in the ray tracer was on ray tracing tori (and cyclides in general), although we also ray traced quadratic surfaces and planes.

In ray tracing, an image is created by casting rays starting at a view point, through each pixel of the desired image, and intersecting the ray with all the objects in the scene. Using the closest intersection, a lighting calculation is performed to determine the colour of the pixel. The ray tracer we implemented performs a diffuse lighting calculation, shadows, and constructive solid geometry operations.

Hierarchical modeling techniques are typically used in creating scenes for a ray tracer. The resulting data structure is a tree of transformations, with primitives (such as spheres and tori) as leaf nodes. These primitives are defined in canonical positions (such as a sphere at the origin). When ray tracing, we traverse this tree structure, and when traversing down the hierarchy, at a node with transformation T, we apply T^{-1} to the point and vector of the ray. At the leaf nodes, this transformed ray is intersected with the canonical primitives. We used such a hierarchical approach in our ray tracer, and thus the ray-intersect-tori computations we computed using the special torus defined as in Eq. 6. However, we also ray traced other cyclides as seen in the picture, intersecting rays with arbitrary cyclides.

Figure 1 shows a scene ray traced with our software.

Fig. 1. Ray traced image.

We note that while ray tracing can be performed in DCGA, the constructions are at times a bit awkward. In particular, there is no concept of a DCGA vector. As such, to construct a ray, one needs to map from DCGA into one of CGA1 or CGA2, and perform various computations in one of these subalgebras. Other than that, implementing the DCGA ray tracer was straightforward.

7 Conclusions

In studying how to use the representation of cyclides in DCGA for ray tracing, we investigated the colinear-point-quadruple resulting from a line-intersect-cyclide, finding how to extract the points from a CPQ, as well as how to construct a CPQ directly from four points. We further investigated some other constructions needed in a ray tracer.

Acknowledgement. This work was funded in part by the National Sciences and Engineering Research Council of Canada and National Natural Science Foundation of China (NSFC) under Grant 51525504.

References

1. Breuils, S., Nozick, V., Fuchs, L., Sugimoto, A.: Efficient representation and manipulation of quadratic surfaces using Geometric Algebras. arXiv:1903.02444v1 [cs.CG] (2019)
2. Dorst, L., Fontijne, D., Mann, S.: Geometric Algebra for Computer Science. Morgan-Kaufmann (2007)
3. Easter, R.B., Hitzer, E.: Double conformal geometric algebra. Adv. Appl. Clifford Algebras **27**, 2175–2199 (2017). https://doi.org/10.1007/s00006-017-0784-0
4. Fontijne, F. Gaigen 2: a geometric algebra implementation generator. In: GPCE 2006. ACM (2006). https://doi.org/10.1145/1173706.1173728
5. Glassner, A.S.: An Introduction to Ray Tracing, Academic Press (1989)
6. Rees, E.L.: Graphics discussion of the roots of a quartic equation. Am. Math. Mon. **29**(2), 51–55 (1922). https://doi.org/10.1080/00029890.1922.11986100
7. Hestenes, D., Li, H., Rockwood, A.: New algebraic tools for classical geometry. In: Sommer, G. (ed.) Geometric Computing with Clifford Algebras. Springer, Heidelberg (2001). https://doi.org/10.1007/978-3-662-04621-0_1

Search for Similarity Transformation Between Image Point Clouds Using Geometric Algebra for Conics

Anna Derevianko(✉) and Petr Vašík

Brno University of Technology, Brno, Czech Republic
Anna.Derevianko@vutbr.cz

Abstract. We introduce a novel way of searching for the similarity transformation of pictures using Geometric Algebra for Conics (GAC). In our approach, we do not represent the image objects by their contour but, instead, by the ellipse fitted into the contour points. Such representation makes the consequent similarity search fast and memory-saving and makes the search for the needed transformation easier. Examples of application on the real object images are also included.

Keywords: geometric algebra · Clifford algebra · transformation · picture comparison · image processing

1 Introduction

The problem of image comparison and recognition is still popular. There are many different methods such as Blob detection technique, template matching, SURF feature extraction etc. [1].

The paper considers the search for a transformation, consisting of translation, rotation and scaling, which allows one object to be as close as possible to another in order to simplify their further comparison. However, existing methods are computationally demanding: neural networks or iterative closest point search require plenty of time and memory [8]. Therefore, we have presented a relatively fast method. As a fundamental contour of our object, we consider ellipses inscribed into the extracted contour points in a specific way, namely, using GAC-based Iterative Conic Fitting Algorithm. This will simplify the search for the required transformation.

2 Perceptual Hashing

Perceptual hashing is widely used for the tasks that require matching of similar images. Generally, perceptual hashing algorithms generate a fingerprint for each image so that similar images will be mapped to the same or similar hash code.

Five of the most used hashing techniques are well known: A-Hash, D-Hash, P-Hash, W-Hash, SVD-Hash. Let us describe the main idea of each of the types.

D. W. Silva et al. (Eds.): ICACGA 2022, LNCS 13771, pp. 15–26, 2024.
https://doi.org/10.1007/978-3-031-34031-4_2

- A-Hash
  The average hashing technique (A-Hash) produces the hash value of the image based on its low frequencies, representing the image structure, and points out the higher frequencies, which correspond to the image details. The goal of the A-Hash is to find the average color of the image by calculating the mean of the image matrix values.
- D-Hash
  The difference hash technique (D-Hash), is similar to the A-Hash. They both focus on the image structure, which is achieved by eliminating the higher frequencies from the image. The main difference lays in generating hashes by computing the difference based on the change of color gradient between the adjacent pixels in the image matrix.
- P-Hash
  The perceptive hash (P-Hash) is a technique extending the A-Hash by adding the Discrete Cosine Transform (DCT) to obtain the most sensitive information of the human vision system (HVS). Instead of using image intensities for the hash generation process, it uses a range of low frequencies obtained after applying the DCT technique.
- W-Hash
  The wavelet hash technique (W-Hash) is using the Discrete Wavelet Transform (DWT) for generating perceptual hashes.
- SVD-Hash
  The Singular Value Decomposition hash(SVD-Hash) was first introduced by Kozat et al. [10]. The general idea of the technique is to derive a secondary image from the original one using a pseudo-randomly (PR) extracting features that approximately demonstrate geometric characteristics. [4]

Perceptual hashing algorithms use perceptual features of images (see [12]) to generate their hashes. The primary goal is to generate hashes that remain unchanged or change slightly when content preserving modifications are made to the image. Given two images I and I',$hI = H(I)$ and $hI' = H(I')$ their corresponding perceptual hashes, and $D(hI, hI')$ is a similarity metric, and τ is an empirically determined threshold, $D(hI, hI') < \tau$ indicates that I and I' are copies of the same image with minor content preserving modifications. The main three steps involved in perceptual hashing algorithms are image pre-processing, perceptual feature extraction, and quantization or compression to generate the final hash string. One of the problems with this approach is that even the same, but rotated images have different hash due to the intrinsic limitation of perceptual hash algorithms.

For the experiment we use the module OpenCV in Python, that brings implementations of different image hashing algorithms, slowing to extract the hash of images and find similar images in huge data set.

3 Geometric Algebra for Conics

Let us briefly describe the basic concepts of GAC.

By geometric algebra we mean a Clifford algebra with a specific embedding of a Euclidean space in such a way that the intrinsic geometric primitives as well as their transformations are viewed as its elements, precisely multivectors. For more details see [3,5,7].

For our goals we will use the algebra for conics, proposed by C. Perwass to generalize the concept of (two–dimensional) conformal geometric algebra $\mathbb{G}_{3,1}$, [14] with the notation of [6]. In the usual basis $\bar{n}, e_1, e_2, n$, embedding of a plane in $\mathbb{G}_{3,1}$ is given by

$$(x, y) \mapsto \bar{n} + xe_1 + ye_2 + \frac{1}{2}(x^2 + y^2)n,$$

where e_1, e_2 form Euclidean basis and $\bar{n}$ and n, defined by specific linear combination of additional basis vectors e_3, e_4 with $e_3^2 = 1$ and $e_4^2 = -1$, are the coordinate origin and infinity, respectively, [14]. Hence the objects representable by vectors in $\mathbb{G}_{3,1}$ are linear combinations of $1, x, y, x^2+y^2$, i.e. circles, lines, point pairs and points. For the general conics, we need to add two terms: $\frac{1}{2}(x^2 - y^2)$ and xy. It turns out that we need two new infinities for that and also their two corresponding counterparts (Witt pairs), [11]. Thus the resulting dimension of the space generating the appropriate geometric algebra is eight.

Let $\mathbb{R}^{5,3}$ denote the eight–dimensional real coordinate space $\mathbb{R}^8$ equipped with a non–degenerate symmetric bilinear form of signature $(5,3)$. The form defines Clifford algebra $\mathbb{G}_{5,3}$ and this is the Geometric Algebra for Conics in the algebraic sense. To add the geometric meaning we have to describe an embedding of the plane into $\mathbb{R}^{5,3}$. To do so, let us choose a basis of $\mathbb{R}^{5,3}$ such that the corresponding bilinear form is

$$B = \begin{pmatrix} 0 & 0 & -1_{3\times3} \\ 0 & 1_{2\times2} & 0 \\ -1_{3\times3} & 0 & 0 \end{pmatrix}, \tag{1}$$

where $1_{2\times2}$ and $1_{3\times3}$ denote the unit matrices. Analogously to CGA and to the notation in [14], the corresponding basis elements are denoted as follows

$$\bar{n}_+, \bar{n}_-, \bar{n}_\times, e_1, e_2, n_+, n_-, n_\times.$$

Note that there are three orthogonal 'origins' $\bar{n}$ and three corresponding orthogonal 'infinities' n. In terms of this basis, a point of the plane $\mathbf{x} \in \mathbb{R}^2$ defined by $\mathbf{x} = xe_1 + ye_2$ is embedded using the operator $\mathcal{C} : \mathbb{R}^2 \to \mathit{Cone} \subset \mathbb{R}^{5,3}$, which is defined by

$$\mathcal{C}(x, y) = \bar{n}_+ + xe_1 + ye_2 + \frac{1}{2}(x^2 + y^2)n_+ + \frac{1}{2}(x^2 - y^2)n_- + xyn_\times. \tag{2}$$

The image *Cone* of the plane in $\mathbb{R}^{5,3}$ is an analogue of the conformal cone. In fact, it is a two–dimensional real projective variety determined by five homogeneous polynomials of degree one and two.

Definition 1. *Geometric Algebra for Conics (GAC) is the Clifford algebra $\mathbb{G}_{5,3}$ together with the embedding $\mathbb{R}^2 \to \mathbb{R}^{5,3}$ given by (2) in the basis determined by matrix (1).*

Note that, up to the last two terms, the embedding (2) is the embedding of the plane into the two–dimensional conformal geometric algebra $\mathbb{G}_{3,1}$. In particular, it is evident that the scalar product of two embedded points is the same as in $\mathbb{G}_{3,1}$, i.e. for two points $\mathbf{x}, \mathbf{y} \in \mathbb{R}^2$ we have

$$\mathcal{C}(\mathbf{x}) \cdot \mathcal{C}(\mathbf{y}) = -\frac{1}{2}\|\mathbf{x} - \mathbf{y}\|^2, \tag{3}$$

where the standard Euclidean norm is considered on the right hand side. This demonstrates linearisation of distance problems. In particular, each point is represented by a null vector. Let us recall that the invertible algebra elements are called versors and they form a group, the Clifford group, and that conjugations with versors give transformations intrinsic to the algebra. Namely, if the conjugation with a $\mathbb{G}_{5,3}$ versor R preserves the set *Cone*, i.e. for each $\mathbf{x} \in \mathbb{R}^2$ there exists such a point $\bar{\mathbf{x}} \in \mathbb{R}^2$ that

$$R\mathcal{C}(\mathbf{x})\tilde{R} = \mathcal{C}(\bar{\mathbf{x}}), \tag{4}$$

where $\tilde{R}$ is the reverse of R, then $\mathbf{x} \to \bar{\mathbf{x}}$ induces a transformation $\mathbb{R}^2 \to \mathbb{R}^2$ which is intrinsic to GAC. See [6] to find that the conformal transformations are intrinsic to GAC.

Let us also recall the outer (wedge) product, inner product and the duality

$$A^* = AI^{-1}. \tag{5}$$

However we use the definitions as in [14]. Note that in GAC the pseudoscalar is given by $I = \bar{n}_+\bar{n}_-\bar{n}_\times e_1 e_2 n_+ n_- n_\times$. For our purposes, we stress that these operations correspond to sums and products only. Indeed, the wedge product is calculated as the outer product of vectors on each vector space of the same grade blades, while the inner product acts on these spaces as the scalar product. The extension of both operations to general multivectors adds no computational complexity due to linearity of both operations. Let us also recall that if a conic C is seen as a wedge of five different points (which determines a conic uniquely), we call the appropriate 5–vector E^* an outer product null space representation (OPNS) and its dual E, indeed a 1–vector, the inner product null space (IPNS) representation.

Let us recall the definition of inner product representation. An element $A_I \in \mathbb{G}_{5,3}$ is the inner product representation of a geometric entity A in the plane if and only if $A = \{\mathbf{x} \in \mathbb{R}^2 : \mathcal{C}(\mathbf{x}) \cdot A_I = 0\}$. Hence, given a fixed geometric algebra, the representable objects can be found by examining the inner product of a vector and an embedded point. A general vector in the conic space $\mathbb{R}^{5,3}$ in terms of our basis is of the form

$$v = \bar{v}^+\bar{n}_+ + \bar{v}^-\bar{n}_- + \bar{v}^\times\bar{n}_\times + v^1 e_1 + v^2 e_2 + v^+ n_+ + v^- n_- + v^\times n_\times$$

Hence a conic is uniquely represented (in a homogeneous sense) by a vector in $\mathbb{R}^{5,3}$ modulo this subspace. This gives the desired dimension six. In other words, the inner representation of a conic in GAC can be defined as a vector

$$Q_I = \bar{v}^+ \bar{n}_+ + \bar{v}^- \bar{n}_- + \bar{v}^\times \bar{n}_\times + v^1 e_1 + v^2 e_2 + v^+ n_+. \tag{6}$$

The classification of conics is well known. Among the non–degenerate conics there are three types, the ellipse, hyperbola, and parabola. Now, we present the vector form (6) appropriate to the simplest case, i.e. an axes–aligned ellipse E_I with its centre in the origin and semi–axes a, b. Correctness may be verified easily by multiplying its vector by an embedded point which means the application of (1) and (2). The corresponding GAC vector is of the form

$$E_I = (a^2 + b^2)\bar{n}_+ + (a^2 - b^2)\bar{n}_- - a^2 b^2 n_+. \tag{7}$$

More generally, an ellipse E with the semi–axes a, b centred in $(u, v) \in \mathbb{R}^2$ rotated by angle θ is in the GAC inner product null space (IPNS) representation given by

$$\begin{aligned} E = {} & \bar{n}_+ - (\alpha \cos 2\theta)\bar{n}_- - (\alpha \sin 2\theta)\bar{n}_\times \\ & + (u + u\alpha \cos 2\theta - v\alpha \sin 2\theta)e_1 + (v + v\alpha \cos 2\theta - u\alpha \sin 2\theta)e_2 \\ & + \tfrac{1}{2}\left(u^2 + v^2 - \beta - (u^2 - v^2)\alpha \cos 2\theta - 2uv\alpha \sin 2\theta\right) n_+. \end{aligned} \tag{8}$$

For proofs and further details about other conics see [6].

3.1 Parameter Extraction

It is well known that the type of a given unknown conic can be read off its matrix representation, which in our case for a conic given by

$$Q = \begin{pmatrix} -\frac{1}{2}(\bar{v}^+ + \bar{v}^-) & -\frac{1}{2}\bar{v}^\times & \frac{1}{2}v^1 \\ -\frac{1}{2}\bar{v}^\times & -\frac{1}{2}(\bar{v}^+ - \bar{v}^-) & \frac{1}{2}v^2 \\ \frac{1}{2}v^1 & \frac{1}{2}v^2 & -v^+ \end{pmatrix}. \tag{9}$$

The entries of (9) can be easily computed by means of the inner product:

$$\begin{aligned} q_{11} &= Q_I \cdot \tfrac{1}{2}(n_+ + n_-), \\ q_{22} &= Q_I \cdot \tfrac{1}{2}(n_+ - n_-), \\ q_{33} &= Q_I \cdot \bar{n}_+, \\ q_{12} = q_{21} &= Q_I \cdot \tfrac{1}{2}n_\times, \\ q_{13} = q_{31} &= Q_I \cdot \tfrac{1}{2}e_1, \\ q_{23} = q_{32} &= Q_I \cdot \tfrac{1}{2}e_2. \end{aligned}$$

It is also well known how to determine the internal parameters of an unknown conic and its position and the orientation in the plane from the matrix (9). Hence all this can be determined from the GAC vector Q_I by means of the inner product.

The parameters of a conic can be obtained from the matrix (9) of its IPNS representation, for example:

– center of an ellipse or hyperbola:

$$x_c = \frac{q_{12}q_{23} - 2q_{22}q_{13}}{4q_{11}q_{22} - q_{12}^2}, \quad y_c = \frac{q_{13}q_{12} - 2q_{11}q_{23}}{4q_{11}q_{22} - q_{12}^2} \tag{10}$$

– semiaxis of an ellipse:

$$a, b = \frac{\sqrt{(2A(q_{11} + q_{22} \pm \sqrt{(q_{11} - q_{22})^2 + q_{12}^2})}}{(4q_{11}q_{22} - q_{12}^2))}, \tag{11}$$

where $A = q_{11}q_{23}^2 + q_{22}q_{13}^2 - q_{12}q_{13}q_{23} + (q_{12}^2 - 4q_{11}q_{22})q_{33}$

– angle of rotation

$$\theta = \begin{cases} -\arctan \frac{q_{22} - q_{11} - \sqrt{(q_{11} - q_{22})^2 + q_{12}^2}}{q_{12}}, & q_{12} \neq 0 \\ 0, & q_{12} = 0, \ q_{11} < q_{22} \\ \frac{\pi}{2}, & q_{12} = 0, \ q_{11} > q_{22} \end{cases} \tag{12}$$

Other parameters can be derived with the help of eigenvalues of the quadratic form matrix. For more details see [9].

3.2 Transformations

The main advantage of GAC compared to other models (for instance, $\mathbb{G}_6$) is that it is fully operational in the sense that it allows all Euclidean transformations, i.e. rotations and translations. But not just that, it also allows scaling in the sense of (4). Hence, like in the case of CGA (or $\mathbb{G}_{3,1}$), one obtains all conformal transformations. The exact form of GAC versor for rotation (rotor), translation (translator), and scaling (scalor) is given as follows.

The rotor for a rotation around the origin by the angle φ is given by $R = R_+(R_1 \wedge R_2)$, where

$$R_+ = \cos(\tfrac{\varphi}{2}) + \sin(\tfrac{\varphi}{2})e_1 \wedge e_2, \tag{13}$$

$$R_1 = \cos(\varphi) + \sin(\varphi)\bar{n}_\times \wedge n_-, \tag{14}$$

$$R_2 = \cos(\varphi) - \sin(\varphi)\bar{n}_- \wedge n_\times. \tag{15}$$

The translator is given by $T = T_+T_-T_\times$, where

$$T_+ = 1 - \tfrac{1}{2}ue_1 \wedge n_+ \tag{16}$$

$$T_- = 1 - \tfrac{1}{2}ue_1 \wedge n_- + \tfrac{1}{4}u^2 n_+ \wedge n_- \tag{17}$$

$$T_\times = 1 - \tfrac{1}{2}ue_2 \wedge n_\times \tag{18}$$

for a translation in the direction e_1 around u. Similarly, for a translation in the direction e_2 around v one has

$$T_+ = 1 - \tfrac{1}{2}ve_2 \wedge n_+ \tag{19}$$

$$T_- = 1 + \tfrac{1}{2}ve_2 \wedge n_- \tfrac{1}{4}v^2 n_+ \wedge n_- \tag{20}$$

$$T_\times = 1 - \tfrac{1}{2}ve_1 \wedge n_\times \tag{21}$$

The scalor for a scaling by $\alpha \in \mathbb{R}^+$ is given by $S = S_+ S_- S_\times$, where

$$S_+ = \tfrac{\alpha+1}{2\sqrt{\alpha}} + \tfrac{\alpha-1}{2\sqrt{\alpha}}\bar{n}_+ \wedge n_+, \quad (22)$$

$$S_- = \tfrac{\alpha+1}{2\sqrt{\alpha}} + \tfrac{\alpha-1}{2\sqrt{\alpha}}\bar{n}_- \wedge n_-, \quad (23)$$

$$S_\times = \tfrac{\alpha+1}{2\sqrt{\alpha}} + \tfrac{\alpha-1}{2\sqrt{\alpha}}\bar{n}_\times \wedge n_\times. \quad (24)$$

For proof see [6]. All transformations apply on a vector in GAC by conjugation (4) of the appropriate versor formal exponential. This holds also for translations and rotations, for their precise form see [6]. Consider python implementation of transformations. Now let us demonstrate transformations by visualization on example.

Example 1. Consider axis-aligned ellipse with the semi-axes $a = 2, b = 4$ centred in $(u, v) = (0, 2)$.

The result of applying the rotation by angle $\phi = \frac{\pi}{6}$, scaling by 2 and translations by vectors $(0, 2), (2, 0), (-3, 2)$ respectively is shown in Fig. 1 [2] (Fig. 2).

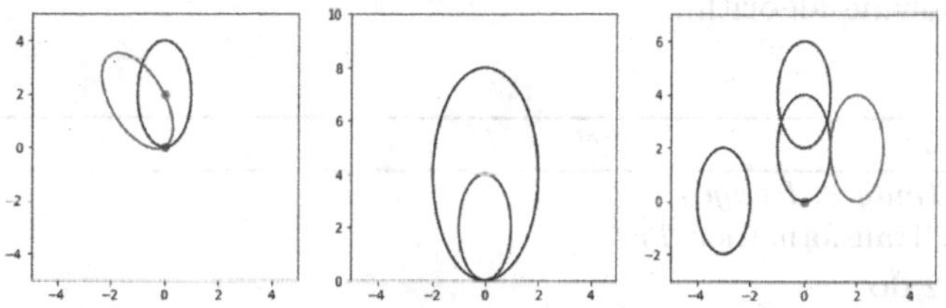

Fig. 1. Transformations

4 Algorithm for Finding the Transformation

Scaling parameter for our transformation can be found by **Algorithm** 1.

Algorithm 1

Inputs:two co–centric ellipses, where one is the scaled copy of another
Output: scaling parameter SP

1: Construct a line l passing through the points $\bar{n}^+$ and e_2:

$$l = e_2 \wedge n_+ \wedge \bar{n}_+ \wedge n_- \wedge n_\times.$$

2: Find the intersection points $C = E_1 \cap l$, $B = E_2 \cap l$ of the line and both ellipses, i.e. solve a quadratic equation in a Euclidean space.
3: The scale parameter between ellipses is

$$SP = \frac{|\bar{n}_+ \cdot \mathcal{C}(B)|}{|\bar{n}_+ \cdot \mathcal{C}(C)|}.$$

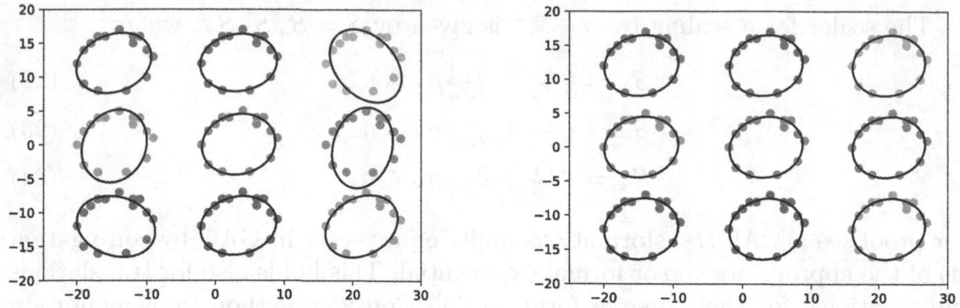

(a) Direct fitting (b) Iterative fitting

Fig. 2. Comparison of conic fitting algorithms applied on a differently translated sample point set. Precision prescribed for the iterative algorithm was $\varepsilon = 10^{-6}$ and, in all cases, 3 or 4 iterations sufficed to achieve the precision.

For more details, see [2]. Therefore, we may find the desired transformation, using the following algorithm.

Algorithm 2

Inputs: $Image_1$, $Image_2$
Output: Transformation Tr

for $i = 1, 2$ **do**
    Upload $Image_i$ and create a binary thresholded $ImageT_i$
    Find the contours in the thresholded $ImageT_i$
    Find the contour ($Contour_i$) enclosing the biggest area
    Apply **Algorithm 1** to $Contour_i$
    Extract parameters $x_c^i, y_c^i, a_i, b_i, \theta_i$ from the fitted conic C_i
    Apply translation T of conic C_i by vector $(-x_c^i, -y_c^i)$
Find the transformation $Tr = SR$ for the conic C_2, where
    R is rotation of a conic around the origin by the angle $\theta_1 - \theta_2$;
    S is scaling of a conic by factor s; which is found in **Algorithm 1**

By applying transformation Tr to the conic C_2 in a way $TrC_2\widetilde{Tr}$ we get the conic, aligned with the C_1. By applying that transformation to the all points in set it is possible to get aligned pictures. To apply the transformation, we need to crop images as circles, co-centered with ellipses and bigger ellipse's semiaxis as circle radius and then to rotate images on the angle difference.

Perceptual hash algorithms describe a class of functions for generating comparable hashes. Image characteristics are used to generate an individual (but not unique) fingerprint, and these fingerprints can be compared with each other. For our research we use perceptual hashes. In cryptography, every hash is random. The data used to generate the hash acts as a source of random numbers, so that the same data will give the same result, but different data will give a different result. If the hashes are different, then the data is different. If the hashes match, then the data is most likely the same (since there is a possibility of collisions, the same hashes do not guarantee that the data will match). In contrast, perceptual hashes can be compared with each other and inferred about the degree of difference between the two datasets [13].

5 Simulation

As a demonstration, let us compare two photos of the same object taken by a drone. The images are shown in the picture Fig. 3a,3b.

(a) Image 1

(b) Image 2

Fig. 3. xx

First let us compare two images using perceptual hashes. The parameter shows the degree of similarity of the images. For similar images the output parameter is up to 15, for different images it is more than 15. Comparing the pictures gives us the result:

```
hash1  1111100111110000111100110011001100011001000110110110111101111111
hash2  0000001111100011111001111111111111110000000111000000111110001111111
result:  30
```

So the hash comparison shows that pictures are very different. Now we will find the transformation and apply it to the image (rotate and translate). Consider the image in Fig. 5. After highlighting the contours, the maximal one was selected, it is shown in Fig. 4b. After fitting the ellipse into the given contour, an ellipse was obtained with the following parameters:

```
S = [285.1305490824287 , 325.1182024501176]
a = 156.15972309402093
b = 89.45721142437789
theta = 139.62977432632437
```

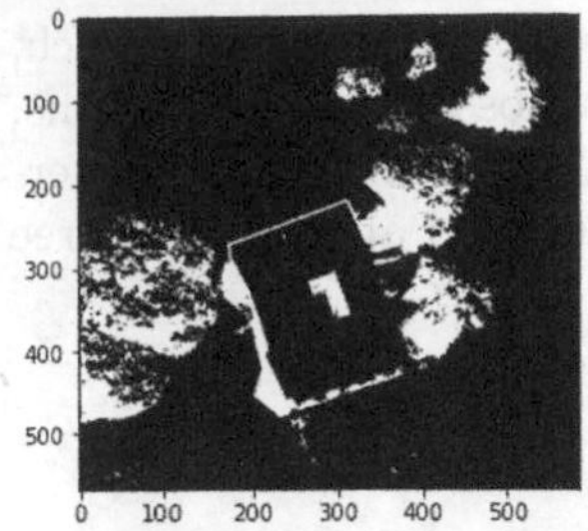

(a) Thresholded Image

(b) Detected Contours

(c) Maximal Contour

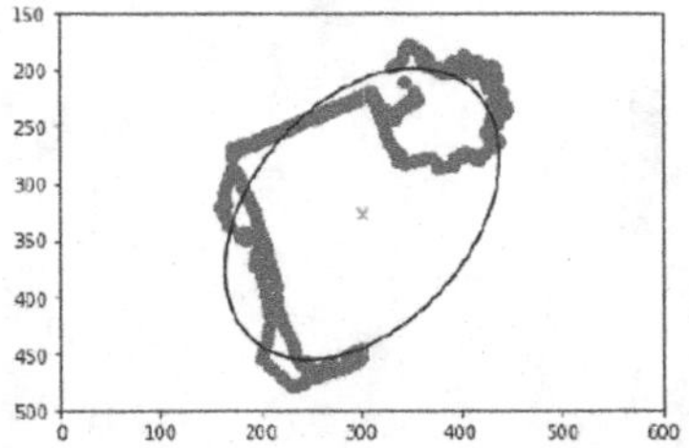

(d) Fitted Conic

Fig. 4. Simulation

We carry out a similar procedure with another image and, as a result, we obtain an ellipse with the following parameters:

```
S = [348.97305779468377 , 275.565547053376]
a = 152.99108015608545
b = 93.58490358200152
theta = 47.82948709748718
```

Therefore, the necessary translation for moving the ellipse center to the origin is $T = T(-209.28270157371304, -219.21051222645468)$.

Note that transferring conics to the origin was necessary for the scaling procedure to work correctly. After cropping images as circles we get the following photos. Now we can rotate images on the angle difference, i.e. 91.8002872283719 degrees.

Now the parameter is already equal to 0, so objects on the pictures are found to be same (Fig. 6).

```
hash1  1111101111111001111110000011010000110100100111001101100111001111
hash2  1111101111111001111110000011010000110100100111001101100111001111
result:  0
```

Therefore objects on the pictures are found to be the same object.

Perceptual hashes method works well in the case when the pictures have different sizes, and it also works when there is a little noise in the image (the

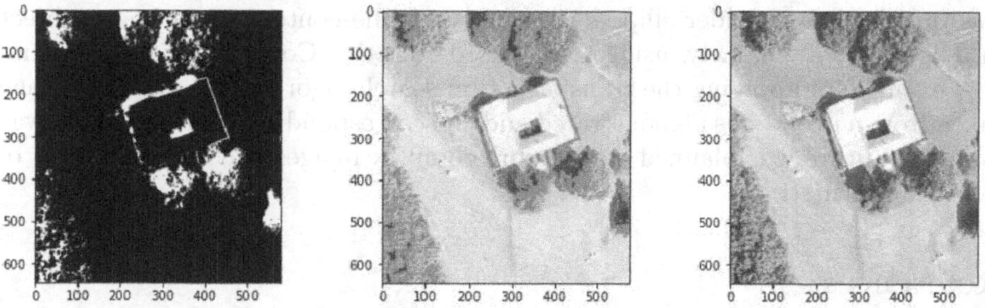

(a) Thresholded Image (b) Detected Contours (c) Maximal Contour

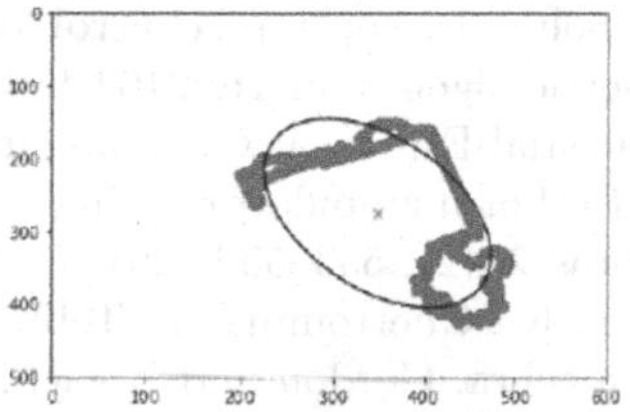

(d) Fitted Conic

Fig. 5. Simulation

(a) Image 1 after object cutting (b) Image 2 after object cutting (c) Image 2 after object cutting and rotation

Fig. 6. xxx

noise disappears when the picture is scaled). Using our transformation (rotation, translation and scaling) we solve the problem of rotated images, that are usually found to be different by this method.

6 Conclusion

The algorithm, searching for transformation, consisting of translation, rotation and scaling, allowing one object to be aligned to another, was presented. It was proposed to consider ellipses inscribed into the contour of an image object in a specific way, namely, using GAC-based Iterative Conic Fitting Algorithm. The example of applying the transformation search algorithm on the real image taken from a drone was demonstrated and the corresponding transformation was found. In future it is planned to use more complex images and extend the list of used transformations.

References

1. Andersson, O., M., S.R.: A comparison of object detection algorithms using unmanipulated testing images: Comparing SIFT, KAZE, AKAZE and ORB (2016)
2. Derevianko, A., Vašík, P.: Solver-free optimal control for linear dynamical switched system by means of geometric algebra. arXiv:2103.13803 [math.OC]
3. Gonzalez-Jimenez, L., Carbajal-Espinosa, O., Loukianov, A., Bayro-Corrochano, E.: Robust pose control of robot manipulators using conformal geometric algebra. Adv. Appl. Clifford Algebras **24**(2), 533–552 (2014)
4. Hamadouche, M., Zebbiche, K., Guerroumi, M., Tebbi, H., Zafoune, Y
5. Hestenes, D.: Space-time algebra. Gordon and Breach, New York (1966)
6. Hrdina, J., Návrat, A., Vašík, P.: Geometric algebra for conics. Adv. Appl. Clifford Algebras **28**(66) (2018). https://doi.org/10.1007/s00006-018-0879-2
7. Hrdina, J., Návrat, A., Vašík, P., Matoušek, R.: Geometric algebras for uniform colour spaces. Math. Meth. Appl. Sci. (2017). https://doi.org/10.1002/mma.4489
8. Khazari, A.E., Que, Y., Sung, T.L., Lee, H.J.: Deep global features for point cloud alignment. Sensors 20(14) (2020). https://doi.org/10.3390/s20144032. https://www.mdpi.com/1424-8220/20/14/4032
9. Korn, G., Korn, T.: Mathematical Handbook for Scientists and Engineers: Definitions, Theorems, and Formulas for Reference and Review. Dover Civil and Mechanical Engineering Series, Dover Publications (2000). https://books.google.cz/books?id=xHNd5zCXt-EC
10. Kozat, S., Venkatesan, R., Mihcak, M.: Robust perceptual image hashing via matrix invariants. In: 2004 International Conference on Image Processing, 2004. ICIP 2004, vol. 5, pp. 3443–3446 (2004). https://doi.org/10.1109/ICIP.2004.1421855
11. Lounesto, P.: Clifford Algebra and Spinors. CUP, 2nd edn., Cambridge (2006)
12. Mojsilovi, R., Gomes, J., Rogowitz, B.: Isee: Perceptual features for image library navigation. Proceedings of SPIE - The International Society for Optical Engineering 4662 (05 2002). DOI: https://doi.org/10.1117/12.469523
13. Niu, X.M., Jiao, Y.H.: Overview of perceptual hashing 36, 1405–1411 (07 2008)
14. Perwass, C.: Geometric Algebra with Applications in Engineering. Springer Verlag (2009)

Line-Bound Vectors, Plane-Bound Bivectors and Tetrahedra in the Conformal Model of Three-Dimensional Space

Timothy F. Havel(✉)

MIT, 77 Massachusetts Avenue, Cambridge, MA 02139, USA
tfhavel@gmail.com

Abstract. The representation of some elementary geometric concepts in the conformal geometric algebra of three dimensions are reviewed, and their connections to a recently discovered extension of Heron's formula for the area of a triangle to the volume of a tetrahedron are discussed.

Keywords: Conformal Model · Heron's Formula · Plücker Identity · Tetrahedron

1 Background and Introduction

The intuitions which inspired Hermann Günther Grassmann to invent the inner and outer products of vectors and points are seldom emphasized in modern presentations of geometric algebra [7]. This is unfortunate because they still have something to offer, both pedagogically and as a source of inspiration for its applications and further developments. I have recently had the rare pleasure of experiencing some of Grassmann's spirit in the course of discovering a natural extension of Heron's formula to the tetrahedron, and envisioning how those same geometric principles can be applied in higher dimensions [6]. The algebra was tough going, but the payoff was in the geometric insights it led to.

In order to emphasize the elementary nature of those results and render them accessible to the widest possible audience, I eschewed the use of modern geometric algebra in that work and confined myself to the better-known vector algebra of Gibbs and Heaviside. Nevertheless, the conformal model would seem to be the ideal framework within which to build upon the geometric insights obtained in three dimensions in order to extend Heron's formula to yet higher dimensions. Doing so, however, will require some combinatorial analysis in addition to algebraic and geometric, and as such is outside the scope of the present paper. Instead, some of the techniques utilized in Ref. [6] will here be reformulated within the conformal model [2,9], with emphasis on their intuitive aspects.

2 Barycentric and Affine Sums in the Conformal Model

One curious feature of the 5D conformal model is that, although it contains the 4D homogeneous model as a subspace, barycentric sums in that subspace do

D. W. Silva et al. (Eds.): ICACGA 2022, LNCS 13771, pp. 27–39, 2024.
https://doi.org/10.1007/978-3-031-34031-4_3

not correspond to weighted sums of the corresponding conformal points.[1] This is because a naive barycentric sum of conformal points $\boldsymbol{a} = \boldsymbol{n}_0 + \mathbf{a} + \boldsymbol{n}_\infty \mathbf{a}^2/2$, $\boldsymbol{b} = \boldsymbol{n}_0 + \mathbf{b} + \boldsymbol{n}_\infty \mathbf{b}^2/2$ ($\boldsymbol{n}_0^2 = \boldsymbol{n}_\infty^2 = 0$, $\boldsymbol{n}_0 \cdot \boldsymbol{n}_\infty = -1$) is not itself null vector:

$$(\beta_\mathsf{a}\, \boldsymbol{a} + \beta_\mathsf{b}\, \boldsymbol{b})^2 = 2\,\beta_\mathsf{a}\,\beta_\mathsf{b}\, \boldsymbol{a} \cdot \boldsymbol{b} = -\,\beta_\mathsf{a}\beta_\mathsf{b}\, \|\mathbf{a} - \mathbf{b}\|^2 \qquad (\beta_\mathsf{a} + \beta_\mathsf{b} = 1) \tag{1}$$

Instead a non-linear correction term has to be added on, specifically:

$$\beta_\mathsf{a}\, \boldsymbol{a} + \beta_\mathsf{b}\, \boldsymbol{b} + \beta_\mathsf{a}\beta_\mathsf{b}\, \boldsymbol{a} \cdot \boldsymbol{b}\, \boldsymbol{n}_\infty = \boldsymbol{n}_0 + \beta_\mathsf{a}\, \mathbf{a} + \beta_\mathsf{b}\, \mathbf{b} + \tfrac{1}{2} \|\beta_\mathsf{a}\, \mathbf{a} + \beta_\mathsf{b}\, \mathbf{b}\|^2\, \boldsymbol{n}_\infty \tag{2}$$

For a general indexed sum $\sum_{i=1}^N \beta_i\, \boldsymbol{a}_i$ of points with $\sum_{i=1}^N \beta_i = 1$, the corresponding formula is:

$$\sum_{i=1}^N \beta_i\, \boldsymbol{a}_i + \boldsymbol{n}_\infty \sum_{j>i=1}^N \beta_i\beta_j\, \boldsymbol{a}_i \cdot \boldsymbol{a}_j = \boldsymbol{n}_0 + \sum_{i=1}^N \beta_i\, \mathbf{a}_i + \frac{1}{2} \left\| \sum_{i=1}^N \beta_i\, \mathbf{a}_i \right\|^2 \boldsymbol{n}_\infty \tag{3}$$

Far from being an encumbrance, this correction contains valuable metrical information. For example, if $\boldsymbol{b}$ is such a corrected barycentric sum of the $\boldsymbol{a}_i$ we can easily derive Lagrange's first identity [4] for the radius of gyration of $\{\mathbf{a}_i\}$ about the barycenter $\mathbf{b}$ thereof as follows:

$$\begin{aligned} 0 = \boldsymbol{b}^2 &= \boldsymbol{b} \cdot \Big(\textstyle\sum_i \beta_i\, \boldsymbol{a}_i + \boldsymbol{n}_\infty \sum_{j>i} \beta_i\beta_j\, \boldsymbol{a}_i \cdot \boldsymbol{a}_j \Big) \\ &= \textstyle\sum_i \beta_i\, \boldsymbol{a}_i \cdot \boldsymbol{b} - \sum_{j>i} \beta_i\beta_j\, \boldsymbol{a}_i \cdot \boldsymbol{a}_j \\ \Longrightarrow\ \textstyle\sum_i \beta_i\, \|\mathbf{a}_i - \mathbf{b}\|^2 &= \textstyle\sum_{j>i} \beta_i\beta_j\, \|\mathbf{a}_i - \mathbf{a}_j\|^2 \end{aligned} \tag{4}$$

We can also derive Lagrange's second identity, which connects these distances with those to an arbitrary third point $\boldsymbol{c}$, namely:

$$\begin{aligned} \boldsymbol{c} \cdot \boldsymbol{b} &= \textstyle\sum_i \beta_i\, \boldsymbol{c} \cdot \boldsymbol{a}_i + \boldsymbol{c} \cdot \boldsymbol{n}_\infty \sum_{j>i} \beta_i\beta_j\, \boldsymbol{a}_i \cdot \boldsymbol{a}_j \\ \Longrightarrow\ \sum_i \beta_i\, \|\mathbf{a}_i - \mathbf{c}\|^2 &= \|\mathbf{b} - \mathbf{c}\|^2 + \textstyle\sum_{j>i} \beta_i\beta_j\, \|\mathbf{a}_i - \mathbf{a}_j\|^2 \end{aligned} \tag{5}$$

Finally, from Lagrange's first and second identities together we get the Huygens-Leibniz identity:

$$\textstyle\sum_i \beta_i\, \|\mathbf{a}_i - \mathbf{c}\|^2 = \|\mathbf{b} - \mathbf{c}\|^2 + \sum_i \beta_i\, \|\mathbf{a}_i - \mathbf{b}\|^2 \tag{6}$$

If we are given a second barycentric sum $\mathbf{b}' = \sum_i \beta'_i\, \mathbf{a}_i$, the squared distance between the two is

$$\|\mathbf{b} - \mathbf{b}'\|^2 = -2\, \boldsymbol{b} \cdot \boldsymbol{b}' = -2\, \boldsymbol{b} \cdot \textstyle\sum_j \beta'_j\, \boldsymbol{a}_j - 2\, \boldsymbol{b} \cdot \boldsymbol{n}_\infty \sum_{j>i} \beta'_i\beta'_j\, \boldsymbol{a}_i \cdot \boldsymbol{a}_j\,, \tag{7}$$

but since $\boldsymbol{n}_\infty \cdot \boldsymbol{a}_k = -1$ for all k and $\sum_k \beta'_k = 1$, we have $\boldsymbol{b} \cdot \sum_j \beta'_j\, \boldsymbol{a}_j =$

$$\begin{aligned} &\textstyle\sum_{i,j} \beta_i\beta'_j\, \boldsymbol{a}_i \cdot \boldsymbol{a}_j + \big(\sum_{j>i} \beta_i\beta_j\, \boldsymbol{a}_i \cdot \boldsymbol{a}_j \big) \big(\sum_k \beta'_k\, \boldsymbol{n}_\infty \cdot \boldsymbol{a}_k \big) \\ &\qquad = \textstyle\sum_{i,j} \big(\beta_i\beta'_j - \tfrac{1}{2}\, \beta_i\beta_j \big)\, \boldsymbol{a}_i \cdot \boldsymbol{a}_j\,. \end{aligned} \tag{8}$$

[1] Such sums can nonetheless be interpreted as a "pencil of coaxial spheres" (cf. e.g. §15.2.4 of Ref. [2]).

Together with Eq. (7), this yields: $\|\mathbf{b}-\mathbf{b}'\|^2 =$

$$\begin{aligned}\textstyle\sum_{i,j}(-2\,\beta_i\beta_j' + \beta_i\beta_j + \beta_i'\beta_j')\,\boldsymbol{a}_i\cdot\boldsymbol{a}_j &= \textstyle\sum_{i,j}(\beta_i-\beta_j)(\beta_i'-\beta_j')\,\boldsymbol{a}_i\cdot\boldsymbol{a}_j \\ &= -\tfrac{1}{2}\textstyle\sum_{i,j}(\beta_i-\beta_i')(\beta_j-\beta_j')\,\|\mathbf{a}_i-\mathbf{a}_j\|^2\end{aligned} \tag{9}$$

This last expression is known as *Schönberg's quadratic form* [1]. It is well known that it is non-negative for all values of the variables $\delta_i = \beta_i - \beta_i'$ with $\sum_i \delta_i = 0$ if and only if the coefficients $\|\mathbf{a}_i - \mathbf{a}_j\|^2$ are indeed the squared distances among a system of points in Euclidean space. The geometric interpretation of this form as being itself a squared distance shows very clearly why this is so.

In the homogeneous model the difference between two points $\boldsymbol{n}_0+\mathbf{a}$ and $\boldsymbol{n}_0+\mathbf{b}$ yields a "free" vector $\mathbf{v} = \mathbf{b} - \mathbf{a}$. It acts as a translation by addition, producing in particular $\boldsymbol{n}_0 + \mathbf{b} = \boldsymbol{n}_0 + \mathbf{a} + \mathbf{v}$. The difference between two *conformal* points $\boldsymbol{a}$ and $\boldsymbol{b}$, however, contains another term proportional to $\boldsymbol{n}_\infty$, namely

$$\boldsymbol{v} := \boldsymbol{b} - \boldsymbol{a} = \mathbf{v} + \tfrac{1}{2}(\mathbf{b}^2 - \mathbf{a}^2)\,\boldsymbol{n}_\infty\,. \tag{10}$$

Even though $\boldsymbol{b} = \boldsymbol{a} + \boldsymbol{v}$ and $\boldsymbol{v}^2 = \|\mathbf{b} - \mathbf{a}\|^2$, the addition of $\boldsymbol{v}$ to an conformal point does *not* yield the translated point in general. Instead, multiplying it from the right by $\boldsymbol{n}_\infty$ yields the bivector $\mathbf{v}\boldsymbol{n}_\infty$, which in turn generates a rotor $\exp(\mathbf{v}\boldsymbol{n}_\infty) = 1 + \mathbf{v}\boldsymbol{n}_\infty$ that translates a point $\boldsymbol{c} = \boldsymbol{n}_0 + \mathbf{c} + \boldsymbol{n}_\infty\mathbf{c}^2/2$ via the usual multiplicative two-sided action:

$$(1 + \mathbf{v}\boldsymbol{n}_\infty)\,\boldsymbol{c}\,(1 + \boldsymbol{n}_\infty\mathbf{v}) = \boldsymbol{n}_0 + \mathbf{c} + \mathbf{v} + \tfrac{1}{2}\boldsymbol{n}_\infty\,(\mathbf{c} + \mathbf{v})^2\,. \tag{11}$$

The difference between two "flat" points $\boldsymbol{n}_\infty \wedge \boldsymbol{b} - \boldsymbol{n}_\infty \wedge \boldsymbol{a} = \boldsymbol{n}_\infty\mathbf{v}$ generates this same rotor $\exp(-\boldsymbol{n}_\infty\mathbf{v})$, and more generally any affine sum of flat points $\sum_i \delta_i(\boldsymbol{n}_\infty \wedge \boldsymbol{a}_i) = \boldsymbol{n}_\infty \sum_i \delta_i \mathbf{a}_i = \boldsymbol{n}_\infty\mathbf{v}$ with $\sum_i \delta_i = 0$ also generates a translation.

3 Line-Bound Vectors and Tetrahedra

The next step up from flat points are *line-bound* vectors, which have the form

$$\begin{aligned}\boldsymbol{n}_\infty \wedge \boldsymbol{a} \wedge \boldsymbol{b} &= \boldsymbol{n}_\infty \wedge (\boldsymbol{n}_0 + \mathbf{a} + \mathbf{a}^2\boldsymbol{n}_\infty/2) \wedge (\boldsymbol{n}_0 + \mathbf{b} + \mathbf{a}^2\boldsymbol{n}_\infty/2) \\ &= \boldsymbol{n}_\infty \wedge (\boldsymbol{n}_0 + \mathbf{a}) \wedge (\boldsymbol{n}_0 + \mathbf{b}) = \boldsymbol{n}_\infty \wedge \big(\boldsymbol{n}_0 \wedge (\mathbf{b} - \mathbf{a}) + \mathbf{a} \wedge \mathbf{b}\big) \\ &= \boldsymbol{N} \wedge (\mathbf{b} - \mathbf{a}) + \boldsymbol{n}_\infty \wedge \mathbf{a} \wedge \mathbf{b} = \boldsymbol{N}(\mathbf{b} - \mathbf{a}) + \boldsymbol{n}_\infty\,\mathbf{a} \wedge \mathbf{b}\end{aligned} \tag{12}$$

with $\boldsymbol{N} := \boldsymbol{n}_\infty \wedge \boldsymbol{n}_0$. The inner product of this with $\boldsymbol{n}_0$ is a bivector $(\mathbf{b} - \mathbf{a})\boldsymbol{n}_\infty - \mathbf{a} \wedge \mathbf{b}$ which generates a Euclidean "screw" motion (translation and rotation about an axis thereof). This line-bound vector itself is determined by an oriented segment $[\mathbf{a}, \mathbf{b}]$ of a line in space, where the line's direction is $(\mathbf{b} - \mathbf{a})/\|\mathbf{b} - \mathbf{a}\|$ and its (minimum) distance from the origin is $\|\mathbf{a} \wedge \mathbf{b}\|/(\|\mathbf{b} - \mathbf{a}\|)$. The position of this oriented line segment on the line is however indeterminate, or alternatively, the line-bound vector only determines the equivalence class of all collinear oriented line segments with length $\|\mathbf{b} - \mathbf{a}\|$.

The length of the oriented line segment may be obtained by squaring the line-bound vector, which yields a simple example of what is widely known as a Cayley-Menger determinant [1,3,5]:

$$(\boldsymbol{n}_\infty \wedge \boldsymbol{a} \wedge \boldsymbol{b})^2 = -(\boldsymbol{b} \wedge \boldsymbol{a} \wedge \boldsymbol{n}_\infty) \cdot (\boldsymbol{n}_\infty \wedge \boldsymbol{a} \wedge \boldsymbol{b}) = \tag{13}$$
$$-\det\begin{bmatrix} \boldsymbol{n}_\infty \cdot \boldsymbol{n}_\infty & \boldsymbol{n}_\infty \cdot \boldsymbol{a} & \boldsymbol{n}_\infty \cdot \boldsymbol{b} \\ \boldsymbol{n}_\infty \cdot \boldsymbol{a} & \boldsymbol{a} \cdot \boldsymbol{a} & \boldsymbol{a} \cdot \boldsymbol{b} \\ \boldsymbol{n}_\infty \cdot \boldsymbol{b} & \boldsymbol{a} \cdot \boldsymbol{b} & \boldsymbol{b} \cdot \boldsymbol{b} \end{bmatrix} = \det\begin{bmatrix} 0 & 1 & 1 \\ 1 & 0 & \frac{1}{2}\|\mathbf{a}-\mathbf{b}\|^2 \\ 1 & \frac{1}{2}\|\mathbf{a}-\mathbf{b}\|^2 & 0 \end{bmatrix}$$
$$= \tfrac{1}{2}\,\|\mathbf{a}-\mathbf{b}\|^2 + \tfrac{1}{2}\,\|\mathbf{a}-\mathbf{b}\|^2 = \|\mathbf{a}-\mathbf{b}\|^2$$

A little more generally, the inner product of two line-bound vectors is the inner product of the corresponding free vectors:

$$-(\boldsymbol{b} \wedge \boldsymbol{a} \wedge \boldsymbol{n}_\infty) \cdot (\boldsymbol{n}_\infty \wedge \boldsymbol{c} \wedge \boldsymbol{d}) =$$
$$-\tfrac{1}{2}\left(\|\mathbf{a}-\mathbf{c}\|^2 - \|\mathbf{a}-\mathbf{d}\|^2 - \|\mathbf{b}-\mathbf{c}\|^2 + \|\mathbf{b}-\mathbf{d}\|^2\right) = (\mathbf{b}-\mathbf{a}) \cdot (\mathbf{d}-\mathbf{c}) \tag{14}$$

In order to obtain the higher-grade parts the geometric product of two such line-bound vectors, we expand it via Eq. (12) to get:

$$\begin{aligned} &(\boldsymbol{n}_\infty \wedge \boldsymbol{a} \wedge \boldsymbol{b})(\boldsymbol{n}_\infty \wedge \boldsymbol{c} \wedge \boldsymbol{d}) \\ &= \big(\boldsymbol{N}(\mathbf{b}-\mathbf{a}) + \boldsymbol{n}_\infty(\mathbf{a} \wedge \mathbf{b})\big)\big(\boldsymbol{N}(\mathbf{d}-\mathbf{c}) + \boldsymbol{n}_\infty(\mathbf{c} \wedge \mathbf{d})\big) \\ &= (\mathbf{b}-\mathbf{a})(\mathbf{d}-\mathbf{c}) + (\boldsymbol{n}_\infty \cdot \boldsymbol{N})(\mathbf{a} \wedge \mathbf{b})(\mathbf{d}-\mathbf{c}) - (\boldsymbol{N} \cdot \boldsymbol{n}_\infty)(\mathbf{b}-\mathbf{a})(\mathbf{c} \wedge \mathbf{d}) \\ &= (\mathbf{b}-\mathbf{a})(\mathbf{d}-\mathbf{c}) + \boldsymbol{n}_\infty(\mathbf{a} \wedge \mathbf{b})(\mathbf{d}-\mathbf{c}) + \boldsymbol{n}_\infty(\mathbf{b}-\mathbf{a})(\mathbf{c} \wedge \mathbf{d}) \end{aligned} \tag{15}$$

The two-vector part of this, which is easily seen to be the same as the commutator product of the line-bound vectors, is:

$$\big\langle(\boldsymbol{n}_\infty \wedge \boldsymbol{a} \wedge \boldsymbol{b})(\boldsymbol{n}_\infty \wedge \boldsymbol{c} \wedge \boldsymbol{d})\big\rangle_2 = \tag{16}$$
$$(\mathbf{b}-\mathbf{a}) \wedge (\mathbf{d}-\mathbf{c}) + \boldsymbol{n}_\infty\big((\mathbf{a} \wedge \mathbf{b}) \cdot (\mathbf{d}-\mathbf{c}) + (\mathbf{b}-\mathbf{a}) \cdot (\mathbf{c} \wedge \mathbf{d})\big)$$

Finally, the four-vector part of this product of line-bound vectors is:

$$\begin{aligned} &\big\langle(\boldsymbol{n}_\infty \wedge \boldsymbol{a} \wedge \boldsymbol{b})(\boldsymbol{n}_\infty \wedge \boldsymbol{c} \wedge \boldsymbol{d})\big\rangle_4 \\ &\quad = \boldsymbol{n}_\infty\,\big((\mathbf{a} \wedge \mathbf{b}) \wedge (\mathbf{d}-\mathbf{c}) + (\mathbf{b}-\mathbf{a}) \wedge (\mathbf{c} \wedge \mathbf{d})\big) \\ &\quad = \boldsymbol{n}_\infty\,\big(\mathbf{a} \wedge \mathbf{b} \wedge \mathbf{d} - \mathbf{a} \wedge \mathbf{b} \wedge \mathbf{c} + \mathbf{b} \wedge \mathbf{c} \wedge \mathbf{d} - \mathbf{a} \wedge \mathbf{c} \wedge \mathbf{d}\big) \\ &\quad = \boldsymbol{n}_\infty\,\big((\mathbf{b}-\mathbf{a}) \wedge (\mathbf{c}-\mathbf{a}) \wedge (\mathbf{d}-\mathbf{a})\big) \end{aligned} \tag{17}$$

This may be interpreted as $3!\,\boldsymbol{n}_\infty$ times the volume of the oriented tetrahedron $[\mathbf{a},\mathbf{b},\mathbf{c},\mathbf{d}]$. Since the line-bound vectors do not change when $[\mathbf{a},\mathbf{b}]$ and $[\mathbf{c},\mathbf{d}]$ are translated along their respective lines, we see that the volume of the tetrahedron is also unchanged by such translations, as well as translations of its other edges along the lines they span.

Letting $\boldsymbol{L} := \boldsymbol{n}_\infty \wedge \boldsymbol{a} \wedge \boldsymbol{b}$ and $\boldsymbol{M} := \boldsymbol{n}_\infty \wedge \boldsymbol{c} \wedge \boldsymbol{d}$, the product of $\boldsymbol{LM}$ with its reverse is easily seen to be just

$$\boldsymbol{L}\boldsymbol{M}\tilde{\boldsymbol{M}}\tilde{\boldsymbol{L}} = \tilde{\boldsymbol{M}}\tilde{\boldsymbol{L}}\boldsymbol{L}\boldsymbol{M} = \|\mathbf{b}-\mathbf{a}\|^2\,\|\mathbf{d}-\mathbf{c}\|^2\,. \tag{18}$$

This product, however, can also be written as

$$\begin{aligned} \boldsymbol{LM\tilde{M}\tilde{L}} \;&=\; \langle \boldsymbol{LM}\rangle_0 \langle \boldsymbol{LM}\rangle_0^\sim + \langle \boldsymbol{LM}\rangle_2 \langle \boldsymbol{LM}\rangle_2^\sim + \langle \boldsymbol{LM}\rangle_4 \langle \boldsymbol{LM}\rangle_4^\sim \\ &\quad + 2\,\langle \boldsymbol{LM}\rangle_0 \langle \boldsymbol{LM}\rangle_2^\sim + 2\,\langle \boldsymbol{LM}\rangle_0 \langle \boldsymbol{LM}\rangle_4^\sim + 2\,\langle \boldsymbol{LM}\rangle_2 \langle \boldsymbol{LM}\rangle_4^\sim \end{aligned} \tag{19}$$

Because $\langle \boldsymbol{LM}\rangle_4$ contains a factor of $\boldsymbol{n}_\infty$ by Eq. (17), $\langle \boldsymbol{LM}\rangle_4 \langle \boldsymbol{LM}\rangle_4^\sim = 0$. In addition, by Eq. (14), $\langle \boldsymbol{LM}\rangle_0 = (\mathbf{b}-\mathbf{a})\cdot(\mathbf{d}-\mathbf{c})$, while the last three terms on the right-hand side of Eq. (19) contain no scalar part. It follows that the magnitude of $\langle \boldsymbol{LM}\rangle_2$ is:

$$\begin{aligned} \left\|\langle \boldsymbol{LM}\rangle_2\right\|^2 = \langle \boldsymbol{LM}\rangle_2 * \langle \boldsymbol{LM}\rangle_2^\sim \;:=\; \left\langle \langle \boldsymbol{LM}\rangle_2 \langle \boldsymbol{LM}\rangle_2^\sim \right\rangle_0 \;=& \\ \|\mathbf{b}-\mathbf{a}\|^2\,\|\mathbf{d}-\mathbf{c}\|^2 \;-\; \left((\mathbf{b}-\mathbf{a})\cdot(\mathbf{d}-\mathbf{c})\right)^2 \;=\; \|(\mathbf{b}-\mathbf{a})\wedge(\mathbf{d}-\mathbf{c})\|^2 & \end{aligned} \tag{20}$$

Geometrically, this quantity is the square of 4 times the area of the *medial parallelogram* $[(\mathbf{a}+\mathbf{c})/2, (\mathbf{c}+\mathbf{b})/2, (\mathbf{b}+\mathbf{d})/2, (\mathbf{d}+\mathbf{a})/2]$ of the tetrahedraon $[\mathbf{a},\mathbf{b},\mathbf{c},\mathbf{d}]$. To see that this is indeed a parallelogram, note first that

$$\begin{aligned} (\mathbf{a}+\mathbf{c})\wedge(\mathbf{c}+\mathbf{b})\wedge(\mathbf{b}+\mathbf{d})\wedge(\mathbf{d}+\mathbf{a}) =& \\ \mathbf{a}\wedge\mathbf{c}\wedge\mathbf{b}\wedge\mathbf{d} + \mathbf{c}\wedge\mathbf{b}\wedge\mathbf{d}\wedge\mathbf{a} \;=\; 0 & \end{aligned} \tag{21}$$

since the other 14 of the 16 blades obtained on expanding the left-hand side contain a repeated factor. Thus the midpoints of the four edges $[\mathbf{a},\mathbf{c}]$, $[\mathbf{c},\mathbf{b}]$, $[\mathbf{b},\mathbf{d}]$, $[\mathbf{d},\mathbf{a}]$ are coplanar. Furthermore, the outer product of the vectors from the midpoint of any edge, say $(\mathbf{a}+\mathbf{c})/2$, to the midpoints of its adjacent edges $(\mathbf{c}+\mathbf{b})/2$ and $(\mathbf{d}+\mathbf{a})/2$, is simply

$$\tfrac{1}{4}\left((\mathbf{c}+\mathbf{b})-(\mathbf{a}+\mathbf{c})\right)\wedge\left((\mathbf{d}+\mathbf{a})-(\mathbf{a}+\mathbf{c})\right) \;=\; \tfrac{1}{4}(\mathbf{b}-\mathbf{a})\wedge(\mathbf{d}-\mathbf{c})\,, \tag{22}$$

whence this paragraph's first assertion follows.

In general, the sum of line-bound vectors is not itself a line-bound vector, but a composite entity which can be interpreted in various ways. Classical interpretations include the result of a system of forces acting at various points on a rigid body, and the result of a system of infinitesimal motions applied to a rigid body. We will not develop these theories here, but refer to interested reader to the extensive literature on the subject [8]. Instead we shall seek to interpret the graded components of the product of such an entity with itself in the context of the conformal model, as above.

Hence consider an arbitrary composite line-bound vector of the form:

$$\boldsymbol{L}+\boldsymbol{M} \;=\; \boldsymbol{n}_\infty\wedge\boldsymbol{a}\wedge\boldsymbol{b} + \boldsymbol{n}_\infty\wedge\boldsymbol{c}\wedge\boldsymbol{d} \;=\; \boldsymbol{n}_\infty\wedge(\boldsymbol{a}\wedge\boldsymbol{b}+\boldsymbol{c}\wedge\boldsymbol{d}) \tag{23}$$

By Eq. (14), the inner square of this expands to:

$$\begin{aligned} (\boldsymbol{n}_\infty\wedge(\boldsymbol{a}\wedge\boldsymbol{b}+\boldsymbol{c}\wedge\boldsymbol{d}))\cdot(\boldsymbol{n}_\infty\wedge(\boldsymbol{a}\wedge\boldsymbol{b}+\boldsymbol{c}\wedge\boldsymbol{d})) \;=& \\ (\mathbf{b}-\mathbf{a})^2 + 2\,(\mathbf{b}-\mathbf{a})\cdot(\mathbf{d}-\mathbf{c}) + (\mathbf{d}-\mathbf{c})^2 \;=\; 4\,\|(\mathbf{b}+\mathbf{d})/2-(\mathbf{a}+\mathbf{c})/2\|^2 & \end{aligned} \tag{24}$$

Thus we see that the length of this diagonal of the medial parallelogram (which is also a bimedian of the tetrahedron) remains unchanged as $[\mathbf{a},\mathbf{b}]$ and $[\mathbf{c},\mathbf{d}]$

are translated along their respective lines. The lengths of its sides $\|\mathbf{b}-\mathbf{a}\|/2$ and $\|\mathbf{d}-\mathbf{c}\|/2$ are of course also invariant under such translations. Finally, Eq. (16) shows that the bivector of the medial parallelogram is likewise invariant. Together, these observations prove that translations of $[\mathbf{a},\mathbf{b}]$ and $[\mathbf{c},\mathbf{d}]$ along their respective lines merely translate the corresponding medial parallelogram in space without changing its shape or aspect.

The anti-symmetry of Eq. (16) under the $(\mathbf{a},\mathbf{b}) \leftrightarrow (\mathbf{c},\mathbf{d})$ swap shows that the 2-vector part of the square of a composite line-bound vector vanishes. The 4-vector part, however, is:

$$\left\langle (\boldsymbol{n}_\infty \wedge (\boldsymbol{a}\wedge\boldsymbol{b}+\boldsymbol{c}\wedge\boldsymbol{d}))^2 \right\rangle_4 = 2\,\boldsymbol{n}_\infty\,((\mathbf{b}-\mathbf{a})\wedge(\mathbf{c}\wedge\mathbf{d}) + (\mathbf{a}\wedge\mathbf{b})\wedge(\mathbf{d}-\mathbf{c})) \quad (25)$$

This clearly does vanish if $\mathbf{b}-\mathbf{a}=\mathbf{d}-\mathbf{c}$ or $\mathbf{a}\wedge\mathbf{b}=\mathbf{c}\wedge\mathbf{d}$, i.e. the line-bound vector in question is simple (non-composite). On taking its dual with respect to the pseudo-scalar $\boldsymbol{\iota}=\mathbf{e}_1\mathbf{e}_2\mathbf{e}_3$ of the free-vector subspace, we obtain:

$$\begin{aligned} \boldsymbol{\iota}\left\langle (\boldsymbol{n}_\infty \wedge (\boldsymbol{a}\wedge\boldsymbol{b}+\boldsymbol{c}\wedge\boldsymbol{d}))^2 \right\rangle_4 &= \\ 2\,\boldsymbol{n}_\infty\,((\mathbf{b}-\mathbf{a})\cdot(\tilde{\boldsymbol{\iota}}\,\mathbf{c}\wedge\mathbf{d}) + (\tilde{\boldsymbol{\iota}}\,\mathbf{a}\wedge\mathbf{b})\cdot(\mathbf{d}-\mathbf{c})) &= \\ 2\,\boldsymbol{n}_\infty\,((\mathbf{b}-\mathbf{a})\cdot(\mathbf{c}\times\mathbf{d}) + (\mathbf{a}\times\mathbf{b})\cdot(\mathbf{d}-\mathbf{c})) & \end{aligned} \quad (26)$$

The vanishing of the coefficient of $2\,\boldsymbol{n}_\infty$ on the right is a well-known condition for the lines spanned by $[\mathbf{a},\mathbf{b}]$ and $[\mathbf{c},\mathbf{d}]$ to either intersect or be parallel [8]. It is in fact just a way of writing the famous Plücker identity for the six Plücker coordinates of a line in projective three-space.

It is seldom noted, nevertheless, that this same expression is equal to $\pm 3!$ times the volume $|[\mathbf{a},\mathbf{b},\mathbf{c},\mathbf{d}]|$ of the tetrahedron. This follows from the fact that the 4 vector parts of $2\,\boldsymbol{LM}$ and $(\boldsymbol{L}+\boldsymbol{M})^2$ are obviously the same (cf. Eq. (17)). Given this observation, it is perhaps no surprise that the real algebraic variety of all degenerate (volume $=0$) tetrahedra should be intimately related to the Klein quadric $\mathcal{K} := \{\mathbf{x}\in\mathbb{R}^6 \,|\, x_1x_6 - x_2x_5 + x_3x_4 = 0\}$ [6]. A thorough study of the representation of the more general Grassmann-Plücker relations in geometric algebra was recently given by Sobczyk [10].

4 Plane-Bound Bivectors and Tetrahedra

We now consider the relations between two plane-bound bivectors, along with the space-bound trivector (oriented volume) which a flat point and plane-bound bivector mutually define. In the conformal model, a plane-bound bivector has the form:

$$\begin{aligned} \boldsymbol{n}_\infty\wedge\boldsymbol{b}\wedge\boldsymbol{c}\wedge\boldsymbol{d} &= \boldsymbol{n}_\infty\wedge(\boldsymbol{n}_0+\mathbf{b})\wedge(\boldsymbol{n}_0+\mathbf{c})\wedge(\boldsymbol{n}_0+\mathbf{d}) \\ &= \boldsymbol{N}\,(\mathbf{c}\wedge\mathbf{d}-\mathbf{b}\wedge\mathbf{d}+\mathbf{b}\wedge\mathbf{c}) + \boldsymbol{n}_\infty\,\mathbf{b}\wedge\mathbf{c}\wedge\mathbf{d} \\ &= \boldsymbol{N}\,(\mathbf{c}-\mathbf{b})\wedge(\mathbf{d}-\mathbf{b}) + \boldsymbol{n}_\infty\,\mathbf{b}\wedge\mathbf{c}\wedge\mathbf{d} \end{aligned} \quad (27)$$

The free bivector $(\mathbf{c}-\mathbf{b})\wedge(\mathbf{d}-\mathbf{b})$ has a squared magnitude (area) of

$$\|(\mathbf{c}-\mathbf{b})\wedge(\mathbf{d}-\mathbf{b})\|^2 \;=\; \|\mathbf{c}-\mathbf{b}\|^2\|\mathbf{d}-\mathbf{b}\|^2 \;-\; \big((\mathbf{c}-\mathbf{b})\cdot(\mathbf{d}-\mathbf{b})\big)^2 \qquad (28)$$

Its normal vector parallel to $(\mathbf{c}-\mathbf{b})\times(\mathbf{d}-\mathbf{b}) = -\iota\,(\mathbf{c}-\mathbf{b})\wedge(\mathbf{d}-\mathbf{b})$ determines the direction of the plane containing $[\mathbf{b},\mathbf{c},\mathbf{d}]$, while the distance of the plane from the origin is $\|\mathbf{b}\wedge\mathbf{c}\wedge\mathbf{d}\|/(\|(\mathbf{c}-\mathbf{b})\wedge(\mathbf{d}-\mathbf{b})\|)$. The same plane-bound bivector, however, is obtained for any other triple of points $\mathbf{b}'$, $\mathbf{c}'$, $\mathbf{d}'$ that lie in the same plane and span a triangle $[\mathbf{b}',\mathbf{c}',\mathbf{d}']$ with the same area.

The three-point Cayley-Menger determinant obtained by multiplying the plane-bound bivector with its reverse is:

$$\begin{aligned}
&(\boldsymbol{d}\wedge\boldsymbol{c}\wedge\boldsymbol{b}\wedge\boldsymbol{n}_\infty)(\boldsymbol{n}_\infty\wedge\boldsymbol{b}\wedge\boldsymbol{c}\wedge\boldsymbol{d})\\
&\quad= 2\big((\boldsymbol{b}\cdot\boldsymbol{c})(\boldsymbol{b}\cdot\boldsymbol{d}) + (\boldsymbol{b}\cdot\boldsymbol{c})(\boldsymbol{c}\cdot\boldsymbol{d}) + (\boldsymbol{b}\cdot\boldsymbol{d})(\boldsymbol{c}\cdot\boldsymbol{d})\big)\\
&\qquad\qquad - \big((\boldsymbol{b}\cdot\boldsymbol{c})^2 + (\boldsymbol{b}\cdot\boldsymbol{d})^2 + (\boldsymbol{c}\cdot\boldsymbol{d})^2\big)\\
&\quad= \tfrac{1}{4}\big(\|\mathbf{b}-\mathbf{c}\| + \|\mathbf{b}-\mathbf{d}\| + \|\mathbf{c}-\mathbf{d}\|\big)\big(\|\mathbf{b}-\mathbf{c}\| + \|\mathbf{b}-\mathbf{d}\| - \|\mathbf{c}-\mathbf{d}\|\big)\\
&\qquad\big(\|\mathbf{b}-\mathbf{c}\| - \|\mathbf{b}-\mathbf{d}\| + \|\mathbf{c}-\mathbf{d}\|\big)\big(-\|\mathbf{b}-\mathbf{c}\| + \|\mathbf{b}-\mathbf{d}\| + \|\mathbf{c}-\mathbf{d}\|\big) \qquad (29)
\end{aligned}$$

The factorized formula on the right-hand side of this equation is of course *Heron's formula* for the squared area of a triangle $[\mathbf{b},\mathbf{c},\mathbf{d}]$ (times 4). It can be written rather more compactly as $\big|[\mathbf{b},\mathbf{c},\mathbf{d}]\big|^2 = suvw$, where s is half the triangle's perimeter (i.e. the first factor above), and u, v, w are half the deviations of the three triangle inequalities from saturation (i.e. the last three factors). By expanding the squared dot product in Eq. (28) via the law of cosines $\mathbf{x}\cdot\mathbf{y} = (\|\mathbf{x}\|^2 + \|\mathbf{y}\|^2 - \|\mathbf{x}-\mathbf{y}\|^2)/2$, it is readily shown that this three-point Cayley-Menger determinant also equals the free bivector's squared magnitude.

The inner product of different plane-bound bivectors gives a non-symmetric Cayley-Menger determinant. Providing the planes are not parallel, we may assume the point triples $\boldsymbol{a},\boldsymbol{b},\boldsymbol{c}$ and $\boldsymbol{a},\boldsymbol{b},\boldsymbol{d}$ have a pair $\boldsymbol{a},\boldsymbol{b}$ in common, so that:

$$\begin{aligned}
&(\boldsymbol{n}_\infty\wedge\boldsymbol{a}\wedge\boldsymbol{b}\wedge\boldsymbol{c})\cdot(\boldsymbol{n}_\infty\wedge\boldsymbol{a}\wedge\boldsymbol{b}\wedge\boldsymbol{d}) =\\
&\big(\boldsymbol{N}\,(\mathbf{b}-\mathbf{a})\wedge(\mathbf{c}-\mathbf{a}) + \boldsymbol{n}_\infty\,\mathbf{a}\wedge\mathbf{b}\wedge\mathbf{c})\big)\cdot\big(\boldsymbol{N}\,(\mathbf{b}-\mathbf{a})\wedge(\mathbf{d}-\mathbf{a}) + \boldsymbol{n}_\infty\,\mathbf{a}\wedge\mathbf{b}\wedge\mathbf{d})\big)\\
&\qquad = \big((\mathbf{b}-\mathbf{a})\wedge(\mathbf{c}-\mathbf{a})\big)\cdot\big((\mathbf{b}-\mathbf{a})\wedge(\mathbf{d}-\mathbf{a})\big) \qquad (30)
\end{aligned}$$

We also have the trivial identity among free bivectors,

$$(\mathbf{b}-\mathbf{a})\wedge(\mathbf{d}-\mathbf{c}) \;=\; (\mathbf{b}-\mathbf{a})\wedge(\mathbf{d}-\mathbf{a}) \;-\; (\mathbf{b}-\mathbf{a})\wedge(\mathbf{c}-\mathbf{a})\;, \qquad (31)$$

which in turn implies

$$\begin{aligned}
\|(\mathbf{b}-\mathbf{a})\wedge(\mathbf{d}-\mathbf{c})\|^2 \;=\;& \|(\mathbf{b}-\mathbf{a})\wedge(\mathbf{d}-\mathbf{a})\|^2 \;+\; \|(\mathbf{b}-\mathbf{a})\wedge(\mathbf{c}-\mathbf{a})\|^2\\
&-2\big((\mathbf{b}-\mathbf{a})\wedge(\mathbf{d}-\mathbf{a})\big)\cdot\big((\mathbf{b}-\mathbf{a})\wedge(\mathbf{c}-\mathbf{a})\big)\;.
\end{aligned} \qquad (32)$$

The symmetry of the inner product of bivectors together with Eqs. (20), (30) and (32) thus gives us the *areal law of cosines*,

$$\begin{aligned}
&(\boldsymbol{n}_\infty \wedge \boldsymbol{a} \wedge \boldsymbol{b} \wedge \boldsymbol{c}) \cdot (\boldsymbol{n}_\infty \wedge \boldsymbol{a} \wedge \boldsymbol{b} \wedge \boldsymbol{d}) \\
&= \|(\mathbf{b}-\mathbf{a}) \wedge (\mathbf{c}-\mathbf{a})\| \, \|(\mathbf{b}-\mathbf{a}) \wedge (\mathbf{d}-\mathbf{a})\| \, \cos(\varphi_{\mathsf{ab}}) \\
&= \tfrac{1}{2}\left(\|(\mathbf{b}-\mathbf{a}) \wedge (\mathbf{c}-\mathbf{a})\|^2 + \|(\mathbf{b}-\mathbf{a}) \wedge (\mathbf{d}-\mathbf{a})\|^2 - \|(\mathbf{b}-\mathbf{a}) \wedge (\mathbf{d}-\mathbf{c})\|^2\right) \\
&= \tfrac{1}{2}\left(\left\|\boldsymbol{n}_\infty \wedge \boldsymbol{a} \wedge \boldsymbol{b} \wedge \boldsymbol{c}\right\|^2 + \left\|\boldsymbol{n}_\infty \wedge \boldsymbol{a} \wedge \boldsymbol{b} \wedge \boldsymbol{d}\right\|^2 - \left\|\langle \boldsymbol{L}\boldsymbol{M}\rangle_2\right\|^2\right), \quad (33)
\end{aligned}$$

where $\boldsymbol{L}$ and $\boldsymbol{M}$ are the line-bound vectors from Eq. (18) and φ_{ab} is the dihedral angle between the planes spanned by $[\mathbf{a}, \mathbf{b}, \mathbf{c}]$ and $[\mathbf{a}, \mathbf{b}, \mathbf{d}]$.

There is also a corresponding *areal law of sines*, which is obtained from a simple case of the dual (Grassmann's regressive) outer product "$\vee$" which corresponds to the meet of the associated subspaces. Letting $\mathbf{X}$ be an arbitrary non-null blade and $\mathbf{y}, \mathbf{z}$ be vectors with $\mathbf{X} \wedge \mathbf{y}, \mathbf{X} \wedge \mathbf{z} \neq 0$, this may be written as

$$(\mathbf{X} \wedge \mathbf{y}) \vee (\mathbf{X} \wedge \mathbf{z}) \;=\; \mathbf{X}(\mathbf{X} \wedge \mathbf{y} \wedge \mathbf{z})^* \;=\; \mathbf{X}\,\|\mathbf{X} \wedge \mathbf{y} \wedge \mathbf{z}\|\,, \qquad (34)$$

where $(\mathbf{X} \wedge \mathbf{y} \wedge \mathbf{z})^*$ is the dual w.r.t. the unit pseudo-scalar of the subspaces' join. Upon taking norms and recalling [2] that the norm of such a dual outer product is $\|\mathbf{X} \wedge \mathbf{y}\|\|\mathbf{X} \wedge \mathbf{z}\| \sin(\varphi)$ where φ is the angle between the subspaces of $\mathbf{X} \wedge \mathbf{y}$ and $\mathbf{X} \wedge \mathbf{z}$, this implies:

$$\begin{aligned}
&\left\|(\boldsymbol{n}_\infty \wedge \boldsymbol{a} \wedge \boldsymbol{b} \wedge \boldsymbol{c}) \vee (\boldsymbol{n}_\infty \wedge \boldsymbol{a} \wedge \boldsymbol{b} \wedge \boldsymbol{d})\right\| \\
&\quad= \|(\mathbf{b}-\mathbf{a}) \wedge (\mathbf{c}-\mathbf{a})\| \, \|(\mathbf{b}-\mathbf{a}) \wedge (\mathbf{d}-\mathbf{a})\| \, \sin(\varphi_{\mathsf{ab}}) \\
&\quad= \|\mathbf{b}-\mathbf{a}\| \, \left\|(\mathbf{b}-\mathbf{a}) \wedge (\mathbf{c}-\mathbf{a}) \wedge (\mathbf{d}-\mathbf{a})\right\| \\
&\quad= \|\boldsymbol{n}_\infty \wedge \boldsymbol{a} \wedge \boldsymbol{b}\| \, \|\boldsymbol{n}_\infty \wedge \boldsymbol{a} \wedge \boldsymbol{b} \wedge \boldsymbol{c} \wedge \boldsymbol{d}\|
\end{aligned} \qquad (35)$$

where φ_{ab} is the dihedral angle as above.

Now consider the product of a flat point with the plane-bound bivector:

$$\begin{aligned}
&(\boldsymbol{n}_\infty \wedge \boldsymbol{a}) \;\; (\boldsymbol{n}_\infty \wedge \boldsymbol{b} \wedge \boldsymbol{c} \wedge \boldsymbol{d}) \;= \\
&\qquad (\boldsymbol{N} + \boldsymbol{n}_\infty \mathbf{a})\left(\boldsymbol{N}\,(\mathbf{c}-\mathbf{b}) \wedge (\mathbf{d}-\mathbf{b}) + \boldsymbol{n}_\infty \mathbf{b} \wedge \mathbf{c} \wedge \mathbf{d}\right) \\
&\qquad\quad = (1 + \boldsymbol{n}_\infty \mathbf{a})\,(\mathbf{c}-\mathbf{b}) \wedge (\mathbf{d}-\mathbf{b}) \;-\; \boldsymbol{n}_\infty \mathbf{b} \wedge \mathbf{c} \wedge \mathbf{d} \\
&\qquad\quad = (\mathbf{c}-\mathbf{b}) \wedge (\mathbf{d}-\mathbf{b}) + \boldsymbol{n}_\infty \mathbf{a} \cdot \left((\mathbf{c}-\mathbf{b}) \wedge (\mathbf{d}-\mathbf{b})\right) \\
&\qquad\qquad + \boldsymbol{n}_\infty (\mathbf{a}-\mathbf{b}) \wedge (\mathbf{c}-\mathbf{b}) \wedge (\mathbf{d}-\mathbf{b}) \qquad (36)
\end{aligned}$$

The $\boldsymbol{n}_\infty \mathbf{a} \cdot \left((\mathbf{c}-\mathbf{b}) \wedge (\mathbf{d}-\mathbf{b})\right)$ term is unfortunately not translation invariant, but the rest of its 2-vector part $(\mathbf{c}-\mathbf{b}) \wedge (\mathbf{d}-\mathbf{b})$ is, as is its 4-vector part $\boldsymbol{n}_\infty (\mathbf{a}-\mathbf{b}) \wedge (\mathbf{c}-\mathbf{b}) \wedge (\mathbf{d}-\mathbf{b})$. The latter shows that the volume $\left|[\mathbf{a}, \mathbf{b}', \mathbf{c}', \mathbf{d}']\right|$ of the tetrahedron is the same for all $\mathbf{b}'$, $\mathbf{c}'$, $\mathbf{d}'$ in the same plane as $[\mathbf{b}, \mathbf{c}, \mathbf{d}]$ and spanning a triangle of the same area. It is also the same, of course, for any translate of $\mathbf{a}$ in the plane through $\mathbf{a}$ parallel to $[\mathbf{b}, \mathbf{c}, \mathbf{d}]$. We can put $\mathbf{a}$ at the origin by translating all the points with the rotor $\tilde{\boldsymbol{T}}_{\mathbf{a}} := 1 - \mathbf{a}\boldsymbol{n}_\infty$, obtaining:

$$\begin{aligned}
&\tilde{\boldsymbol{T}}_{\mathbf{a}}\,(\boldsymbol{n}_\infty \wedge \boldsymbol{a})\,\boldsymbol{T}_{\mathbf{a}}\,\tilde{\boldsymbol{T}}_{\mathbf{a}}\,(\boldsymbol{n}_\infty \wedge \boldsymbol{b} \wedge \boldsymbol{c} \wedge \boldsymbol{d})\,\boldsymbol{T}_{\mathbf{a}} \qquad (37) \\
&\quad= \;(\mathbf{c}-\mathbf{b}) \wedge (\mathbf{d}-\mathbf{b}) \;-\; \boldsymbol{n}_\infty \mathbf{b} \wedge \mathbf{c} \wedge \mathbf{d}
\end{aligned}$$

Thus our remarks following Eq. (28) show that the height of $\boldsymbol{n}_\infty \wedge \boldsymbol{a}$ above the plane of $\boldsymbol{n}_\infty \wedge \boldsymbol{b} \wedge \boldsymbol{c} \wedge \boldsymbol{d}$ is

$$h_\mathsf{a} = \frac{\|(\mathbf{b}-\mathbf{a}) \wedge (\mathbf{c}-\mathbf{a}) \wedge (\mathbf{d}-\mathbf{a})\|}{\|(\mathbf{c}-\mathbf{b}) \wedge (\mathbf{d}-\mathbf{b})\|} = \frac{\|\boldsymbol{n}_\infty \wedge \boldsymbol{a} \wedge \boldsymbol{b} \wedge \boldsymbol{c} \wedge \boldsymbol{d}\|}{\|\boldsymbol{n}_\infty \wedge \boldsymbol{b} \wedge \boldsymbol{c} \wedge \boldsymbol{d}\|} \tag{38}$$

which is the same as the ratio of the norms of the corresponding translation-independent terms in Eq. (36).

Finally, we have the following discrete corollary of Stoke's theorem amongst the plane-bound bivectors of the four faces of the tetrahedron:

$$\begin{aligned} \boldsymbol{n}_\infty \wedge (\boldsymbol{b} \wedge \boldsymbol{c} \wedge \boldsymbol{d} - \boldsymbol{a} \wedge \boldsymbol{c} \wedge \mathbf{d} + \boldsymbol{a} \wedge \boldsymbol{b} \wedge \boldsymbol{d} - \boldsymbol{a} \wedge \boldsymbol{b} \wedge \boldsymbol{c}) &= \\ \boldsymbol{n}_\infty \wedge (\boldsymbol{b}-\boldsymbol{a}) \wedge (\boldsymbol{c}-\boldsymbol{a}) \wedge (\boldsymbol{d}-\boldsymbol{a}) &= \boldsymbol{n}_\infty (\mathbf{b}-\mathbf{a}) \wedge (\mathbf{c}-\mathbf{a}) \wedge (\mathbf{d}-\mathbf{a}) \end{aligned} \tag{39}$$

This implies, in particular, that the sum of the free bivector parts of these plane-bound bivectors vanishes, a result usually attributed to Hermann Minkowski although it must have been known to Grassmann. The bivectors of the four faces determine that of the medial parallelogram $(\mathbf{b}-\mathbf{a}) \wedge (\mathbf{d}-\mathbf{c})$ via Eq. (31) and its analogues for the other two medial parallelogram bivectors $(\mathbf{c}-\mathbf{a}) \wedge (\mathbf{d}-\mathbf{b})$ and $(\mathbf{c}-\mathbf{b}) \wedge (\mathbf{d}-\mathbf{a})$. This system of linear relations among these areal bivectors can be inverted to express the areal bivectors of each face as a signed sum of those of the three medial parallelograms. These analoges of Minkowski's identity justify the heterodox point-of-view that the tetrahedron actually has *seven* faces, where its three medial parallelograms qualify as *interior* faces. Further justification derives from the fact that the areas of these seven faces mutually determine a non-degenerate tetrahedron up to isometry [6].

5 Heron's Formula for Tetrahedra, and Their In-Spheres

The author's extension of Heron's formula to the tetrahedron is based on the above linear relations amongst the bivectors of its seven faces, along with their connections to the in-sphere thereof. This section summarizes these results using the language of conformal geometric algebra, and discusses how they might be extended to yet-higher dimensions within that framework.

The in-radius ρ of a tetrahedron is $\rho = \tau/\sigma$, where $\tau := \|\boldsymbol{n}_\infty \wedge \boldsymbol{a} \wedge \boldsymbol{b} \wedge \boldsymbol{c} \wedge \boldsymbol{d}\|$ is 3! times its volume and σ is twice its surface area:

$$\begin{aligned} \sigma := \alpha_\mathsf{d} + \alpha_\mathsf{c} + \alpha_\mathsf{b} + \alpha_\mathsf{a} &:= \|\boldsymbol{n}_\infty \wedge \boldsymbol{a} \wedge \boldsymbol{b} \wedge \boldsymbol{c}\| \\ &+ \|\boldsymbol{n}_\infty \wedge \boldsymbol{a} \wedge \boldsymbol{b} \wedge \boldsymbol{d}\| + \|\boldsymbol{n}_\infty \wedge \boldsymbol{a} \wedge \boldsymbol{c} \wedge \boldsymbol{d}\| + \|\boldsymbol{n}_\infty \wedge \boldsymbol{b} \wedge \boldsymbol{c} \wedge \boldsymbol{d}\| \end{aligned} \tag{40}$$

As is likewise well known, the barycentric coordinates of the in-center $\boldsymbol{i}$ are just the ratios of these four areas to their sum σ, whence it follows from Eq. (3) that the dual representation of the in-sphere itself is:

$$\boldsymbol{s} := (\alpha_\mathsf{a} \boldsymbol{a} + \cdots + \alpha_\mathsf{d} \boldsymbol{d})/\sigma + \boldsymbol{n}_\infty(\alpha_\mathsf{a}\alpha_\mathsf{b}\, \boldsymbol{a} \cdot \boldsymbol{b} + \cdots + \alpha_\mathsf{c}\alpha_\mathsf{d}\, \boldsymbol{c} \cdot \boldsymbol{d} - \tau^2/2)/\sigma^2 \tag{41}$$

The in-sphere touches the (exterior) faces of the tetrahedron at four points known as its *in-touch points* $\boldsymbol{j}$, $\boldsymbol{k}$, $\boldsymbol{\ell}$, $\boldsymbol{m}$, where their free vectors relative to $\boldsymbol{n}_0$ satisfy

$\mathbf{j} \in [\mathbf{b}, \mathbf{c}, \mathbf{d}]$, $\mathbf{k} \in [\mathbf{a}, \mathbf{c}, \mathbf{d}]$, $\mathbf{l} \in [\mathbf{a}, \mathbf{b}, \mathbf{d}]$, $\mathbf{m} \in [\mathbf{a}, \mathbf{b}, \mathbf{c}]$. These divide the exterior faces into three triangles each, and the six pairs of triangles each sharing a common edge of the tetrahedron can easily be shown to be congruent. The areas (times 2) of these triangles also determine the areas of its seven faces (*vide infra*) and hence the tetrahedron itself (up to isometry if non-degenerate). Accordingly, these six areas have been named the *natural parameters* of the tetrahedron. Any assumed non-negative values for the areas of the seven faces, in turn, determine the natural parameters of a tetrahedron, providing they satisfy certain consistency relations.

To formulate these consistency relations, suppose the areas of the interior faces (times 4) are $\alpha_{\mathsf{ab}} = \alpha_{\mathsf{cd}} = \|(\mathbf{b} - \mathbf{a}) \wedge (\mathbf{d} - \mathbf{c})\|$, and similarly for $\alpha_{\mathsf{ac}} = \alpha_{\mathsf{bd}}$ and $\alpha_{\mathsf{ad}} = \alpha_{\mathsf{bc}}$. Then the triangle inequality for vectors $\mathbf{v}_1 = \mathbf{v}_2 + \mathbf{v}_3 \implies \|\mathbf{v}_1\| \le \|\mathbf{v}_2\| + \|\mathbf{v}_3\|$, applied to the three bivectors in Eq. (31), implies:[2]

$$\alpha_{\mathsf{cd}} \le \alpha_{\mathsf{c}} + \alpha_{\mathsf{d}}\,, \qquad \alpha_{\mathsf{d}} \le \alpha_{\mathsf{c}} + \alpha_{\mathsf{cd}}\,, \qquad \alpha_{\mathsf{c}} \le \alpha_{\mathsf{d}} + \alpha_{\mathsf{cd}} \tag{42}$$

The deviations of these inequalities from saturation will be denoted by $\mathcal{T}^1_{\mathsf{cd}} := \alpha_{\mathsf{c}} + \alpha_{\mathsf{d}} - \alpha_{\mathsf{cd}}$, $\mathcal{T}^2_{\mathsf{cd}} := \alpha_{\mathsf{cd}} + \alpha_{\mathsf{c}} - \alpha_{\mathsf{d}}$ and $\mathcal{T}^3_{\mathsf{cd}} := \alpha_{\mathsf{cd}} + \alpha_{\mathsf{d}} - \alpha_{\mathsf{c}}$, while $\mathcal{T}^0_{\mathsf{cd}} := \alpha_{\mathsf{c}} + \alpha_{\mathsf{d}} + \alpha_{\mathsf{cd}}$ is a non-degeneracy factor that vanishes if and only if $\alpha_{\mathsf{c}} = \alpha_{\mathsf{d}} = \alpha_{\mathsf{cd}} = 0$. More generally, $\mathcal{T}^k_{\mathsf{xy}}$ will be the corresponding quantities for the other edges of the tetrahedron with $k = 0, 1, 2, 3$ and $\mathsf{x}, \mathsf{y} \in \{\mathsf{a}, \mathsf{b}, \mathsf{c}, \mathsf{d}\}$ $(\mathsf{x} \ne \mathsf{y})$. There are 18 such inequalities in all (6 triples), which will be called the *tetrahedron inequalities*. Finally, the linear dependencies among the seven areal bivectors can be used to show that they also satisfy a quadratic relation known as Yetter's identity [12]:

$$\alpha_{\mathsf{a}}^2 + \alpha_{\mathsf{b}}^2 + \alpha_{\mathsf{c}}^2 + \alpha_{\mathsf{d}}^2 \;=\; \alpha_{\mathsf{ab}}^2 + \alpha_{\mathsf{ac}}^2 + \alpha_{\mathsf{ad}}^2 \quad (=\; \alpha_{\mathsf{cd}}^2 + \alpha_{\mathsf{bd}}^2 + \alpha_{\mathsf{bc}}^2) \tag{43}$$

Using the areal laws of sines and cosines plus a little trigonometry, one can show that the natural parameters of the tetrahedron $[\mathbf{a}, \mathbf{b}, \mathbf{c}, \mathbf{d}]$ are given by the following simple rational functions of the seven areas:

$$\begin{aligned}
u &:= \|\boldsymbol{n}_\infty \wedge \boldsymbol{a} \wedge \boldsymbol{b} \wedge \boldsymbol{\ell}\| = \|\boldsymbol{n}_\infty \wedge \boldsymbol{a} \wedge \boldsymbol{b} \wedge \boldsymbol{m}\| = \mathcal{T}^0_{\mathsf{cd}}\mathcal{T}^1_{\mathsf{cd}}/(2\sigma) \\
v &:= \|\boldsymbol{n}_\infty \wedge \boldsymbol{a} \wedge \boldsymbol{c} \wedge \boldsymbol{k}\| = \|\boldsymbol{n}_\infty \wedge \boldsymbol{a} \wedge \boldsymbol{c} \wedge \boldsymbol{m}\| = \mathcal{T}^0_{\mathsf{bd}}\mathcal{T}^1_{\mathsf{bd}}/(2\sigma) \\
w &:= \|\boldsymbol{n}_\infty \wedge \boldsymbol{a} \wedge \boldsymbol{d} \wedge \boldsymbol{k}\| = \|\boldsymbol{n}_\infty \wedge \boldsymbol{a} \wedge \boldsymbol{d} \wedge \boldsymbol{\ell}\| = \mathcal{T}^0_{\mathsf{bc}}\mathcal{T}^1_{\mathsf{bc}}/(2\sigma) \\
x &:= \|\boldsymbol{n}_\infty \wedge \boldsymbol{b} \wedge \boldsymbol{c} \wedge \boldsymbol{j}\| = \|\boldsymbol{n}_\infty \wedge \boldsymbol{b} \wedge \boldsymbol{c} \wedge \boldsymbol{m}\| = \mathcal{T}^0_{\mathsf{ad}}\mathcal{T}^1_{\mathsf{ad}}/(2\sigma) \\
y &:= \|\boldsymbol{n}_\infty \wedge \boldsymbol{b} \wedge \boldsymbol{d} \wedge \boldsymbol{j}\| = \|\boldsymbol{n}_\infty \wedge \boldsymbol{b} \wedge \boldsymbol{d} \wedge \boldsymbol{\ell}\| = \mathcal{T}^0_{\mathsf{ac}}\mathcal{T}^1_{\mathsf{ac}}/(2\sigma) \\
z &:= \|\boldsymbol{n}_\infty \wedge \boldsymbol{c} \wedge \boldsymbol{d} \wedge \boldsymbol{j}\| = \|\boldsymbol{n}_\infty \wedge \boldsymbol{c} \wedge \boldsymbol{d} \wedge \boldsymbol{k}\| = \mathcal{T}^0_{\mathsf{ab}}\mathcal{T}^1_{\mathsf{ab}}/(2\sigma)
\end{aligned} \tag{44}$$

The fact that each exterior face is divided into three subfaces by its in-touch point allows their areas to be expressed in terms of the natural parameters as:

$$\alpha_{\mathsf{d}} = u+v+x\,, \quad \alpha_{\mathsf{c}} = u+w+y\,, \quad \alpha_{\mathsf{b}} = v+w+z\,, \quad \alpha_{\mathsf{a}} = x+y+z \tag{45}$$

[2] The notation here is a bit confusing in that α_{d} signifies the face $[\mathbf{a}, \mathbf{b}, \mathbf{c}]$ opposite the vertex $\mathbf{d}$, etc., but it is standard and the alternative α_{abc} is cumbersome.

These relations in turn imply $\sigma = 2(u+v+w+x+y+z)$. The squared interior areas, however, are quadratic in the natural parameters:

$$\begin{aligned} \alpha_{\mathsf{cd}}^2 &= (v+w+x+y)^2 - 4\,uz = \alpha_{\mathsf{ab}}^2\,, \quad \alpha_{\mathsf{bd}}^2 = (u+w+x+z)^2 - 4\,vy = \alpha_{\mathsf{ac}}^2\,, \\ \alpha_{\mathsf{bc}}^2 &= (u+v+y+z)^2 - 4\,wx = \alpha_{\mathsf{ad}}^2 \end{aligned} \tag{46}$$

Since the natural parameters determine the areas, they likewise determine a non-degenerate tetrahedron up to isometry.

With these definitions, the extension of Heron's formula to tetrahedra is:

$$\begin{aligned} \tau^4 &= \sigma^2\left(2\,vwxy + 2\,uwxz + 2\,uvyz - u^2z^2 - v^2y^2 - w^2x^2\right) \\ &=: \sigma^2\,\Omega(u,v,w,x,y,z) \;=\; -4\,(u+v+w+x+y+z)^2 \det\begin{bmatrix} 0 & u & v & w \\ u & 0 & x & y \\ v & x & 0 & z \\ w & y & z & 0 \end{bmatrix} \end{aligned} \tag{47}$$

Thus $\Omega > 0$ is a necessary (and, it turns out, sufficient [6]) condition for any $u, \ldots, z > 0$ to determine a non-degenerate tetrahedron. Note also that the parameters u, v, w in the compact version of Heron's formula $suvw$ (given after Eq. (29)) equal the lengths of the line segments into which the triangle's sides are divided by their in-touch points, and that their product uvw likewise equals the analogous 3×3 determinant. Together with Eq. (47), this leads to a rather obvious conjecture as to how the formula should extend to higher dimensions (see Conjecture 4.9 in Ref. [6]).

The observations made in this paper should inform efforts to prove that conjecture. Nevertheless, it is not so obvious how the expression $\|\langle \boldsymbol{LM}\rangle_2\|$ for the area $\alpha_{\mathsf{ab}} = \alpha_{\mathsf{cd}}$ of that interior face should be extended to the "medial sections" (as they are known) of general n-simplicies, and this expression is already a bit unwieldy even for $n = 3$. A clue as to what might be a better approach may be found in a generalization of Cayley-Menger determinants to the "hyper-areas" of general medial sections discovered by István Talata [11]. In particular, Talata's determinant expresses the squared area of an interior face as e.g.

$$\alpha_{\mathsf{ab}}^2 \;=\; \alpha_{\mathsf{cd}}^2 \;=\; \frac{1}{4}\det\begin{bmatrix} 0 & 0 & 1 & 1 & 0 & 0 \\ 0 & 0 & 0 & 0 & 1 & 1 \\ 1 & 0 & 0 & \Delta_{\mathsf{ab}} & \Delta_{\mathsf{ac}} & \Delta_{\mathsf{ad}} \\ 1 & 0 & \Delta_{\mathsf{ab}} & 0 & \Delta_{\mathsf{bc}} & \Delta_{\mathsf{bd}} \\ 0 & 1 & \Delta_{\mathsf{ac}} & \Delta_{\mathsf{bc}} & 0 & \Delta_{\mathsf{cd}} \\ 0 & 1 & \Delta_{\mathsf{ad}} & \Delta_{\mathsf{bd}} & \Delta_{\mathsf{cd}} & 0 \end{bmatrix}, \tag{48}$$

where $\Delta_{\mathsf{ab}} := \|\mathrm{a} - \mathrm{b}\|^2$ etc. Having a double border of 0's and 1's, however, it seems unlikely that this determinant admits a geometric interpretation within the conformal model. Therefore an extension of the conformal model will be proposed that allows this to be done.

This extension posits two pairs of null vectors, $\boldsymbol{n}_0, \boldsymbol{n}_\infty$ and $\boldsymbol{n}_0', \boldsymbol{n}_\infty'$ such that $\boldsymbol{n}_0 \cdot \boldsymbol{n}_\infty = \boldsymbol{n}_0' \cdot \boldsymbol{n}_\infty' = -1$ and $\boldsymbol{n}_0 \cdot \boldsymbol{n}_0' = \boldsymbol{n}_0 \cdot \boldsymbol{n}_\infty' = \boldsymbol{n}_\infty \cdot \boldsymbol{n}_0' = \boldsymbol{n}_\infty \cdot \boldsymbol{n}_\infty' = 0$. The vertices of the tetrahedron are then represented by null vectors of the form:

$$\begin{aligned} \boldsymbol{a} &= \boldsymbol{n}_0 + \boldsymbol{n}_0' + \mathrm{a} + \boldsymbol{n}_\infty \mathrm{a}^2/2\,, \quad \boldsymbol{b} = \boldsymbol{n}_0 + \boldsymbol{n}_0' + \mathrm{b} + \boldsymbol{n}_\infty \mathrm{b}^2/2 \\ \boldsymbol{c} &= \boldsymbol{n}_0 + \boldsymbol{n}_0' + \mathrm{c} + \boldsymbol{n}_\infty' \mathrm{c}^2/2\,, \quad \boldsymbol{d} = \boldsymbol{n}_0 + \boldsymbol{n}_0' + \mathrm{d} + \boldsymbol{n}_\infty' \mathrm{d}^2/2 \end{aligned} \tag{49}$$

It is readily verified that the inner products among $\boldsymbol{n}_0, \boldsymbol{n}_\infty, \boldsymbol{n}'_0, \boldsymbol{n}'_\infty$ reproduce the double border in the above Talata determinant, while the inner products among $\boldsymbol{a}, \boldsymbol{b}, \boldsymbol{c}, \boldsymbol{d}$ are still equal to half the negative squared distances. It will, of course, be necessary to use *four* such pairs of null vectors, and four copies of each vertex each with its own point at infinity, in order to represent the Talata determinants of all three $(2,2)$-medial sections (parallelograms).

Although this algebraic trick clearly works, its geometric implications are not entirely clear. It would be particularly interesting if those led to a simple geometric interpretation of the zeros of the polynomial Ω in Eq. (47), which were studied in depth in the original reference [6] but remain rather mysterious. Almost all these zeros, in fact, are the limits of sequences of non-degenerate tetrahedra the vertices of which go off to plus or minus infinity along a line while the ratios of the distances amongst them remain finite. This "areal" boundary of the set of non-degenerate tetrahedra is wildly different from the usual boundary, which consists of all quadruples of points in the (finite) Euclidean plane.

Note Added in Proof: A greatly expanded version of this paper is currently in press at "Advances in Applied Clifford Algebras."

References

1. Blumenthal, L.M.: Theory and Applications of Distance Geometry. Oxford Univ. Press,: reprinted by Chelsea Publ, p. 1970. Co, Bronx, NY (1953)
2. Dorst, L., Fontijne, D., Mann, S.: The Conformal Model: Operational Euclidean Geometry, chap. 13 of "Geometric Algebra for Computer Science: An Object-Oriented Approach to Geometry." Elsevier, Amsterdam, NL (2009). https://doi.org/10.1016/B978-0-12-374942-0.00018-X
3. Dress, A.W.M., Havel, T.F.: Distance geometry and geometric algebra. Found. Phys. **23**, 1357–1374 (1993). https://doi.org/10.1007/bf01883783
4. Gidea, M., Niculescu, C.: A brief account on Lagrange's algebraic identity. Math. Intell. **34**(3), 55–61 (2012). https://doi.org/10.1007/s00283-012-9305-0
5. Havel, T.F.: Geometric algebra and Möbius sphere geometry as a basis for Euclidean invariant theory. In: White, N.L. (ed.) Invariant Methods in Discrete and Computational Geometry: Proc. Curaçao Conf., 13–17 June, 1994, pp. 245–256. Springer, Netherlands (1995). https://doi.org/10.1007/978-94-015-8402-9_11
6. Havel, T.F.: An extension of Heron's formula to tetrahedra, and the projective nature of its zeros. Cornell Preprint Archive arXiv:2204.08089 (2022)
7. Hestenes, D.: Grassmann's vision. In: Schubring, G. (ed.) Hermann Günther Graßmann (1809–1877): Visionary Mathematician, Scientist and Neohumanist Scholar, pp. 243–254. Springer, Netherlands (1996). https://doi.org/10.1007/978-94-015-8753-2_20
8. Hestenes, D.: New Tools for Computational Geometry and Rejuvenation of Screw Theory. In: Bayro-Corrochano, E., Scheuermann, G. (ed.) Geometric Algebra Computing, pp. 3–33. Springer, London (2010). https://doi.org/10.1007/978-1-84996-108-0_1

9. Li, Hongbo., Hestenes, D., Rockwood, A.: Generalized homogeneous coordinates for computational geometry. In: Sommer, G. (ed.) Geometric Computing with Clifford Algebras, pp. 27–59. Springer, Heidelberg (2001). https://doi.org/10.1007/978-3-662-04621-0_2
10. Sobczyk, G.: Notes on Plücker's relations in geometric algebra. Adv. Math. **363**, 106959 (2020). https://doi.org/10.1016/j.aim.2019.106959
11. Talata, I.: A volume formula for medial sections of simplices. Discrete Comput. Geom. **30**, 343–353 (2003). https://doi.org/10.1007/s00454-003-0015-6
12. Yetter, D.N.: On a formula relating volumes of medial sections of simplices. Discrete Comput. Geom. **43**, 339–345 (2010). https://doi.org/10.1007/s00454-008-9105-7

Geometric Algebras of Compatible Null Vectors

Garret Sobczyk(✉)

Universidad de las Américas-Puebla, 72820 Puebla, Pue., Mexico
garretudla@gmail.com
https://www.garretstar.com

Abstract. A (Clifford) geometric algebra is usually defined in terms of a quadratic form. A null vector v is an algebraic quantity with the property that $v^2 = 0$. The universal algebra generated by taking the sums and products of null vectors over the real or complex numbers is denoted by $\mathcal{N}$. The rules of addition and multiplication are taken to be the familiar rules of addition and multiplication of real or complex square matrices. In a series of ten definitions, the concepts of a Grassmann algebra, its dual Grassmann algebra, the associated real and complex geometric algebras, and their isomorphic real or complex coordinate matrix algebras are set down. This is followed by a discussion of affine transformations, the horosphere and conformal transformations on pseudoeuclidean spaces.

Keywords: affine plane · geometric algebra · Grassmann algebra · horosphere

1 Introduction

The development of the concept of duality in mathematics has a robust history dating back more than 100 years, and involving 20th Century mathematicians of the first rank such as F. Reisz and S. Banach, but also encompassing first rank 19th Century mathematicians such as Gauss, Lobachevsky and Bolyai. A fascinating history of the seminal Hahn-Banach Theorem, and all its ramifications regarding the issue of duality in finite and infinite dimensional Hilbert spaces is given in [1].

Considering infinite dimensional vector spaces broadens and immensely deepens the mathematical issues involved in the concept of duality [2,3]. In this paper we consider duality only in regard to a finite dimensional vector space, where it is well known that duality is equivalent to defining a Euclidean inner product [4]. The purpose of this paper is to show how a new concept of *compatibility* of a pair of null vectors, over the real or complex numbers, not only captures the notion of duality but nails down the corresponding isomorphic real or complex *coordinate matrix algebra* of a *Clifford geometric algebra*.

A series of 10 definitions, given in Sect. 2, is used to define a Grassmann algebra, its compatible dual Grassmann algebra, and their associated geometric algebra. A pair of compatible null vectors is used to define the affine plane and the

D. W. Silva et al. (Eds.): ICACGA 2022, LNCS 13771, pp. 40–47, 2024.
https://doi.org/10.1007/978-3-031-34031-4_4

horosphere of a general pseudoeuclidean geometric algebra. In Sect. 3, conformal transformations in pseudoeuclidean space is defined in terms of Ahlfors-Vahlen matrices, and their corresponding linear fractional transformations in geometric algebra.

2 Ten Definitions

1. Null vectors are algebraic quantities $v \neq 0$ with the property that $v^2 = 0$. They are to be added and multiplied together using the same rules as the addition and multiplication of real or complex square matrices. The trivial null vector is denoted by 0, and the *universal algebra* generated by taking the sums and products of null vectors is denoted by $\mathcal{N}$.
2. Two null vectors $a_1, a_2 \in \mathcal{N}$ are said to be *anticommutative* if
$$a_1a_2 + a_2a_1 = 0.$$
3. A set of mutually anticommuting null vectors $\{a_1, \ldots, a_n\}_{\mathbb{F}}$ is said to be *linearly independent* over $\mathbb{F} = \mathbb{R}$ or $\mathbb{C}$, if $a_1 \cdots a_n \neq 0$.[1] In this case they *generate* over $\mathbb{F}$ the 2^n-dimensional Grassmann algebra
$$\mathcal{G}_n(\mathbb{F}) := gen_{\mathbb{F}}\{a_1, \ldots, a_n\},$$
[6].
4. Let $A_i := \begin{pmatrix} 1 \\ a_i \end{pmatrix}$ for $i = 1, \ldots, n$. The *right directed Kronecker* product,
$$\mathcal{G}_2(\mathbb{F}) := A_1 \overrightarrow{\otimes} A_2 = \begin{pmatrix} \begin{pmatrix} 1 \\ a_1 \end{pmatrix} \otimes 1 \\ \begin{pmatrix} 1 \\ a_1 \end{pmatrix} \otimes a_2 \end{pmatrix} = \begin{pmatrix} 1 \\ a_1 \\ a_2 \\ a_{12} \end{pmatrix},$$
gives the ordered basis elements defining the Grassmann algebra $\mathcal{G}_2(\mathbb{F})$, written in a column matrix. More generally, the right directed Kronecker product
$$\mathcal{G}_n(\mathbb{F}) := A_1 \overrightarrow{\otimes} \cdots \overrightarrow{\otimes} A_n$$
gives the ordered 2^n-*column* matrix of the basis elements
$$\{1; a_1, \ldots, a_n; \ldots; a_{\lambda_1} \cdots a_{\lambda_k}; \ldots; a_1 \cdots a_n\}_{\mathbb{F}}^T$$
of the 2^n-dimensional Grassmann algebra $\mathcal{G}_n(\mathbb{F}) \subset \mathcal{N}$. The $\begin{pmatrix} n \\ k \end{pmatrix}$ elements
$$a_{\lambda_1 \cdots \lambda_k} := a_{\lambda_1} \cdots a_{\lambda_k}$$

[1] More general fields $\mathbb{F}$ can be considered as long as *characteristic* $\mathbb{F} \neq 2$. For an interesting discussion of this issue see [5].

for $1 \le \lambda_1 < \cdots < \lambda_k \le n$ are called k-vectors. Similarly, the *left directed Kronecker* product

$$\mathcal{G}_2(\mathbb{F}) := A_2^T \overleftarrow{\otimes} A_1^T = \left(1 \overleftarrow{\otimes} (1\ a_1)\ a_2 \overleftarrow{\otimes} (1\ a_1)\right) = (1\ a_1\ a_2\ a_{21}),$$

gives the ordered basis elements defining the Grassmann algebra $\mathcal{G}_2(\mathbb{F})$, written in a row matrix. More generally,

$$\mathcal{G}_n(\mathbb{F}) := (A_n^T \overleftarrow{\otimes} \cdots \overleftarrow{\otimes} A_1^T),$$

gives the ordered 2^n-dimensional *row matrix* of the basis elements of the Grassmann algebra $\mathcal{G}_n(\mathbb{R}) \subset \mathcal{N}$, [7, p.82].

5. A pair of null vectors $a, b \in \mathcal{N}$ are said to be *algebraically dual*, or *compatible* if

$$ab + ba = 1. \tag{1}$$

The dual null vectors a and b satisfy the easy to remember, and easily verified, Multiplication Table 1:

Table 1. Multiplication Table.

$\cdot$	a	b	ab	ba
a	0	ab	0	a
b	ba	0	b	0
ab	a	0	ab	0
ba	0	b	0	ba

In this case, we define $a^* := b$ and $b^* := a$, from which it follows that

$$(a^*)^* = b^* = a.$$

Particularly noteworthy is the fact that ab and ba are *idempotents*.

The corresponding anticommuting *pseudoeuclidean* vectors $e := a + b$ and $f := a - b$, satisfy $e^2 = 1 = -f^2$, and define the bivector $u := ef$ in the geometric algebra $\mathbb{G}_{1,1} := \mathbb{R}(e, f) \subset \mathcal{N}$, with the matrix of basis elements

$$\mathbb{G}_{1,1} := AbaB^T = \begin{pmatrix} 1 \\ a \end{pmatrix} ba\,(1\ b) = \begin{pmatrix} ba & b \\ a & ab \end{pmatrix}. \tag{2}$$

Note, we have slightly abused notation by defining the geometric algebra $\mathbb{G}_{1,1}$ both as an extension of the real number system to include the pseudoeuclidean vectors e and f, and in terms of its basis elements given in (2).

A geometric number $g \in \mathbb{G}_{1,1}$ is determined by its *coordinate matrix* $[g] := [g_{ij}]$, for $g_{ij} \in \mathbb{R}$, by

$$g = A^T ba[g]B = (1\ a)\,ba \begin{pmatrix} g_{11} & g_{12} \\ g_{21} & g_{22} \end{pmatrix} \begin{pmatrix} 1 \\ b \end{pmatrix} = g_{11}ba + g_{12}b + g_{21}a + g_{22}ab,$$

see [7, p.67].

6. More generally, let

$$\mathcal{G}_n(\mathbb{R}) := A_1 \overrightarrow{\otimes} \cdots \overrightarrow{\otimes} A_n \subset \mathcal{N} \quad \text{and} \quad \mathcal{G}_n^{\#}(\mathbb{R}) = B_n^T \overleftarrow{\otimes} \cdots \overleftarrow{\otimes} B_1^T \subset \mathcal{N}$$

be two 2^n-dimensional real Grassmann algebras. The Grassmann algebras $\mathcal{G}_n(\mathbb{R})$ and $\mathcal{G}_n^{\#}(\mathbb{R})$ are said to be *compatible Grassmann algebras* if there exists generating bases

$$\mathcal{G}_n(\mathbb{R}) := gen_{\mathbb{R}}\{a_1, \ldots, a_n\} \quad \text{and} \quad \mathcal{G}_n^{\#}(\mathbb{R}) := gen_{\mathbb{R}}\{b_1, \ldots, b_n\}$$

such that

$$2a_i \cdot b_j := a_i b_j + b_j a_i = \delta_{ij}. \tag{3}$$

In this case, $\mathcal{G}_n^*(\mathbb{R}) := \mathcal{G}_n^{\#}(\mathbb{R})$. Of course, nothing is surprising because it is well known that any standard vector space V, and it dual space V^* can be represented in terms of an equivalent *inner product*, [4].
7. For the compatible Grassmann algebras, defined in Definition 6, the *real geometric algebra* $\mathbb{G}_{n,n}(\mathbb{R})$ is defined by

$$\mathbb{G}_{n,n}(\mathbb{R}) := \mathcal{G}_n(\mathbb{R}) \otimes \mathcal{G}_n^*(\mathbb{R}) = gen_{\mathbb{R}}\{a_1, \ldots, a_n, b_1, \ldots, b_n\},$$

where for $i, j = 1, \ldots, n$, the null vectors a_i and b_j satisfy (3).
8. Defining the *idempotents* $u_i := b_i a_i$, the quantity

$$u_{1\cdots n} := u_1 \cdots u_n = \prod_{i=1}^{n} b_i a_i = b_1 a_1 \cdots b_n a_n$$

is a *primitive idempotent* in the geometric algebra $\mathbb{G}_{n,n}$. The *spectral basis* of null vectors of the geometric algebra $\mathbb{G}_{n,n}$ is specified by

$$\mathbb{G}_{n,n} := A_1 \overrightarrow{\otimes} \cdots \overrightarrow{\otimes} A_n u_{1\cdots n} B_n^T \overleftarrow{\otimes} \cdots \overleftarrow{\otimes} B_1^T = \left(\overrightarrow{\otimes}_{i=1}^n A_i\right) u_{1\cdots n} \left(\overrightarrow{\otimes}_{i=1}^n A_i\right)^*,$$

where $A_i^* := B_i^T$ for $i = 1, \ldots, n$. In the spectral basis, any $g \in \mathbb{G}_{n,n}$ is explicitly expressed in terms of its *coordinate matrix* $[g] := [g_{ij}]$ for $g_{ij} \in \mathbb{R}$, by

$$g = \left(\overrightarrow{\otimes}_{i=1}^n A_i^T\right) u_{1\cdots n} [g] \left(\overleftarrow{\otimes}_{i=n}^1 B_i\right),$$

[7, Chapter 5].

In Definition 5, we derived the geometric algebra $\mathbb{G}_{1,1}$, see equation (2). Referring back to Definition 4, for the geometric algebra $\mathbb{G}_{2,2}$, define $u_1 = b_1 a_1$ and $u_2 = b_2 a_2$. The spectral basis for $\mathbb{G}_{2,2}$ is

$$\begin{pmatrix} 1 \\ a_1 \\ a_2 \\ a_{12} \end{pmatrix} u_1 u_2 \begin{pmatrix} 1 & b_1 & b_2 & b_{21} \end{pmatrix} = \begin{pmatrix} u_1 u_2 & b_1 u_2 & b_2 u_1 & b_{21} \\ a_1 u_2 & u_1^\dagger u_2 & a_1 b_2 & -b_2 u_1^\dagger \\ a_2 u_1 & a_2 b_1 & u_1 u_2^\dagger & b_1 u_2^\dagger \\ a_{12} & -a_2 u_1^\dagger & a_1 u_2^\dagger & u_1^\dagger u_2^\dagger \end{pmatrix}, \tag{4}$$

where the *reverses* of u_1 and u_2 are defined by $u_1^\dagger := a_1 b_1$ and $u_2^\dagger := a_2 b_2$. The geometric number $g \in \mathbb{G}_{2,2}$ is defined by its coordinate matrix $[g_{ij}] \in \mathcal{M}_4(\mathbb{R})$ by

$$g = (1 \; a_1 \; a_2 \; a_{12}) \, u_1 u_2 [g_{ij}] \begin{pmatrix} 1 \\ b_1 \\ b_2 \\ b_{21} \end{pmatrix},$$

[7, p. 84].

9. The *standard basis* of $\mathbb{G}_{p,q} := \mathbb{G}(\mathbb{R}^{p,q})$ is specified by

$$\mathbb{G}_{p,q} := \mathbb{R}(e_1, \ldots, e_p, f_1, \ldots, f_q) = gen_{\mathbb{R}}\{e_1, \ldots, e_p, f_1, \ldots, f_q\}$$

where $e_i := a_i + b_i$ and $f_j := a_j - b_j$ for $i = 1, \ldots, p$ and $j = 1, \ldots, q$. The basis vectors are mutually anticommutative and satisfy the basic property

$$e_i^2 = 1, \text{ and } f_j^2 = -1,$$

as can be easily verified. For $p, q > 0$, let $n = p + q$. The *position vector* $x \in \mathbb{R}^{p,q}$ is defined by

$$x = (x^1, x^2, \ldots, x^{p+q}) := \sum_{i=1}^{p} x^i e_i + \sum_{j=1}^{q} x^{p+j} f_j.$$

More details in the construction of the standard basis can be found in [7, p. 71].

10. The real geometric algebra

$$\mathbb{G}_{n,n+1} := \mathbb{R}(e_1, \ldots, e_n, f_1, \ldots, f_{n+1}) = \mathbb{G}_n(\mathbb{C}),$$

where the imaginary i has the geometric interpretation of the oriented unit pseudoscalar in $\mathbb{G}_{n,n+1}$:

$$i = \sqrt{-1} := (e_1 f_1) \cdots (e_n f_n) f_{n+1} \iff f_{n+1} := i(e_1 f_1) \cdots (e_n f_n).$$

How geometric matrices arise as algebraically isomorphic coordinate matrices, and a practical application to the classical Plücker relations, is explored in [8]. A general introduction to geometric algebras and their coordinate matrices is given in [7]. A periodic table of all of the classical Clifford geometric algebras is derived from three *Fundamental Structure Theorems* in [9].

3 Affine Plane and Horosphere

Many ideas of projective geometry have been efficiently formulated in geometric algebra [5,10,11]. Following the approach given in [12, p.321-326], the basic ideas of the affine plane and horosphere are presented in terms of null vectors.

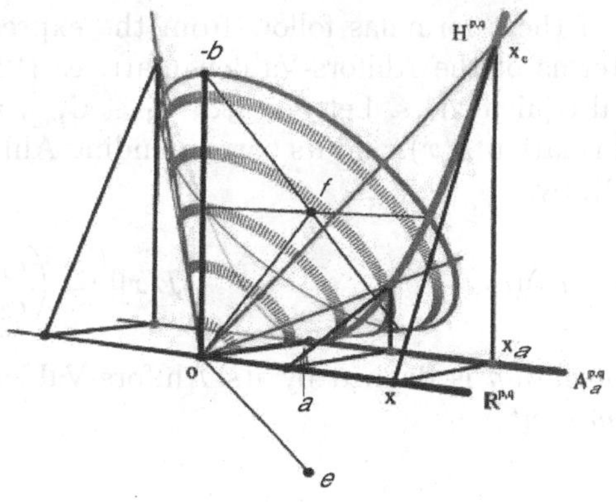

Fig. 1. The affine plane $\mathcal{A}_a^{p,q}$ and horosphere $\mathcal{H}^{p,q}$.

The *affine plane* $\mathcal{A}_a^{p,q} \subset \mathbb{R}^{p+1,q+1}$ is defined by

$$\mathcal{A}_a^{p,q} := \{x_a = x + a| \ \ x \in \mathbb{R}^{p,q} \subset \mathbb{R}^{p+1,q+1}\},$$

and the (p,q)-horosphere

$$\mathcal{H}^{p,q} := \{x_c = x_a b x_a| \ \ x \in \mathbb{R}^{p,q} \subset \mathbb{R}^{p+1,q+1}\},$$

where $a, b \in \mathbb{G}_{1,1}$ are compatible null vectors (see Definition 5.) in $\mathbb{R}^{p+1,q+1}$, orthogonal to the subspace $\mathbb{R}^{p,q}$. A drawing of the affine plane $\mathcal{A}^{p,q}$ and horosphere $\mathcal{H}^{p,q}$ is given in Fig. 1. We now easily establish basic relationships between these fundamental constructions [12, pp.321-325].

For the point $x_a = x + a \in \mathcal{A}^{p,q}$, $x_a^2 = x^2$, and for $x_c = x_a b x_a \in \mathcal{H}^{p,q}$,

$$x_c = (x_a \cdot b + x_a \wedge b)x_a = \frac{1}{2}x_a + (x_a \wedge b)x_a = x_a - x^2 b$$

$$= (a + xba + xab - x^2 b) = (1 + xb)a(1 - xb) = e^{xb} a e^{-xb}.$$

We also calculate

$$x_c \cdot y_c = (x_a - x^2 b) \cdot (y_a - y^2 b) = x \cdot y - \frac{1}{2}(x^2 + y^2) = -\frac{1}{2}(x - y)^2,$$

giving the relationship between the inner product of the points $x_c, y_c \in \mathcal{H}^{p,q}$ and the pseudoeuclidean distance between x and y. The point $x_a \in \mathcal{A}^{p,q}$ can be recovered from $x_c \in \mathcal{H}^{p,q}$,

$$x_a = 2(x_c \wedge b) \cdot a,$$

and the point $x \in \mathbb{R}^{p,q}$ can be recovered from x_c, by

$$x = 2(x \wedge b) \cdot a = 4(x_c \wedge b \wedge a) \cdot (b \wedge a).$$

The importance of these formulas follow from the expression of conformal transformations in terms of the Ahlfors-Vahlen matrices [12, p.287], [13], and their linear fractional equivalences. Let $c_1, c_2, c_3, c_4 \in \mathbb{G}_{p,q}$, and define the linear fractional transformation $L(x)$, and its corresponding Ahlfors-Vahlen matrix transformation $[L(x)]$, by

$$L(x) := (c_1 x + c_2)(c_3 x + c_4)^{-1} \quad \longleftrightarrow \quad [L[x]] := \begin{pmatrix} c_1 & c_2 \\ c_3 & c_4 \end{pmatrix} \begin{pmatrix} x \\ 1 \end{pmatrix}.$$

A *conformal transformation* is defined by its Ahlfors-Vahlen matrix provided that its *pseudodeterminant*

$$pdet \begin{pmatrix} c_1 & c_2 \\ c_3 & c_4 \end{pmatrix} := c_1 c_4^\dagger - c_2 c_3^\dagger \neq 0,$$

where the $\dagger$ denotes the operation of *reverse*, [13, p.271]. When restricted to the values given in the table below, the mapping $L(x)$ defines a conformal transformation $L : \mathbb{R}^{p,q} \rightarrow \mathbb{R}^{p,q}$, mapping the horosphere $\mathcal{H}^{p,q}$ onto itself [12, p.325], [13].

	G	$[G]$	$(c_1 x + c_2)(c_3 x + c_4)^{-1}$
Translation	e^{yb}	$\begin{pmatrix} 1 & y \\ 0 & 1 \end{pmatrix}$	$x + y$
Inversion	$a + b$	$\begin{pmatrix} 0 & 1 \\ 1 & 0 \end{pmatrix}$	$\frac{1}{x}$
Dilation	$e^{\frac{1}{2}\phi u}$	$\begin{pmatrix} e^{\frac{1}{2}\phi} & 0 \\ 0 & e^{-\frac{1}{2}\phi} \end{pmatrix}$	$e^{\phi} x$
Reflection	y	$\begin{pmatrix} y & 0 \\ 0 & -y \end{pmatrix}$	$-yxy^{-1}$
Transversion	e^{ca}	$\begin{pmatrix} 1 & 0 \\ -c & 1 \end{pmatrix}$	$x(1 - cx)^{-1}$

Acknowledgements. The seeds of this note were planted almost 40 years ago in discussions with Professor Zbigniew Oziewicz, a distinguished colleague, about the fundamental role played by *duality* in its many different guises in mathematics and physics [14]. The discussion resurfaced recently in an exchange of emails with Information Scientist Dr. Manfred von Willich. I hope that my treatment here will open up the discussion to the wider scientific community. The author thanks the *Zbigniew Oziewicz Seminar on Fundamental Problems in Physics* group for many fruitful discussions of the ideas herein, which greatly helped sharpen their exposition, [15]. He also thanks the ICACGA Reviewers of this paper for helpful suggestions leading to its improvement.

References

1. Narici, L.: On the Hahn-Banach Theorem. In: Proceedings of the Second International Course of Mathematical Analysis in Andalucia, pp. 87–123, Sept. 20–24, Granada, Editors: M.V. Velasco & Al., World Scientific (2004). http://at.yorku.ca/p/a/a/o/58.pdf
2. De La Harpe, P.: The Clifford algebra and the Spinor group of a Hilbert space. Compositio Mathematica, tome **25**(3), 245–261 (1972)
3. Shale, D., Stinespring, W.: Spinor representations of infinite orthogonal groups. J. Math. Mech. **14**(2), 315–322 (1965)
4. Doran, C., Hestenes, D., Sommen, F., Van Acker, N.: Lie groups as spin groups. J. Math. Phys. **34**, 3642–3669 (1993)
5. Lounesto, P.: Crumeyrolle's Bivectors and Spinors. In: Ablamowicz, R., Lounesto, P. (eds.) Clifford Algebras and Spinor Structures, pp. 137–166, Kluwer (1995)
6. Dieudonné, J.D.: The tragedy of grassmann. Linear Multilinear Algebra **8**, 1–14 (1979)
7. Sobczyk, G.: Matrix Gateway to Geometric Algebra, Spacetime and Spinors, Independent Publisher, Nov. 7, 2019
8. Sobczyk, G.: Notes on Plücker's relations in geometric algebra. Adv. Math. **363**, 106959 (2020)
9. Sobczyk, G.: Periodic Table of Geometric Numbers, 12 March 2020. https://arxiv.org/pdf/2003.07159.pdf
10. Hestenes, D.: The design of linear algebra and geometry. Acta Appl. Math. **23**, 65–93 (1991). Kluwer Academic, Dordrecht
11. Hestenes, D., Ziegler, R.: Projective geometry with Clifford algebra. Acta Appl. Math. Vol. 23, pp. 25–63. Kluwer Academic, Dordrecht (1991)
12. Sobczyk, G.: New Foundations in Mathematics: The Geometric Concept of Number. Birkhäuser, New York (2013)
13. Sobczyk, G.: Conformal mappings in geometric algebra. Notices AMS **59**(2), 264–273 (2012)
14. Oziewicz, Z.: From Grassmann to Clifford, p. 245–256 in Clifford Algebras and Their Applications in Mathematical Physics, eds. J.S.R. Chisholm, A.K. Common, NATO ASI Series C: Mathematical and Physical Sciences, vol. 183 (1986)
15. Cruz Guzman, J., Page, B.: Zbigniew Oziewicz Seminar on Fundamental Problems in Physics. FESC-Cuautitlan Izcalli UNAM, Mexico. https://www.youtube.com/channel/UCBcXAdMO3q6JBNyvBBVLmQg

Inner Product of Two Oriented Points in Conformal Geometric Algebra

Eckhard Hitzer(✉)

International Christian University, 181-8585 Mitaka, Tokyo, Japan
hitzer@icu.ac.jp
https://geometricalgebrajp.wordpress.com/

Abstract. We study the inner product of oriented points in conformal geometric algebra and its geometric meaning. The notion of oriented point is introduced and the inner product of two general oriented points is computed, and analyzed (including symmetry) in terms of point to point distance, and angles between the distance vector and the local orientation planes of the two points. Seven examples illustrate the results obtained. Finally, the results are extended from dimension three to arbitrary dimensions n.

Keywords: Conformal geometric algebra · oriented points · point geometry

1 Introduction

In this work we apply conformal geometric algebra (CGA) to the description of points, including a planar orientation. An excellent general reference on Clifford's geometric algebras is [13], a short engineering oriented tutorial is [10], and [14] describes a free software extension for a standard industrial computer algebra system (MATLAB). Alternatively, all computations could be done in the optimized geometric algebra algorithm software GAALOP [6]. Introductions to CGA are given in [2,4] and efficient computational implementations are described in [6]. CGA has found wide ranging applications in physics, quantum computing, molecular geometry, engineering, signal and image processing, neural networks, computer graphics and vision, encryption, robotics, electronic and power engineering, etc. Up to date surveys are [1,8,12]. An introduction to the notion of oriented point can be found in [5]. A prominent application could be to LIDAR terrain strip adjustment [11].

In the current work, we begin with the CGA expression for oriented points in three Euclidean dimensions and compute their inner products (Sect. 2). We study the geometric information included in this inner product with the help of a wide range of representative examples (Sect. 3), analyze the most important

Dedicated to the truth. Please note that this research is subject to the Creative Peace License [9].

D. W. Silva et al. (Eds.): ICACGA 2022, LNCS 13771, pp. 48–59, 2024.
https://doi.org/10.1007/978-3-031-34031-4_5

term that includes the direction of the line segment connecting the two points and their two point orientations in detail (Sect. 4), and study the symmetries of the inner product of oriented points (Sect. 5). Finally, we extend our framework from three to n Euclidean dimensions (Sect. 6).

2 Computation of Inner Product of Oriented Points

We consider the inner product of two oriented points in conformal geometric algebra [5], as reference for practical CGA computations in this section we recommend [7]. Note that inner product and wedge product have priority over the geometric product, e.g., $\mathbf{i}_q \cdot \boldsymbol{q}E = (\mathbf{i}_q \cdot \boldsymbol{q})E$, etc. An *oriented point* is given by the multivector expression of a *circle with radius zero* ($r = 0$) in CGA,

$$Q = \mathbf{i}_q \wedge \boldsymbol{q} + [\frac{1}{2}\boldsymbol{q}^2\mathbf{i}_q - \boldsymbol{q}(\boldsymbol{q} \cdot \mathbf{i}_q)]\boldsymbol{e}_\infty + \mathbf{i}_q\boldsymbol{e}_0 + \mathbf{i}_q \cdot \boldsymbol{q}E, \tag{1}$$

where the three-dimensional position vector of Q is the vector $\boldsymbol{q} \in \mathbb{R}^3$, the unit oriented bivector of the plane (orthogonal to the normal vector $\boldsymbol{n}_q$ of the plane) is $\mathbf{i}_q \in Cl^2(3,0)$, $\boldsymbol{e}_0$ is the vector for the origin dimension, $\boldsymbol{e}_\infty$ is the vector for the infinity dimension, and the origin-infinity bivector is $E = \boldsymbol{e}_\infty \wedge \boldsymbol{e}_0$, with

$$\boldsymbol{e}_0^2 = \boldsymbol{e}_\infty^2 = 0, \quad \boldsymbol{e}_0 \cdot \boldsymbol{e}_\infty = -1, \tag{2}$$

and $\boldsymbol{e}_0$ and $\boldsymbol{e}_\infty$ are both orthogonal to $\mathbb{R}^3$. For comparison we also state the expression of a conformal point (without orientation: *no*) and circle[1] in CGA:

$$Q_{no} = \boldsymbol{q} + \frac{1}{2}\boldsymbol{q}^2\boldsymbol{e}_\infty + \boldsymbol{e}_0, \qquad C = Q + \frac{1}{2}r^2\mathbf{i}_q\boldsymbol{e}_\infty, \tag{3}$$

where Q_{no} is simply given by the three-dimensional position vector $\boldsymbol{q} \in \mathbb{R}^3$ plus two terms in $\boldsymbol{e}_\infty$ and $\boldsymbol{e}_0$, while the conformal expression for the circle is the same as the oriented point (1), albeit with finite radius $r > 0$.

The Euclidean bivector $\mathbf{i}_q$ specifying the Euclidean carrier of the circle, respectively the orientation (local plane information) of the oriented point, can be obtained as (right contraction: $\lfloor$)

$$\mathbf{i}_q = -(C \wedge \boldsymbol{e}_\infty)\lfloor E = -(Q \wedge \boldsymbol{e}_\infty)\lfloor E. \tag{4}$$

The point Q_{no}, geometrically at the center of the circle C, can be directly obtained from[2]

$$Q_{no} = \widehat{C}\boldsymbol{e}_\infty C = \widehat{Q}\boldsymbol{e}_\infty Q, \tag{5}$$

[1] Two ways to obtain a circle in CGA are: (1) by the outer product of any three conformal points on the circle, (2) by combining center vector $\boldsymbol{q}$, carrier bivector $\mathbf{i}_q$ and radius r as specified by (1) and (3).

[2] For comparison one can norm the result, such that the $\boldsymbol{e}_0$-component becomes one: $Q_{no}/(-Q_{no} \cdot \boldsymbol{e}_\infty)$.

the three-dimensional position vector $\boldsymbol{q} \in \mathbb{R}^3$ from

$$\boldsymbol{q} = \frac{(Q_{no} \wedge E)\lfloor E}{-Q_{no} \cdot \boldsymbol{e}_\infty}, \tag{6}$$

and the radius of the circle as

$$r^2 = \frac{C\widehat{C}}{\mathbf{i}_q^2}. \tag{7}$$

We take a second oriented point P positioned at the origin $\boldsymbol{p} = 0$ with plane orientation bivector $\mathbf{i}_p$,

$$P = \mathbf{i}_p \boldsymbol{e}_0. \tag{8}$$

Now we compute the inner product of P and Q by taking the scalar part of their geometric product

$$\begin{aligned}
P \cdot Q = \langle PQ \rangle &= \Big\langle (\mathbf{i}_p \boldsymbol{e}_0)\, \{\mathbf{i}_q \wedge \boldsymbol{q} + [\tfrac{1}{2}\boldsymbol{q}^2 \mathbf{i}_q - \boldsymbol{q}(\boldsymbol{q} \cdot \mathbf{i}_q)]\, \boldsymbol{e}_\infty + \mathbf{i}_q \boldsymbol{e}_0 + \mathbf{i}_q \cdot \boldsymbol{q} E\} \Big\rangle \\
&= \Big\langle \mathbf{i}_p \boldsymbol{e}_0 \, [\tfrac{1}{2}\boldsymbol{q}^2 \mathbf{i}_q - \boldsymbol{q}(\boldsymbol{q} \cdot \mathbf{i}_q)]\, \boldsymbol{e}_\infty \Big\rangle = -\Big\{ \tfrac{1}{2}\boldsymbol{q}^2 \langle \mathbf{i}_p \mathbf{i}_q \rangle - \langle \mathbf{i}_p \boldsymbol{q}\, (\boldsymbol{q} \cdot \mathbf{i}_q) \rangle \Big\} \\
&= -\frac{1}{2}\boldsymbol{q}^2 \mathbf{i}_p \cdot \mathbf{i}_q + \Big\langle (\mathbf{i}_p \cdot \boldsymbol{q} + \mathbf{i}_p \wedge \boldsymbol{q})\, (\boldsymbol{q} \cdot \mathbf{i}_q) \Big\rangle \\
&= -\frac{1}{2}\boldsymbol{q}^2 \mathbf{i}_p \cdot \mathbf{i}_q + \Big\langle (\mathbf{i}_p \cdot \boldsymbol{q})(\boldsymbol{q} \cdot \mathbf{i}_q) \Big\rangle = -\frac{1}{2}\boldsymbol{q}^2 \mathbf{i}_p \cdot \mathbf{i}_q - \Big\langle (\boldsymbol{q} \cdot \mathbf{i}_p)(\boldsymbol{q} \cdot \mathbf{i}_q) \Big\rangle.
\end{aligned} \tag{9}$$

Note that in this situation $\boldsymbol{q}$ becomes the Euclidean distance vector from P to Q.

We now use the fact that the unit oriented bivector $\mathbf{i}_q$ of the plane is dual to the unit normal vector $\boldsymbol{n}_q$ via multiplication with the three-dimensional Euclidean volume pseudoscalar $i_3 = \boldsymbol{e}_1\boldsymbol{e}_2\boldsymbol{e}_3$,

$$\mathbf{i}_q = \boldsymbol{n}_q i_3, \qquad \mathbf{i}_p = \boldsymbol{n}_p i_3. \tag{10}$$

This gives by (70) and (67) in [10], where $\times$ is the standard cross product of three-dimensional vector algebra,

$$\boldsymbol{q} \cdot \mathbf{i}_p = \boldsymbol{q} \cdot (\boldsymbol{n}_p i_3) = (\boldsymbol{q} \wedge \boldsymbol{n}_p) i_3 = -\boldsymbol{q} \times \boldsymbol{n}_p, \quad \boldsymbol{q} \cdot \mathbf{i}_q = -\boldsymbol{q} \times \boldsymbol{n}_q. \tag{11}$$

Therefore

$$-\langle (\boldsymbol{q} \cdot \mathbf{i}_p)(\boldsymbol{q} \cdot \mathbf{i}_q) \rangle = -\langle (\boldsymbol{q} \times \boldsymbol{n}_p)(\boldsymbol{q} \times \boldsymbol{n}_q) \rangle = -(\boldsymbol{q} \times \boldsymbol{n}_p) \cdot (\boldsymbol{q} \times \boldsymbol{n}_q). \tag{12}$$

Note: The resulting quadruple product appears in the proof of the spherical law of cosines [15]. The quadruple product can be expanded to

$$\begin{aligned}
-(\boldsymbol{q} \times \boldsymbol{n}_p) \cdot (\boldsymbol{q} \times \boldsymbol{n}_q) &= -[\boldsymbol{q}^2 \boldsymbol{n}_p \cdot \boldsymbol{n}_q - (\boldsymbol{q} \cdot \boldsymbol{n}_q)(\boldsymbol{q} \cdot \boldsymbol{n}_p)] \\
&= -\boldsymbol{q}^2 [\boldsymbol{n}_p \cdot \boldsymbol{n}_q - (\hat{\boldsymbol{q}} \cdot \boldsymbol{n}_q)(\hat{\boldsymbol{q}} \cdot \boldsymbol{n}_p)],
\end{aligned} \tag{13}$$

with unit distance direction vector $\hat{\boldsymbol{q}}$, such that $\boldsymbol{q} = |\boldsymbol{q}|\hat{\boldsymbol{q}}$. Note also that from (10)

$$\mathbf{i}_p \cdot \mathbf{i}_q = -\boldsymbol{n}_p \cdot \boldsymbol{n}_q. \tag{14}$$

Then we can write the *full inner product of two oriented points* as

$$\begin{aligned} P\cdot Q &= \frac{1}{2}\boldsymbol{q}^2\boldsymbol{n}_p\cdot\boldsymbol{n}_q - [\boldsymbol{q}^2\boldsymbol{n}_p\cdot\boldsymbol{n}_q - (\boldsymbol{q}\cdot\boldsymbol{n}_q)(\boldsymbol{q}\cdot\boldsymbol{n}_p)] \\ &= -\frac{1}{2}\boldsymbol{q}^2\boldsymbol{n}_p\cdot\boldsymbol{n}_q + (\boldsymbol{q}\cdot\boldsymbol{n}_q)(\boldsymbol{q}\cdot\boldsymbol{n}_p) \\ &= \boldsymbol{q}^2[-\frac{1}{2}\boldsymbol{n}_p\cdot\boldsymbol{n}_q + (\hat{\boldsymbol{q}}\cdot\boldsymbol{n}_q)(\hat{\boldsymbol{q}}\cdot\boldsymbol{n}_p)] \\ &= \boldsymbol{q}^2[-\frac{1}{2}\cos\alpha_{pq} + \cos\Theta_q\cos\Theta_p], \end{aligned} \tag{15}$$

if we define $\cos\alpha_{pq} = \boldsymbol{n}_p\cdot\boldsymbol{n}_q$, $\cos\Theta_q = \hat{\boldsymbol{q}}\cdot\boldsymbol{n}_q$, and $\cos\Theta_p = \hat{\boldsymbol{q}}\cdot\boldsymbol{n}_p$, where α_{pq} is the dihedral angle between the two planes, and Θ_q is the angle between the distance vector $\boldsymbol{q}$ and $\boldsymbol{n}_q$, while Θ_p is the angle between $\boldsymbol{q}$ and $\boldsymbol{n}_p$, respectively. See Fig. 1 for illustration, with P at the origin, and $\boldsymbol{q}$ replaced by $\boldsymbol{d}$.

Remark 1. Note that the above relation is fully general, even if P is a point in general position. Because our special situation, with P at the origin, is only different from the general situation by a global translation, which will not change the inner product $P\cdot Q = \langle PQ\rangle$. In the general case, the vector $\boldsymbol{q}$ will simply be replaced by the Euclidean distance vector between the two positions $\boldsymbol{d} = \boldsymbol{q} - \boldsymbol{p}$, see Fig. 1.

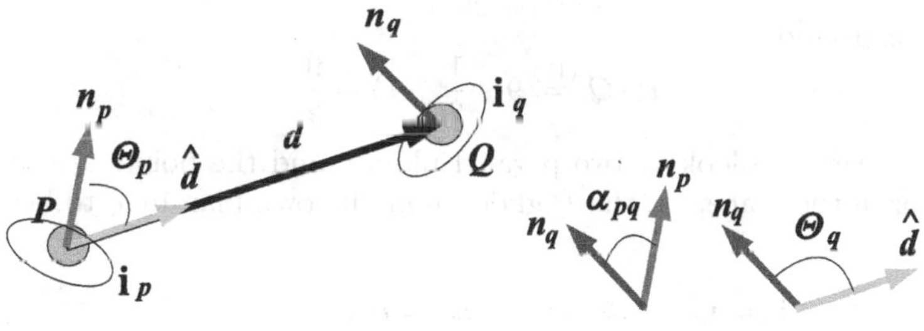

Fig. 1. Illustration of inner product of two oriented points P and Q, with Euclidean distance vector $\boldsymbol{d} = \boldsymbol{q} - \boldsymbol{p}$, α_{pq} dihedral angle between the two orientation planes $\mathbf{i}_p$ and $\mathbf{i}_q$, Θ_q angle between $\boldsymbol{d}$ and $\boldsymbol{n}_q$, and Θ_p angle between $\boldsymbol{d}$ and $\boldsymbol{n}_p$, respectively.

For the special case that the two planes are parallel[3], i.e. $\boldsymbol{n}_p = \boldsymbol{n}_q$, $\boldsymbol{n}_p\cdot\boldsymbol{n}_q = 1$, we have with the consequence $\Theta = \Theta_q = \Theta_p$ that

$$P\cdot Q = \boldsymbol{q}^2(-\frac{1}{2} + \cos^2\Theta) = -\frac{1}{2}\boldsymbol{q}^2(1 - 2\cos^2\Theta) = \frac{1}{2}\boldsymbol{q}^2\cos 2\Theta, \tag{16}$$

using the trigonometric identity $1 - 2\cos^2\Theta = -\cos 2\Theta$.

[3] This will also approximately be the case, if two matching oriented points are compared.

For the special case that additionally $\Theta = 0$, i.e. both planes are parallel and the distance $\boldsymbol{q}$ perpendicular to the planes we have

$$P \cdot Q = \frac{1}{2}\boldsymbol{q}^2. \tag{17}$$

3 Examples

To gain some intuition of what the inner product of two oriented points in CGA (15) means, we compute several examples, always assuming for simplicity that the first point P is positioned at the origin: $\boldsymbol{p} = 0$.

Example 1. First we look at two parallel planes at orthogonal distance three.

$$\mathbf{i}_p = \mathbf{i}_q = \boldsymbol{e}_{12}, \quad \boldsymbol{n}_p = \boldsymbol{n}_q = \boldsymbol{e}_3, \quad \boldsymbol{q} = 3\boldsymbol{e}_3, \quad \boldsymbol{q}^2 = 9, \quad \hat{\boldsymbol{q}} = \boldsymbol{e}_3. \tag{18}$$

Then we can compute directly

$$P \cdot Q = \left\langle \boldsymbol{e}_{12}\,\boldsymbol{e}_0\,[\frac{1}{2}\,9\,\boldsymbol{e}_{12} - 3\,\boldsymbol{e}_3\,(3\,\boldsymbol{e}_3\cdot\boldsymbol{e}_{12})]\,\boldsymbol{e}_\infty \right\rangle = -\frac{1}{2}\,9\,\boldsymbol{e}_{12}\,\boldsymbol{e}_{12} = \frac{9}{2}, \tag{19}$$

because $\boldsymbol{e}_0[\frac{1}{2}9\boldsymbol{e}_{12} - 3\boldsymbol{e}_3(3\boldsymbol{e}_3\cdot\boldsymbol{e}_{12})] = [\frac{1}{2}9\boldsymbol{e}_{12} - 3\boldsymbol{e}_3(3\boldsymbol{e}_3\cdot\boldsymbol{e}_{12})]\boldsymbol{e}_0$, $\boldsymbol{e}_0\cdot\boldsymbol{e}_\infty = -1$, $\boldsymbol{e}_3\cdot\boldsymbol{e}_{12} = 0$, and $\boldsymbol{e}_{12}^2 = -1$. The result also confirms (17).

We obtain the same result, if we apply (15) instead. Toward this we compute

$$\cos\alpha_{pq} = \boldsymbol{e}_3\cdot\boldsymbol{e}_3 = 1, \quad \cos\Theta_q = \boldsymbol{e}_3\cdot\boldsymbol{e}_3 = 1, \quad \cos\Theta_p = \boldsymbol{e}_3\cdot\boldsymbol{e}_3 = 1. \tag{20}$$

Hence, as expected

$$P\cdot Q \overset{(15)}{=} 9(-\frac{1}{2}+1) = \frac{9}{2}. \tag{21}$$

Example 2. Next we look at two parallel planes, and the points are separated by a vector in the plane, i.e. the P glides along its own plan by $\boldsymbol{q}$ to become Q. Assuming

$$\begin{aligned} &\mathbf{i}_p = \mathbf{i}_q = \boldsymbol{e}_{12}, \quad \boldsymbol{n}_p = \boldsymbol{n}_q = \boldsymbol{e}_3, \\ &\boldsymbol{q} = \boldsymbol{e}_1 + \boldsymbol{e}_2, \quad \boldsymbol{q}^2 = 2, \quad \hat{\boldsymbol{q}} = \frac{1}{\sqrt{2}}(\boldsymbol{e}_1+\boldsymbol{e}_2), \end{aligned} \tag{22}$$

we obtain

$$\begin{aligned} &\cos\alpha_{pq} = \boldsymbol{e}_3\cdot\boldsymbol{e}_3 = 1, \quad \cos\Theta_q = \frac{1}{\sqrt{2}}(\boldsymbol{e}_1+\boldsymbol{e}_2)\cdot\boldsymbol{e}_3 = 0, \\ &\cos\Theta_p = \frac{1}{\sqrt{2}}(\boldsymbol{e}_1+\boldsymbol{e}_2)\cdot\boldsymbol{e}_3 = 0. \end{aligned} \tag{23}$$

Applying (15) the inner product becomes

$$P\cdot Q = 2(-\frac{1}{2}+0) = -1. \tag{24}$$

In this case only the first term in (15) proportional to $\boldsymbol{n}_p\cdot\boldsymbol{n}_p$ contributes.

Example 3. We now look again at two parallel planes, but the Euclidean distance vector $\boldsymbol{q}$ is at angle $\pi/4$ with the planes. We assume

$$\begin{aligned} \mathbf{i}_p &= \mathbf{i}_q = \boldsymbol{e}_{12}, \quad \boldsymbol{n}_p = \boldsymbol{n}_q = \boldsymbol{e}_3, \\ \boldsymbol{q} &= \boldsymbol{e}_1 + \boldsymbol{e}_3, \quad \boldsymbol{q}^2 = 2, \quad \hat{\boldsymbol{q}} = \frac{1}{\sqrt{2}}(\boldsymbol{e}_1 + \boldsymbol{e}_3), \end{aligned} \tag{25}$$

and obtain

$$\begin{aligned} &\cos \alpha_{pq} = \boldsymbol{e}_3 \cdot \boldsymbol{e}_3 = 1, \quad \cos \Theta_q = \frac{1}{\sqrt{2}}(\boldsymbol{e}_1 + \boldsymbol{e}_3) \cdot \boldsymbol{e}_3 = \frac{1}{\sqrt{2}}, \\ &\cos \Theta_p = \frac{1}{\sqrt{2}}(\boldsymbol{e}_1 + \boldsymbol{e}_3) \cdot \boldsymbol{e}_3 = \frac{1}{\sqrt{2}}. \end{aligned} \tag{26}$$

Applying (15) the inner product becomes

$$P \cdot Q = 2(-\frac{1}{2}1 + \frac{1}{\sqrt{2}}\frac{1}{\sqrt{2}}) = 0. \tag{27}$$

This is a special case, where both terms in (15) are non-zero, but happen to cancel each other.

Example 4. Now we take two planes vertical to each other, and the distance vector is perpendicular to the first and parallel to the second. We assume

$$\begin{aligned} \mathbf{i}_p &= \boldsymbol{e}_{12}, \quad \mathbf{i}_q = \boldsymbol{e}_{23}, \quad \boldsymbol{n}_p = \boldsymbol{e}_3, \quad \boldsymbol{n}_q = \boldsymbol{e}_1, \\ \boldsymbol{q} &= 3\boldsymbol{e}_3, \quad \boldsymbol{q}^2 = 9, \quad \hat{\boldsymbol{q}} = \boldsymbol{e}_3, \end{aligned} \tag{28}$$

and obtain

$$\cos \alpha_{pq} = \boldsymbol{e}_3 \cdot \boldsymbol{e}_1 = 0, \quad \cos \Theta_q = \boldsymbol{e}_3 \cdot \boldsymbol{e}_1 = 0, \quad \cos \Theta_p = \boldsymbol{e}_3 \cdot \boldsymbol{e}_3 = 1. \tag{29}$$

Applying (15) the inner product becomes

$$P \cdot Q = 9(-\frac{1}{2}0 + 0) = 0. \tag{30}$$

Obviously, if the two planes are vertical to each other, and the Euclidean distance vector is parallel to one of the planes, the result is always zero.

Example 5. This example is simply a variation of the previous one, with a different orientation of $\mathbf{i}_q$. We assume

$$\begin{aligned} &\mathbf{i}_p = \boldsymbol{e}_{12}, \quad \mathbf{i}_q = \frac{1}{\sqrt{2}}(\boldsymbol{e}_{13} + \boldsymbol{e}_{23}), \quad \boldsymbol{n}_p = \boldsymbol{e}_3, \quad \boldsymbol{n}_q = \frac{1}{\sqrt{2}}(-\boldsymbol{e}_2 + \boldsymbol{e}_1), \\ &\boldsymbol{q} = 3\boldsymbol{e}_3, \quad \boldsymbol{q}^2 = 9, \quad \hat{\boldsymbol{q}} = \boldsymbol{e}_3. \end{aligned} \tag{31}$$

and obtain

$$\begin{aligned} &\cos \alpha_{pq} = \boldsymbol{e}_3 \cdot \frac{1}{\sqrt{2}}(-\boldsymbol{e}_2 + \boldsymbol{e}_1) = 0, \quad \cos \Theta_q = \boldsymbol{e}_3 \cdot \frac{1}{\sqrt{2}}(-\boldsymbol{e}_2 + \boldsymbol{e}_1) = 0, \\ &\cos \Theta_p = \boldsymbol{e}_3 \cdot \boldsymbol{e}_3 = 1. \end{aligned} \tag{32}$$

Applying (15) the inner product becomes zero again

$$P \cdot Q = 9(-\frac{1}{2}0 + 0) = 0. \tag{33}$$

Example 6. In this example the second plane is tilted with respect to the first by a dihedral angle of $\pi/4$. The distance vector is perpendicular to the first plane and at angle $\pi/4$ with the second:

$$\begin{aligned} &\mathbf{i}_p = \boldsymbol{e}_{12}, \quad \mathbf{i}_q = \frac{1}{\sqrt{2}}(\boldsymbol{e}_{12} + \boldsymbol{e}_{23}), \quad \boldsymbol{n}_p = \boldsymbol{e}_3, \quad \boldsymbol{n}_q = \frac{1}{\sqrt{2}}(\boldsymbol{e}_3 + \boldsymbol{e}_1), \\ &\boldsymbol{q} = 3\boldsymbol{e}_3, \quad \boldsymbol{q}^2 = 9, \quad \hat{\boldsymbol{q}} = \boldsymbol{e}_3. \end{aligned} \tag{34}$$

We obtain

$$\begin{aligned} &\cos\alpha_{pq} = \boldsymbol{e}_3 \cdot \frac{1}{\sqrt{2}}(\boldsymbol{e}_3 + \boldsymbol{e}_1) = \frac{1}{\sqrt{2}}, \quad \cos\Theta_q = \boldsymbol{e}_3 \cdot \frac{1}{\sqrt{2}}(\boldsymbol{e}_3 + \boldsymbol{e}_1) = \frac{1}{\sqrt{2}}, \\ &\cos\Theta_p = \boldsymbol{e}_3 \cdot \boldsymbol{e}_3 = 1. \end{aligned} \tag{35}$$

Applying (15) the inner product becomes

$$P \cdot Q = 9(-\frac{1}{2}\frac{1}{\sqrt{2}} + \frac{1}{\sqrt{2}}) = \frac{9}{2\sqrt{2}}. \tag{36}$$

Here both terms in (15) contribute and the second term $\cos\Theta_q \cos\Theta_p$ dominates.

Example 7. Here we take the two planes to be parallel, and a more general Euclidean distance vector:

$$\begin{aligned} &\mathbf{i}_p = \boldsymbol{e}_{12}, \quad \mathbf{i}_q = \boldsymbol{e}_{12}, \quad \boldsymbol{n}_p = \boldsymbol{e}_3, \quad \boldsymbol{n}_q = \boldsymbol{e}_3, \\ &\boldsymbol{q} = 2\boldsymbol{e}_2 + 3\boldsymbol{e}_3, \quad \boldsymbol{q}^2 = 13, \quad \hat{\boldsymbol{q}} = \frac{1}{\sqrt{13}}(2\boldsymbol{e}_2 + 3\boldsymbol{e}_3). \end{aligned} \tag{37}$$

We obtain

$$\begin{aligned} &\cos\alpha_{pq} = \boldsymbol{e}_3 \cdot \boldsymbol{e}_3 = 1, \quad \cos\Theta_q = \boldsymbol{e}_3 \cdot \frac{1}{\sqrt{13}}(2\boldsymbol{e}_2 + 3\boldsymbol{e}_3) = \frac{3}{\sqrt{13}}, \\ &\cos\Theta_p = \boldsymbol{e}_3 \cdot \frac{1}{\sqrt{13}}(2\boldsymbol{e}_2 + 3\boldsymbol{e}_3) = \frac{3}{\sqrt{13}}. \end{aligned} \tag{38}$$

Applying (15) the inner product becomes

$$P \cdot Q = 13(-\frac{1}{2} + \frac{9}{13}) = \frac{5}{2}. \tag{39}$$

4 About the Term $(\hat{\boldsymbol{q}} \cdot \boldsymbol{n}_q)(\hat{\boldsymbol{q}} \cdot \boldsymbol{n}_p)$ in $P \cdot Q$

- For $\boldsymbol{n}_q \nparallel \boldsymbol{n}_q$ the two plane normal vectors together define a plane that can be specified by the bivector $\boldsymbol{n}_q \wedge \boldsymbol{n}_q$. This allows to split the Euclidean distance vector $\boldsymbol{q}$ into parts parallel $\boldsymbol{q}_{||}$ and perpendicular $\boldsymbol{q}_{\perp}$ to the $\boldsymbol{n}_q \wedge \boldsymbol{n}_q$-plane. In the inner products of $(\hat{\boldsymbol{q}} \cdot \boldsymbol{n}_q)(\hat{\boldsymbol{q}} \cdot \boldsymbol{n}_p)$ of (15), the perpendicular $\boldsymbol{q}_{\perp}$ part will not contribute, because it is perpendicular to both $\boldsymbol{n}_q$ and $\boldsymbol{n}_q$. So we get

$$(\hat{\boldsymbol{q}} \cdot \boldsymbol{n}_q)(\hat{\boldsymbol{q}} \cdot \boldsymbol{n}_p) = (\hat{\boldsymbol{q}}_{||} \cdot \boldsymbol{n}_q)(\hat{\boldsymbol{q}}_{||} \cdot \boldsymbol{n}_p). \tag{40}$$

- For $\boldsymbol{n}_q = \boldsymbol{n}_p$, the part $\boldsymbol{q}_{\perp}$ perpendicular to $\boldsymbol{n}_q$ drops out, and only the part $\hat{\boldsymbol{q}}_{||}$ parallel to $\boldsymbol{n}_q$ contributes.

$$(\hat{\boldsymbol{q}} \cdot \boldsymbol{n}_q)(\hat{\boldsymbol{q}} \cdot \boldsymbol{n}_p) = (\hat{\boldsymbol{q}}_{||} \cdot \boldsymbol{n}_p)^2. \tag{41}$$

- If $\hat{\boldsymbol{q}} \perp \boldsymbol{n}_q$ or $\hat{\boldsymbol{q}} \perp \boldsymbol{n}_p$, then

$$(\hat{\boldsymbol{q}} \cdot \boldsymbol{n}_q)(\hat{\boldsymbol{q}} \cdot \boldsymbol{n}_p) = 0. \tag{42}$$

- For $\hat{\boldsymbol{q}} = \boldsymbol{n}_q = \boldsymbol{n}_p$ or $\hat{\boldsymbol{q}} = -\boldsymbol{n}_q = -\boldsymbol{n}_p$ we the get a maximal contribution of $(\hat{\boldsymbol{q}} \cdot \boldsymbol{n}_q)(\hat{\boldsymbol{q}} \cdot \boldsymbol{n}_p)$ to the inner product. Then

$$P \cdot Q = \frac{1}{2}\boldsymbol{q}^2. \tag{43}$$

- For $\hat{\boldsymbol{q}} = \boldsymbol{n}_q = -\boldsymbol{n}_p$ or $\hat{\boldsymbol{q}} = -\boldsymbol{n}_q = \boldsymbol{n}_p$ we the get a minimal contribution of $(\hat{\boldsymbol{q}} \cdot \boldsymbol{n}_q)(\hat{\boldsymbol{q}} \cdot \boldsymbol{n}_p)$ to the inner product. Then

$$P \cdot Q = -\frac{1}{2}\boldsymbol{q}^2. \tag{44}$$

5 Symmetries of $P \cdot Q$

The inner product of two oriented points in CGA of (15) is a function of the three unit vectors $\hat{\boldsymbol{q}}$, $\boldsymbol{n}_q$, and $\boldsymbol{n}_p$, i.e. the unit direction of the Euclidean distance, and the two unit normal vectors of the two planes.

$$P \cdot Q = \boldsymbol{q}^2[-\frac{1}{2}\boldsymbol{n}_p \cdot \boldsymbol{n}_q + (\hat{\boldsymbol{q}} \cdot \boldsymbol{n}_q)(\hat{\boldsymbol{q}} \cdot \boldsymbol{n}_p)] = f(\boldsymbol{n}_p, \boldsymbol{n}_q, \hat{\boldsymbol{q}}). \tag{45}$$

The function $f(\boldsymbol{n}_p, \boldsymbol{n}_q, \hat{\boldsymbol{q}})$ has the following symmetries

$$\begin{aligned} f(-\boldsymbol{n}_p, \boldsymbol{n}_q, \hat{\boldsymbol{q}}) &= f(\boldsymbol{n}_p, -\boldsymbol{n}_q, \hat{\boldsymbol{q}}) = -f(\boldsymbol{n}_p, \boldsymbol{n}_q, \hat{\boldsymbol{q}}), \\ f(\boldsymbol{n}_p, \boldsymbol{n}_q, -\hat{\boldsymbol{q}}) &= f(\boldsymbol{n}_p, \boldsymbol{n}_q, \hat{\boldsymbol{q}}). \end{aligned} \tag{46}$$

That is changing the sign of any one of the two plane normal vectors changes the sign of $P \cdot Q$, while changing the sign of the Euclidean distance vector leaves $P \cdot Q$ invariant.

6 Inner Product of Oriented Points for n-Dimensional Euclidean Space

In this section we aim to show that the inner product relationship (15) of oriented points in CGA, applies in any dimension $n \geq 2$, up to an overall sign.

We are now working with CGA $Cl(n+1,1)$ of n-dimensional Euclidean space $\mathbb{R}^n$. Its pseudoscalar is

$$I = I_{n+1,1} = I_n E, \quad I_n = e_1 e_2 \cdots e_n, \quad E = e_\infty \wedge e_0, \tag{47}$$

with squares

$$E^2 = 1, \quad I^2 = (I_n E)^2 = I_n^2 E^2 = I_n^2 = \begin{cases} +1, n \bmod 4 = 1, 0 \\ -1, n \bmod 4 = 2, 3. \end{cases} \tag{48}$$

Depending on the dimension n we therefore have the inverse of the pseudoscalar to be

$$I^{-1} = \begin{cases} +I, n \bmod 4 = 1, 0 \\ -I, n \bmod 4 = 2, 3. \end{cases} \tag{49}$$

The dual of a multivector $M \in Cl(n+1,1)$ is given by

$$M^* = MI^{-1}, \quad M = M^* I. \tag{50}$$

Especially for two bivectors M_b and N_b we have the inner product relationship with the duals of the bivectors to be

$$\begin{aligned} \langle M_b^* N_b^* \rangle &= \langle M_b(\pm I) N_b(\pm I) \rangle = \langle M_b N_b I^2 \rangle \\ &= \langle M_b N_b \rangle \begin{Bmatrix} +1, n \bmod 4 = 1, 0 \\ -1, n \bmod 4 = 2, 3 \end{Bmatrix}, \end{aligned} \tag{51}$$

where we used the commutation of $IN_b = N_b I$ for bivectors N_b. For example in the case of $n = 3$ we have

$$\langle M_b^* N_b^* \rangle = -\langle M_b N_b \rangle. \tag{52}$$

We now construct an oriented point in $Cl(n+1,n)$ by taking a dual sphere vector centered at the Euclidean position of the point $\boldsymbol{p} \in \mathbb{R}^n$ intersected with a dual equator plane[4] orthogonal to normal unit vector $\boldsymbol{n}_p \in \mathbb{R}^n$, $\boldsymbol{n}_p^2 = 1$, and take the limit of the sphere radius $r \to 0$. The dual sphere is

$$\sigma = S^* = C_p - \frac{1}{2} r^2 e_\infty, \tag{53}$$

with conformal center point

$$C_p = \boldsymbol{p} + \frac{1}{2}\boldsymbol{p}^2 e_\infty + e_0. \tag{54}$$

[4] Strictly speaking this is a hyper-plane of dimension $n - 1$.

The dual equator plane that has to include the center C_p and be normal to $\boldsymbol{n}_p$ is

$$\mu = Plane^* = \boldsymbol{n}_p + de_\infty = \boldsymbol{n}_p + (\boldsymbol{p}\cdot\boldsymbol{n}_p)e_\infty, \tag{55}$$

using oriented distance $d \in \mathbb{R}$ from the origin

$$d = C_p \cdot \boldsymbol{n}_p = (\boldsymbol{p} + \frac{1}{2}\boldsymbol{p}^2 e_\infty + e_0)\cdot \boldsymbol{n}_p = \boldsymbol{p}\cdot\boldsymbol{n}_p. \tag{56}$$

The dual of the equator circle[5] is given by the outer product of dual equator plane and dual sphere

$$Circle^* = \mu \wedge \sigma = [\boldsymbol{n}_p + (\boldsymbol{p}\cdot\boldsymbol{n}_p)e_\infty]\wedge(C_p - \frac{1}{2}r^2 e_\infty). \tag{57}$$

Taking the limit of sphere radius $r \to 0$, and inserting the expression for C_p, we get the dual of an oriented point in CGA $Cl(n+1,1)$ located at $\boldsymbol{p}\in\mathbb{R}^n$ and oriented normal to $\boldsymbol{n}_p$

$$\begin{aligned} P^* &= (\boldsymbol{n}_p + \boldsymbol{p}\cdot\boldsymbol{n}_p\, e_\infty)\wedge(\boldsymbol{p} + \frac{1}{2}\boldsymbol{p}^2 e_\infty + e_0) \\ &= \boldsymbol{n}_p\wedge\boldsymbol{p} + \frac{1}{2}\boldsymbol{p}^2\boldsymbol{n}_p e_\infty + (\boldsymbol{p}\cdot\boldsymbol{n}_p)e_\infty\boldsymbol{p} + \boldsymbol{n}_p e_0 + (\boldsymbol{p}\cdot\boldsymbol{n}_p)(e_\infty\wedge e_0) \\ &= \boldsymbol{n}_p\wedge\boldsymbol{p} + [\frac{1}{2}\boldsymbol{p}^2\boldsymbol{n}_p - \boldsymbol{p}(\boldsymbol{p}\cdot\boldsymbol{n}_p)]e_\infty + \boldsymbol{n}_p e_0 + (\boldsymbol{p}\cdot\boldsymbol{n}_p)E, \end{aligned} \tag{58}$$

where we used the anti-commutation $e_\infty\boldsymbol{p} = -\boldsymbol{p}e_\infty$. A second dual oriented point located at $\boldsymbol{q}\in\mathbb{R}^n$ and oriented normal to $\boldsymbol{n}_q$ is then given by

$$Q^* = \boldsymbol{n}_q\wedge\boldsymbol{q} + [\frac{1}{2}\boldsymbol{q}^2\boldsymbol{n}_q - \boldsymbol{q}(\boldsymbol{q}\cdot\boldsymbol{n}_q)]e_\infty + \boldsymbol{n}_q e_0 + (\boldsymbol{q}\cdot\boldsymbol{n}_q)E. \tag{59}$$

Locating the first dual oriented point at the origin, i.e. $\boldsymbol{p}=0$, it becomes

$$P^* = \boldsymbol{n}_p e_0. \tag{60}$$

The inner product with the second dual oriented point in general position $\boldsymbol{q}$, $\boldsymbol{q}$ therefore marking the oriented distance vector of the two points, becomes

$$\begin{aligned} \langle P^*Q^*\rangle &= \langle \boldsymbol{n}_p e_0\{\boldsymbol{n}_q\wedge\boldsymbol{q} + [\frac{1}{2}\boldsymbol{q}^2 - \boldsymbol{q}(\boldsymbol{q}\cdot\boldsymbol{n}_q)]e_\infty + \boldsymbol{n}_q e_0 + (\boldsymbol{q}\cdot\boldsymbol{n}_q)E\}\rangle \\ &= \langle \boldsymbol{n}_p e_0[\frac{1}{2}\boldsymbol{q}^2\boldsymbol{n}_q - \boldsymbol{q}(\boldsymbol{q}\cdot\boldsymbol{n}_q)]e_\infty\rangle = \langle \boldsymbol{n}_p[\frac{1}{2}\boldsymbol{q}^2\boldsymbol{n}_q - \boldsymbol{q}(\boldsymbol{q}\cdot\boldsymbol{n}_q)]\rangle \\ &= \frac{1}{2}\boldsymbol{q}^2(\boldsymbol{n}_p\cdot\boldsymbol{n}_q) - (\boldsymbol{q}\cdot\boldsymbol{n}_p)(\boldsymbol{q}\cdot\boldsymbol{n}_q), \end{aligned} \tag{61}$$

[5] Note that in general dimensions this is a hyper-circle in the sense that for $n=2$ it is a point pair, for $n=3$ a normal circle, for $n=4$ the circle is itself a three-dimensional sphere embedded in four dimensions, etc.

using $e_0\boldsymbol{v} = -\boldsymbol{v}e_0$ for any vector $\boldsymbol{v} \in \mathbb{R}^n$, especially for $\boldsymbol{v} = [\frac{1}{2}\boldsymbol{q}^2\boldsymbol{n}_q - \boldsymbol{q}(\boldsymbol{q}\cdot\boldsymbol{n}_q)]$, and $-\langle e_0 e_\infty\rangle = 1$. Because P^* and Q^* are bivectors, the inner product of P and Q becomes by (51)

$$\langle PQ\rangle = \langle P^*Q^*\rangle \begin{Bmatrix} +1, n \bmod 4 = 1,0 \\ -1, n \bmod 4 = 2,3 \end{Bmatrix}$$
$$= \begin{Bmatrix} +1, n \bmod 4 = 1,0 \\ -1, n \bmod 4 = 2,3 \end{Bmatrix} [\frac{1}{2}\boldsymbol{q}^2(\boldsymbol{n}_p\cdot\boldsymbol{n}_q) - (\boldsymbol{q}\cdot\boldsymbol{n}_p)(\boldsymbol{q}\cdot\boldsymbol{n}_q)], \quad (62)$$

and obviously agrees by (52) in three dimensions ($n = 3$) with (15).

The analysis of the preceding Sects. 2, 4 and 5 therefore fully applies in general dimensions $n \geq 2$, up to an overall sign[6] due to the value of I^2, which is easy to take into account. And examples analogous to Sect. 3 are obviously easy to construct.

7 Conclusion

In this work we have reviewed the formulation of oriented points in conformal geometric algebra (CGA), and computed the inner product of two oriented points in terms of their distance vector (its direction and length) and their two point orientations. The geometric meaning of this inner product is elucidated based on a set of representative examples, analysis of the key term in the inner product, and symmetry analysis. Finally, the approach is extended from three to n Euclidean dimensions. Our new results may find application in LIDAR terrain strip adjustment computations, where points on overlapping strips need to be compared together with the local plane orientation of the respective strip, see e.g. [11].

Acknowledgments. The author wishes to thank God: *In the beginning God created the heavens and the earth. The earth was without form, and void; and darkness was on the face of the deep. And the Spirit of God was hovering over the face of the waters. Then God said, "Let there be light"; and there was light.* (NKJV, Biblegateway [3]). He further thanks his colleagues W. Benger, D. Hildenbrand and M. Niederwieser, as well as everybody involved organizing ICACGA 2022. The author finally thanks all anonymous reviewers for providing excellent comments and advice.

References

1. Breuils, S., Tachibana, K., Hitzer, E.: New applications of Clifford's geometric algebra. Adv. Appl. Clifford Algebras **32**, 17 (2022). https://doi.org/10.1007/s00006-021-01196-7
2. Dorst, L., Fontijne, D., Mann, S.: Geometric Algebra for Computer Science, an Object-Oriented Approach to Geometry. Morgan Kaufmann, Burlington (2007)

[6] Taking the absolute value $|\langle PQ\rangle| = |\langle P^*Q^*\rangle|$, completely removes this overall dimension dependent sign, and may be all that is needed in many applications.

3. Genesis chapter 1 verse 1, in The Holy Bible, English Standard Version. Wheaton (Illinois): Crossway Bibles, Good News Publishers (2001)
4. Hestenes, D., Li, H., Rockwood, A.: New algebraic tools for classical geometry. In: Sommer, G. (ed.) Geometric Computing with Clifford Algebras. Springer, Berlin (2001). https://doi.org/10.1007/978-3-662-04621-0_1
5. Hildenbrand, D., Charrier, P.: Conformal geometric objects with focus on oriented points. In: Gürlebeck, K. (ed.) Proceedings of 9th International Conference on Clifford Algebras and their Applications in Mathematical Physics, Weimar, Germany, 15–20 July 2011, 10 p. (2011). http://www.gaalop.de/wp-content/uploads/LongConformalEntities_ICCA91.pdf
6. Hildenbrand, D.: Foundations of Geometric Algebra Computing, Springer, Berlin (2013). Introduction to Geometric Algebra Computing. CRC Press, Taylor & Francis Group, Boca Raton (2019). https://doi.org/10.1007/978-3-642-31794-1
7. Hitzer, E., Tachibana, K., Buchholz, S., Yu, I.: Carrier method for the general evaluation and control of pose, molecular conformation, tracking, and the like. Adv. Appl. Clifford Algebras **19**(2), 339–364 (2009). https://doi.org/10.1007/s00006-009-0160-9, https://www.researchgate.net/publication/226288320_Carrier_Method_for_the_General_Evaluation_and_Control_of_PoseMolecular_Conformation_Tracking_and_the_Like
8. Hitzer, E., Nitta, T., Kuroe, Y.: Applications of Clifford's geometric algebra. Adv. Appl. Clifford Algebras **23**, 377–404 (2013). https://doi.org/10.1007/s00006-013-0378-4
9. Hitzer, E.: Creative Peace License. http://gaupdate.wordpress.com/2011/12/14/the-creative-peace-license-14-dec-2011/. Accessed 12 June 2020
10. Hitzer, E.: Introduction to Clifford's geometric algebra. SICE J. Control Meas. Syst. Integr. **51**(4), 338–350 (2012). http://arxiv.org/abs/1306.1660, https://doi.org/10.48550/arXiv.1306.1660. Accessed 12 June 2020
11. Hitzer, E., Benger, W., Niederwieser, M., Baran, R., Steinbacher, F.: Foundations for strip adjustment of airborne laserscanning data with conformal geometric algebra. Adv. Appl. Clifford Algebras **32**(1), 1–34 (2021). https://doi.org/10.1007/s00006-021-01184-x
12. Hitzer, E., Lavor, C., Hildenbrand, D.: Current survey of Clifford geometric algebra applications. Math. Meth. Appl. Sci. 1–31 (2022). https://onlinelibrary.wiley.com/doi/10.1002/mma.8316
13. Lounesto, P.: Clifford Algebras and Spinors, 2nd edn. CUP, Cambridge (2006)
14. Sangwine, S.J., Hitzer, E.: Clifford multivector toolbox (for MATLAB). Adv. Appl. Clifford Algebras **27**(1), 539–558 (2017). https://doi.org/10.1007/s00006-016-0666-x, http://repository.essex.ac.uk/16434/1/authorfinal.pdf
15. "Spherical law of cosines," Wikipedia. https://en.wikipedia.org/wiki/Spherical_law_of_cosines. Accessed 04 Sept 2021

Computer Science Applications

Geometric Algebra Models of Proteins for Three-Dimensional Structure Prediction

Alberto Pepe[1(✉)], Joan Lasenby[1], and Pablo Chacón[2]

[1] Signal Processing and Communications Group, Cambridge University Engineering Department, Trumpington Street, Cambridge CB2 1PZ, UK
{ap2219,jl221}@cam.ac.uk

[2] Chacon Lab, Rocasolano Physical Chemistry Institute, 28006 Madrid, Spain

Abstract. A protein can be regarded as a chain of amino acids with unique folding in the three-dimensional (3D) space. Knowing the folding of a protein is desirable since the folding controls the protein properties. However, determining it experimentally is expensive and time consuming: estimating the 3D structure of a protein computationally - known as protein structure prediction (PSP) - can overcome these issues. In this paper, we explore the advantage of using Geometric Algebra (GA) to model proteins for PSP applications. In particular, we employ GA to define a metric of the orientation of the amino acids in the chain. We then encode this metric in matrix form and show how patterns in these images mirror folding patterns of proteins. Lastly, we prove that this metric is predictable through a standard deep learning (DL) architecture for the inference of pairwise amino acids distances. We demonstrate that GA is a powerful tool to obtain a compact representation of the protein geometry with potential to improve the prediction accuracy of standard PSP pipelines.

Keywords: Protein Structure Prediction · Deep Learning · Geometric Algebra

1 Introduction

The 3D structure of a protein - known as tertiary structure - is the arrangement in space of its amino acid chain - the primary structure - and it determines the protein behaviour and cellular function. Determining the structure experimentally, however, is expensive and time consuming.

For this reason, there has been a great deal of recent interest in deep learning (DL) algorithms to predict the protein structure starting from the amino acid sequence [1,2]. By cutting time and cost and achieving unprecedented accuracies, protein structure prediction (PSP) has a huge potential impact on medicine and biotechnologies. The state of the art in PSP is represented by [3]: the AlphaFold2

D. W. Silva et al. (Eds.): ICACGA 2022, LNCS 13771, pp. 63–74, 2024.
https://doi.org/10.1007/978-3-031-34031-4_6

pipeline can directly predict the 3D coordinates of heavy atoms and reach a median backbone accuracy of 0.96 Å (as the interresidue distance is of the order of Å) on the CASP14 dataset [4]. From a geometrical point of view, proteins are represented as a residue gas: each amino acid - also called a residue - is associated with a rigid body (triangles) for the backbone and an angle for the sidechain. Similar processing strategies are found in [5], where 1D, 2D and 3D data are combined in a pipeline of several neural networks producing mutual predictions.

Most PSP pipelines based on DL have contact and distance maps as their end goal, which are then used to predict the protein structure. However, in [6], it has been demonstrated that adding orientational information improves the accuracy of the structure prediction: adding angle maps (three in total, one for each dihedral angle associated with a residue) can improve the precision of the top L long-range contacts of up to 2.2% on the CASP13 dataset.

In this paper, we propose a single map based on a Geometric Algebra (GA) description of the protein geometry. This has two main advantages compared to common angle maps: (1) it has a clear correspondence to the protein's secondary structure and (2) can be represented as a single, symmetric map instead of three asymmetric ones.

The rest of the paper is structured as follows: in Sect. 2, the fundamentals of Conformal GA are introduced. In Sect. 3, the proposed protein model is presented and the GA cost and cost maps are introduced. In Sect. 4, the prediction algorithm and strategy are presented, while in Sect. 5 the prediction results are shown. Lastly, in Sect. 6, conclusions are drawn.

2 Conformal Geometric Algebra

Conformal Geometric Algebra (CGA) maps GA $\mathcal{G}_{p,q,r}$ of dimension $n = p+q+r$ to $\mathcal{G}_{p+1,q+1,r}$ by introducing two basis vectors, e and $\bar{e}$, with $e^2 = +1$ and $\bar{e}^2 = -1$. Having introduced e and $\bar{e}$, we can compose the vectors

$$\begin{aligned} n_\infty &= e + \bar{e} \\ n_0 &= \frac{1}{2}(\bar{e} - e) \end{aligned} \tag{1}$$

which help define a mapping of the kind

$$x \in \mathcal{G}_{p,q,r} \longrightarrow F(x) \in \mathcal{G}_{p+1,q+1,r} \tag{2}$$

in which $F(x)$ is defined as

$$\begin{aligned} F(x) &= -\frac{1}{2}(x - e)n_\infty(x - e) \\ F(x) &= \frac{1}{2}(x^2 n_\infty + 2x - n_0) \end{aligned} \tag{3}$$

In the case in which we are dealing with a 3D space (i.e. $\mathcal{G}_{3,0,0}$), the equivalent CGA will be $\mathcal{G}_{4,1,0}$. When working in CGA, point pairs, lines, planes, circles and spheres are all conveniently represented by blades in the 5D CGA. A summary is provided in Table 1.

Table 1. Objects in CGA

Grade	Symbol	Object
1	A	point
2	A ∧ B	point pair
3	A ∧ B ∧ C	circle ($\mathcal{C}$)
3	A ∧ B ∧ n_∞	line (L)
4	A ∧ B ∧ C ∧ D	sphere (Σ)
4	A ∧ B ∧ C ∧ n_∞	plane (Π)

3 CGA in Protein Geometry

3.1 Cost Function

A protein can be simplified into a backbone chain and several side chains. The backbone is responsible for the 3D shape of the protein, and it is composed of a series of carbon, nitrogen, and oxygen atoms. The α-carbons are the main feature of the backbone, to which the side chains that differentiate each amino acid are bonded. Each α-carbon is preceded by a nitrogen atom and followed by a carbon atom. Hence, to each amino acid i we can associate a triplet of atoms $\{N, C_\alpha, C\}_i$.

Each $\{N, C_\alpha, C\}$ triplet lies on a plane, constraining the protein folding (see Fig. 1). We can hence conveniently model a protein backbone in CGA so any three $\{N, C_\alpha, C\}$ atoms will lie on a plane (not too dissimilar to the residue gas of [3]): let A_i, B_i and C_i be the Euclidean coordinates expressed in Conformal space of the atoms $\{N, C_\alpha, C\}_i$, respectively. The plane associated with residue i can be expressed as the 4-blade:

$$\Pi_i = A_i \wedge B_i \wedge C_i \wedge n_\infty \tag{4}$$

Given two planes Π_i, Π_j corresponding to the amino acids i, j, we can compute the rotor that brings one to the other as described in [7]:

$$R_{ij} = \frac{1}{\sqrt{\langle K \rangle_0}}(1 - \Pi_i \Pi_j) \tag{5}$$

where $K = 2 - (\Pi_i \Pi_j + \Pi_j \Pi_i)$ and $\langle \cdot \rangle$ is the grade projector operator. We now use the cost function $C_\lambda(R)$ that measures how much the rotor R varies from the identity, as defined in [8]. $C_\lambda(R)$ is a weighted sum of a translational and a rotational term:

$$C_{\lambda_1 \lambda_2}(R) = \lambda_1 \langle R_{\|} \tilde{R}_{\|} \rangle_0 + \lambda_2 \langle (R_\perp - 1)(\tilde{R}_\perp - 1) \rangle_0 \tag{6}$$

in which the translational error is represented by $R_{\|} = R \cdot e$, and the rotational error by $\langle (R_\perp - 1)(\tilde{R}_\perp - 1) \rangle_0 = \langle (R - 1)(\tilde{R} - 1) \rangle_0$. As we are interested in an orientational feature, we will focus exclusively on the rotational part (case $\lambda_1 = 0, \lambda_2 = 1$).

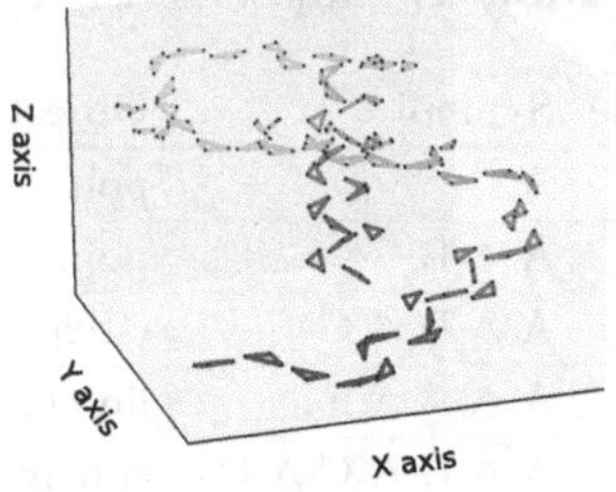

Fig. 1. First 70 $\{N, C_\alpha, C\}$ planar triplets of the haemoglobin backbone.

3.2 Cost Maps

Inter-residue interactions are commonly represented as matrices - also called maps. A contact map $\mathbf{C}$ of a protein consisting of M residues, for example, is a binary $M \times M$ matrix of the type:

$$\mathbf{C}_{ij} = \begin{cases} 1 & \text{if } d_{ij} < 15\ \text{Å} \\ 0 & \text{otherwise} \end{cases} \tag{7}$$

where d_{ij} is the distance between residues i, j expressed in Å measured as the Euclidean distance between the C_α coordinates of residues i and j. A cost map can be interpreted as: two residues are in contact if they are within a certain distance from each other. A more informative metric, usually real valued, is given by distance maps, which are similarly defined as:

$$\mathbf{D}_{ij} = d_{ij} \tag{8}$$

From either or both contact and distance maps it is possible to obtain accurate 3D shape estimation. However, when contact or distance maps are predicted and not exact, errors are introduced into the 3D reconstruction step. Having an additional map grasping the orientation between residues can help to further constrain the search space for the protein folding. We can hence employ our cost function to produce a cost map which contains orientational information as follows:

$$\mathbf{M}_{ij} = \begin{cases} C_{\lambda_1\lambda_2}(R_{ij}) & \text{if } d_{ij} < 15\ \text{Å} \\ 0 & \text{otherwise} \end{cases} \tag{9}$$

3.3 Examples

A comparison between contact map $\mathbf{C}$, distance map $\mathbf{D}$ and cost map $\mathbf{M}$ is given in Fig. 2 for an example protein. We label protein according to their 4-character alphanumeric PDB identifier [10]

It is possible to establish a relation between patterns in cost maps and the protein secondary structure. By secondary structure we refer to the local folding

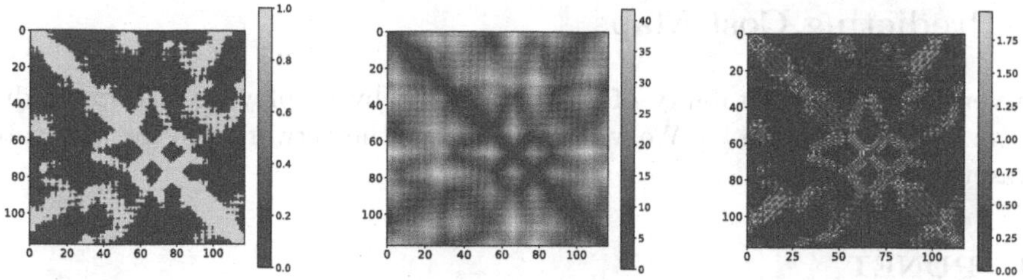

Fig. 2. From left to right: contact, distance and cost map for protein 2hc5a.

of a segment of a protein, e.g. α-helices, β-sheets or turns. Secondary structure information is a common feature in PSP pipelines and one of the most important in predicting distance and contact maps, as shown in [5,9].

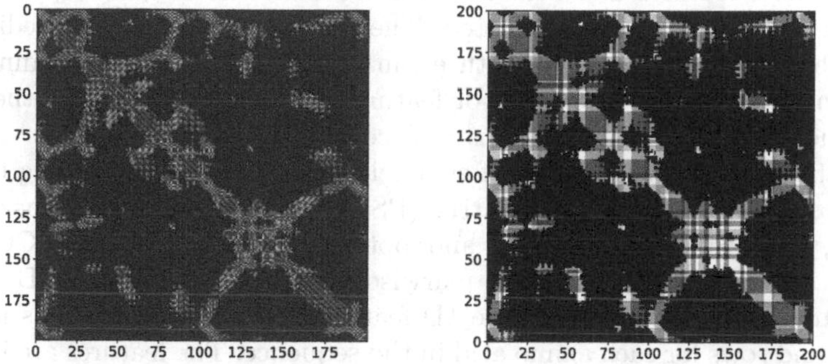

By assigning a colour to each secondary structure, it is possible visualize the secondary structure of each amino acid pair. We arbitrarily assigned red to α-helices, green to β-sheets, blue to turns and white to all the others. Any combination of these four colours gives the possible secondary structures of the pair, for a total of 10 different colour combinations. As shown in Sect. 3.3, it is possible to find a clear correspondence between secondary structures and patterns in the cost maps. To the best of our knowledge, this is the first example of an orientational map that also encodes the secondary structure of the protein.

4 Predicting Cost Maps

We verified the predictability of our cost maps by employing a deep residual network as presented in [9]. We will refer to both the network and the associated dataset as PDNET.

4.1 PDNET

PDNET is residual neural network composed of 128 blocks. Each residual block consists of a batch normalization layer, a ReLU activation function, a 2D convolutional layer with 3×3 kernel, a dropout layer with $\alpha = 0.3$, a ReLU activation function, and a 2D convolutional layer, for a total of ~ 9.5M tunable parameters.

PDNET was originally designed to predict either: (i) contact maps, (ii) binned distance maps or (iii) real-valued distance maps. We demonstrate that from the same features and with the same architecture originally presented in [11], cost maps can also be estimated. The task of distance map prediction is comparable to the problem of depth estimation: the three RGB channels of a colour image are replaced by tens of feature matrices derived from the amino acid sequence, and the depth map is replaced by the distance map.

Specifically, the total number of channels is $N = 57$, corresponding to 7 features: position specific scoring matrix (PSSM), secondary structure, entropy, FreeCon, CCMPred, surface area and potential energy. Of these CCMpred, FreeCon and potential energy are pairwise features, the rest are 1D features relative to a single amino acid. The 1D features are encoded twice as identical columns and rows for each amino acid in the sequence. The features are identical to those of the PDNET dataset of [9], which includes a more detailed description of their biochemical meaning. They are either derived from previous DL based prediction or multiple sequence alignment queries.

When PDNET is employed to predict real valued distances, it employs the reciprocal logcosh as a loss function:

$$L_{\mathbf{D}}^{(i)} = \log\left(\cosh\left(\frac{K}{\mathbf{D}_P^{(i)} + \epsilon} - \frac{K}{\mathbf{D}_T^{(i)} + \epsilon}\right)\right) \tag{10}$$

where $\mathbf{D}_P^{(i)}$ is the predicted distance matrix, $\mathbf{D}_T^{(i)}$ the true distance matrix, ϵ a small positive number and K is a scalar set equal to 100. The inverse of the maps is taken in order to prioritize short-range interaction, for which higher accuracy is desirable, over long-range interaction, which is less relevant in terms of the overall 3D structure. The loss is evaluated pixel by pixel and summed over the total number of pixels.

4.2 Training Details

The GA-based cost maps are also real valued and bounded in the range [0, 2], as we verified empirically by evaluating $C_{\lambda_1,\lambda_2}(R)$. However, since the cost does not increase for residues further away as it is a purely orientational measure, we changed the loss to be:

$$L_{\mathbf{M}}(i) = \log\left(\cosh\left(\mathbf{M}_P^{(i)} - \mathbf{M}_T^{(i)}\right)\right) \tag{11}$$

where $\mathbf{M}_P^{(i)}, \mathbf{M}_T^{(i)}$ are the predicted and true cost maps for protein (i) in the training set, respectively.

For training the network, we kept the features unchanged from those of PDNET, namely a stack of images of the type $\{\mathbf{X}^{(i)}\}_{i=1}^{N}$, with $N = 57$ and $\mathbf{X}^{(i)} \in \mathbb{R}^{M\times M}$, in which M is the length of the protein sequence. The change comes in substituting the target $\mathbf{D}_T \in \mathbb{R}^{M\times M}$ - the true, real-valued distance maps, with $\mathbf{M}_T \in \mathbb{R}^{M\times M}$ - the true, real-valued cost maps, obtained from the protein coordinates in the protein database [10]. Again, the loss is evaluated per pixel.

The training set has been kept to 1000 proteins from the DEEPCOV dataset, and the testing set to 150 proteins from the PSICOV dataset, as in the original PDNET pipeline. The code has been implemented using the Keras API of Tensorflow for the Machine Learning modules, the Clifford library for operations in Geometric Algebra and the PDB Module of the Biopython library for handling protein data. The code was written in the form of Jupyter Notebooks on Google Colaboratory and all the experiments have been run on an NVIDIA Tesla K80 GPU. All the scripts and data are available upon request to the authors.

We considered scenarios (see Fig. 3): (a) predicting cost maps with 57 feature channels (standard PDNET), (b) predicting cost maps with 57 feature channels + 1 (real) distance channel (ideal case, as distance maps would not be available), (c) predicting cost maps with 57 feature channels + 1 (predicted) distance channel also via PDNET (realistic case, as distance maps also need to be predicted in PSP).

5 Results

We evaluated two metrics, namely: (i) mean absolute error (MAE), as in common regression problems, and (ii) structural similarity index (SSIM) between $\mathbf{M}_P, \mathbf{M}_T$, since a low MAE does not necessarily mean that the patterns in the cost maps are captured successfully. The MAE is measured in Å, while the SSIM ranges between [0, 1], with SSIM = 1 meaning fully similar matrices and SSIM = 0 fully dissimilar matrices. They are defined as follows:

$$MAE(\mathbf{M}_P, \mathbf{M}_T) : \frac{1}{M^2}\sum_{i=1}^{M}\sum_{j=1}^{M}|\mathbf{M}_{Pij} - \mathbf{M}_{Tij}| \tag{12}$$

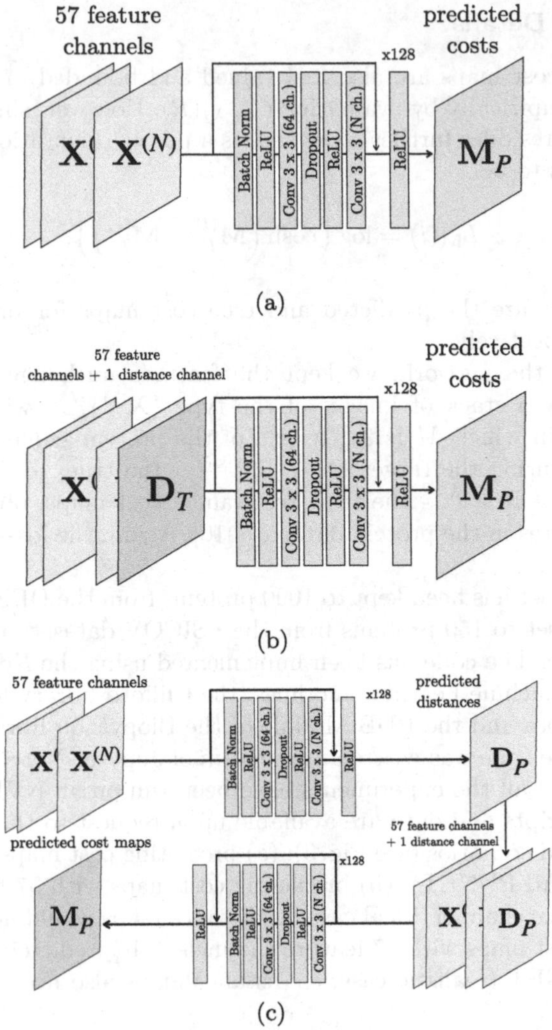

Fig. 3. The three processing schemes: (a) predicting costs from PDNET; (b) predicting costs from PDNET + true distance maps; (c) predicting costs from PDNET + predicted distances, themselves predicted from PDNET.

$$SSIM(\mathbf{M}_P, \mathbf{M}_T) : \frac{(2\mu_{\mathbf{M}_P}\mu_{\mathbf{M}_T} + c_1)(2\sigma_{\mathbf{M}_P\mathbf{M}_T} + c_2)}{(\mu^2_{\mathbf{M}_P} + \mu^2_{\mathbf{M}_T} + c_1)(\sigma^2_{\mathbf{M}_P} + \sigma^2_{\mathbf{M}_T} + c_2)} \tag{13}$$

with $\mu_{\mathbf{M}_T}$ being the mean of $\mathbf{M}_T$, $\mu_{\mathbf{M}_P}$ the mean of $\mathbf{M}_P$, $\sigma_{\mathbf{M}_P\mathbf{M}_T}$ the covariance of $\mathbf{M}_P$ and $\mathbf{M}_T$, $\sigma^2_{\mathbf{M}_P}$ the variance of $\mathbf{M}_P$, $\sigma^2_{\mathbf{M}_T}$ the variance of $\mathbf{M}_T$, $c_1 = (k_1L)^2, c_2 = (k_2L)^2$ with $k_1 = 0.01, k_2 = 0.03$ and L being the dynamic range, set to $L = 255$.

Results are summarized in Tables 2 and 3.

Table 2. MAE between original and predicted cost maps (Å)

	no distance			with distance			with pred. distance		
	Max	Mean	Min	Max	Mean	Min	Max	Mean	Min
DEEPCOV (val)	0.1080	0.0218	0.0009	0.0418	0.0108	0.0005	0.0607	0.01825	0.0005
PSICOV (test)	0.0342	0.0158	0.0029	0.0275	0.0125	0.0028	0.0327	0.01490	0.0029

Table 3. SSIM between original and predicted cost maps.

	no distance			with distance			with pred. distance		
	Max	Mean	Min	Max	Mean	Min	Max	Mean	Min
DEEPCOV (val)	0.9946	0.9041	0.4990	0.9986	0.9652	0.8387	0.9991	0.9360	0.7442
PSICOV (test)	0.9937	0.9431	0.8592	0.9941	0.9632	0.9130	0.9936	0.9519	0.8851

It can be noticed that cost maps are indeed predictable based on features commonly used to predict distances. However, when predicting cost maps without distance information, only close range contacts (i.e. the pixels close to the diagonal) are predicted accurately. Adding predicted distance information, on the other hand, allows us to significantly improve the prediction of the patterns in cost maps, with a mean MAE decrease by 16.3% for the training set and by 5.7% for the testing set. The average SSIM increased by 3.5% and by 1% for the training and testing sets, respectively. The better the prediction of the distance information (i.e., the closer the predicted distance maps are to the original ones), the higher the improvement on cost prediction.

Examples of the predicted cost maps in comparison with the original cost maps over the testing set are given in Fig. 4.

Lastly, we evaluated the which is intuitive feature importance (PFI) to rank the most relevant features in the prediction of cost maps. We did so by training the network by permuting one feature at a time and then taking the ratio of our metric with and without permutation of that feature. By permutation we refer to the shuffling of a single feature across the training set, meaning that when we evaluate the PFI for feature n, each protein will have associated an erroneous feature n belonging to a different protein during training, while leaving the testing set unchanged. We then measured the PFI of feature n as:

$$PFI^{(n)}_{MAE} = \frac{MAE(\mathbf{M}_P, \mathbf{M}_T)}{MAE^{(n)}(\mathbf{M}_P, \mathbf{M}_T)} \tag{14}$$

$$PFI^{(n)}_{SSIM} = \frac{SSIM^{(n)}(\mathbf{M}_P, \mathbf{M}_T)}{SSIM(\mathbf{M}_P, \mathbf{M}_T)} \tag{15}$$

In which $f(\mathbf{M}_P, \mathbf{M}_T)$ is the metric f measured with standard training procedure, and $f(\mathbf{M}_P, \mathbf{M}_T)^{(n)}$ is the metric f measured when permuting feature n during training.

Results for the validation set (DEEPCOV) and for two testing set (PSICOV and CAMEO HARD) are shown in Fig. 5. The PSSM and secondary structures

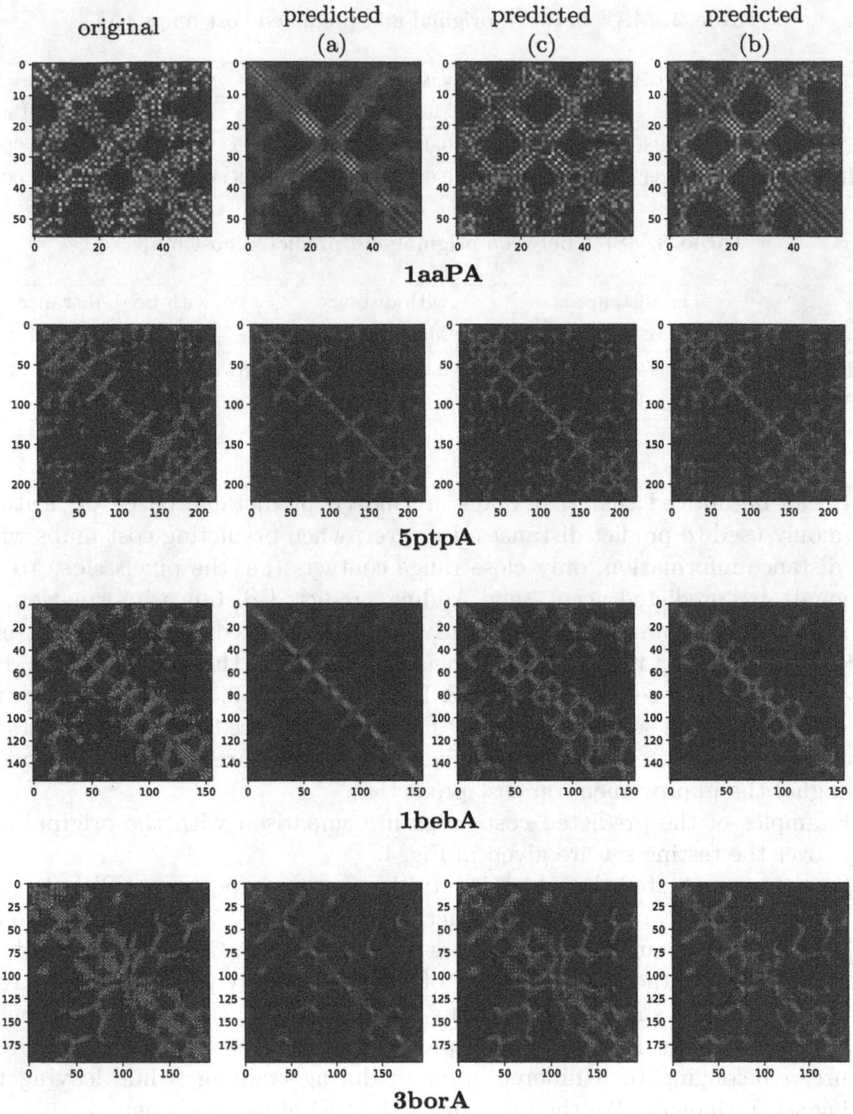

Fig. 4. Examples of the GA cost map for four protein chains predicted with the three approaches (a,b,c) of Fig. 3. Note how adding distances significantly improves the quality of the prediction.

appear to be the two most relevant features, a result which mirrors that found for distance maps in [9]. This is in agreement with the findings of Sect. 3, where we saw the close relationship between cost patterns and secondary structures.

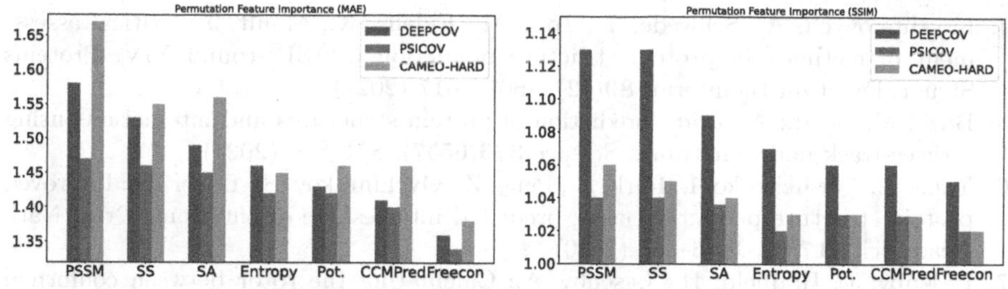

Fig. 5. Permutation Feature Importance over MAE and SSIM for each of the 7 features across validation and test sets.

6 Conclusions

In this paper, we have introduced a new feature based on GA describing the relative amino acid orientation for PSP. We firstly presented the criterion behind the modeling of a protein backbone as a collection of planes. We then evaluated the rotor between each pair of planes and associated a cost to it. The pairwise costs were then arranged in matrix form to produce cost maps. We proceeded to show how patterns in cost maps can be directly associated to the protein secondary structure and verified how features and algorithms employed in PSP to predict distance maps can also be used to predict our proposed GA cost maps. Adding distance information - even if only predicted - can further improve the predicted cost maps in terms of MAE and SSIM.

Our cost maps therefore constitute a useful tool for protein modelling and may provide new orientation-based features that could improve the 3D structure prediction. We believe that GA could hence constitute a successful tool to model proteins and provide new orientational features that can improve the precision of the 3D structure prediction and reduce the number of required features.

Future work might include employing predicted costs, along with feature and distance maps, to predict the 3D coordinates of C_α atoms in the protein backbone on the basis of [5,12] and verify whether the cost maps can further constrain the search space and improve the accuracy of the 3D coordinates, or employing different GA modeling choices.

References

1. AlQuraishi, M.: Machine learning in protein structure prediction. Curr. Opin. Chem. Biol. **1**(65), 1–8 (2021)
2. Pearce, R., Zhang, Y.: Deep learning techniques have significantly impacted protein structure prediction and protein design. Curr. Opin. Struct. Biol. **1**(68), 194–207 (2021)
3. Jumper, J., et al.: Highly accurate protein structure prediction with AlphaFold. Nature **596**(7873), 583–9 (2021)

4. Kryshtafovych, A., Schwede, T., Topf, M., Fidelis, K., Moult, J.: Critical assessment of methods of protein structure prediction (CASP)-round XIV. Proteins Struct. Function Bioinform. **89**(12), 1607–1617 (2021)
5. Baek, M., et al.: Accurate prediction of protein structures and interactions using a three-track neural network. Science **373**(6557), 871–876 (2021)
6. Yang, J., Anishchenko, I., Park, H., Peng, Z., Ovchinnikov, S., Baker, D.: Improved protein structure prediction using predicted interresidue orientations. Proc. Natl. Acad. Sci. **117**(3), 1496–503 (2020)
7. Lasenby, J., Hadfield, H., Lasenby, A.: Calculating the rotor between conformal objects. Adv. Appl. Clifford Algebras **29**(5), 1–9 (2019)
8. Eide, E.R.: Camera calibration using conformal geometric algebra. Master's degree thesis. University of Cambridge (2018)
9. Adhikari, B.: A fully open-source framework for deep learning protein real-valued distances. Sci. Rep. **10**(1), 1 (2020)
10. Burley, S.K., Berman, H.M., Kleywegt, G.J., Markley, J.L., Nakamura, H., Velankar, S.: Protein data bank (PDB): the single global macromolecular structure archive. Protein Crystallogr. 627–641 (2017)
11. Adhikari, B.: DEEPCON: protein contact prediction using dilated convolutional neural networks with dropout. Bioinformatics **36**(2), 470–477 (2020)
12. Costa, A., Ponnapati, M., Jacobson, J.M., Chatterjee, P.: Distillation of MSA embeddings to folded protein structures with graph transformers. bioRxiv (2021)

Clifford Convolutional Neural Networks for Lymphoblast Image Classification

Guilherme Vieira[1], Marcos Eduardo Valle[1(✉)], and Wilder Lopes[2]

[1] Universidade Estadual de Campinas, Campinas, Brazil
vieirant.gui@gmail.com, valle@ime.unicamp.br
[2] Ogarantia, Paris, France
wilder@ogarantia.com

Abstract. This paper features convolutional neural network (CNN) models on Clifford algebras applied to a medical image classification task, namely the diagnosis of acute lymphoblastic leukemia (ALL). ALL is a type of cancer identified by malformed lymphocytes, known as lymphoblasts, in the bloodstream. The image classification task aims to discriminate healthy cells from lymphoblasts. This work shows that CNNs featuring parameters in Clifford algebras significantly outperform real-valued networks of equivalent size in this application. Indeed, the real-valued and a Clifford CNN achieved an average accuracy of 94.60% and 97.02%, respectively, in the ALL-IDB dataset with a 50% train-test split. Moreover, we present smaller versions of Clifford CNNs with roughly 75% fewer parameters that yielded a 96.50% average accuracy. The results reported in this work are comparable to high-end models in the literature despite having several orders of magnitude fewer parameters.

Keywords: Clifford algebra · convolutional neural network · deep learning · acute lymphoblastic leukemia · computer assisted diagnosis

1 Introduction

Neural networks (NNs) are artificial intelligence models inspired by the human nervous systems [12]. NNs seek to mimic the synapse process, responsible for carrying information between neurons in the brain. Learning is achieved by strengthening synaptic connections that are frequently activated. Likewise, NNs learn by processing examples and adjusting their synaptic weights to match the expected output. Apart from biological motivation, an NN can be interpreted as a non-linear parametric function in mathematical terms. The parameters correspond to the synaptic weights and are adjusted by minimizing a loss function built on a set of examples called the training set.

A branch of NNs that flourished in the recent decades is the so-called deep learning (DL) [12]. DL uses networks with numerous layers and a high number

This work was supported in part by CNPq under grant no. 315820/2021-7, FAPESP under grant no. 2022/01831-2, and Coordenação de Aperfeiçoamento de Pessoal de Nível Superior - Brasil (CAPES) - Finance Code 001.

D. W. Silva et al. (Eds.): ICACGA 2022, LNCS 13771, pp. 75–87, 2024.
https://doi.org/10.1007/978-3-031-34031-4_7

of free parameters to learn alternate representations of data. These deep neural networks (DNNs) rose in popularity due to an exponential increase in computational power in the last decades. One such type of DNN is the convolutional neural network (CNN). CNNs feature a particular type of operation based on the convolution of filters with the inputs, and are incredibly well-suited at learning and representing local patterns. Consequently, CNNs are often regarded as the basis of state-of-the-art models in image processing and pattern recognition.

Traditional NNs are based on real-valued inputs, outputs, and synaptic weights. In contrast, hypercomplex-valued NNs (HvNNs) such as Clifford neural networks use hypercomplex values instead of real numbers [1,15]. Because hypercomplex values can be interpreted as vector space elements, HvNNs are adequate to process multidimensional data. Furthermore, they can benefit from the geometric properties of the hypercomplex algebras. Several works showcase the advantage of hypercomplex networks over their real-valued counterparts, especially on tasks involving multi-channel data such as image processing [5,17,21]. Recent works also show that CNNs in hypercomplex algebras excel at reducing computational complexity while delivering similar or higher performance when compared to real-valued models [8,13,17].

This paper addresses an application of HvNN for acute lymphoblast leukemia (ALL) diagnosis. ALL is a type of blood cancer that appears and multiplies rapidly. It is characterized by the presence of many lymphoblasts in the blood and also in the bone marrow. A common diagnosis technique is the peripheral blood smear, in which a hematologist counts the number of lymphoblasts in a blood sample with a microscope. However, manually counting lymphoblasts is a rather monotonous task that is prone to error and takes the time of a professional who could be more productive in other matters. For this and several other reasons, computer models to perform automatic lymphoblast counts have been proposed in the literature [4,7,18,19,23]. In particular, successful automated ALL diagnosis has also been achieved by combining deep learning models with transfer learning and unsharpening techniques [10,11,23]. Besides the results using real-valued models, a quaternion-valued CNN exhibited robust performance for computer-aided ALL diagnosis with only 34% of the total number of parameters of the corresponding real-valued NN [13]. In fact, CNNs on multiple hypercomplex algebras were applied for ALL detection in [22], noticeably outperforming an equivalent real network. In this paper, we investigate further the performance of HvNN models for ALL diagnoses by considering Clifford algebras besides quaternions. On top of extending the quaternion-valued CNN to other Clifford algebras, this paper improves the HvNN proposed in [13] by reducing the number of parameters without compromising the network performance.

This work is organized as follows: Sect. 2 provides an overview of basic concepts on Clifford algebras; the core concepts of Clifford neural networks are discussed in Sect. 3; Sect. 4 introduces the application of the proposed Clifford networks in a lymphoblast classification task; lastly, Sect. 5 provides concluding remarks regarding the attained results.

2 A Brief on Clifford Algebras

In a very simplified manner, this section presents the basic concepts of Clifford algebras. The readers interested in further details are invited to read [14,20].

Let $\mathcal{V}$ be a finite-dimensional vector space over the real numbers, that is, $\mathcal{V} = \mathbb{R}^n$. A Clifford algebra is obtained by enriching the vector space $\mathcal{V}$ with scalars and multivectors. Multivectors generalize the concept of vectors. From a geometric point of view, while vectors have length, multivectors are associated with properties like area and volumes. Besides their geometric properties, multivectors are derived algebraically by the product of vectors as follows.

Consider an orthonormal basis $\{\gamma_1, \ldots, \gamma_d\}$ of $\mathcal{V}$. We define a multivector γ_{ij} as the product of two distinct basis vectors γ_i and γ_j, that is, $\gamma_{ij} \equiv \gamma_i \gamma_j$, for $i, j \in \{1, \ldots, d\}$ with $i \neq j$. Moreover, the square of vectors and multivectors are scalars, and we denote the scalar unit by $\gamma_0 \equiv 1$.

Let us denote by Γ the set of products of all combinations of up to d vectors as well as the scalar unit 1 and the d vector basis. Note that Γ has the same number of elements as the power set of $\{1, \ldots, d\}$, that is, $\text{Card}(\Gamma) = 2^d$. For example, if $\{\gamma_1, \gamma_2, \gamma_3\}$ is an orthonormal basis for a vector space $\mathcal{V}$, we have

$$\Gamma = \{1, \gamma_1, \gamma_2, \gamma_3, \gamma_{12}, \gamma_{13}, \gamma_{23}, \gamma_{123}\}. \tag{1}$$

Alternatively, we can write $\Gamma = \{\gamma_\lambda : \lambda \in \Lambda\}$, where Λ denotes the set of all 2^d ordered indexes defined by

$$\Lambda = \{i_1 i_2 \cdots i_k : 1 \leq i_1 < i_2 < \cdots < i_k \leq d, \quad 1 \leq k \leq d\} \cup \{0\}. \tag{2}$$

A Clifford algebra is defined on the set $\mathcal{G}(\mathcal{V})$ of all linear combinations of scalars, vectors, and multivectors derived from $\mathcal{V}$. Formally, $\mathcal{G}(\mathcal{V})$ denotes the vector space spanned by the set Γ given by (1), that is,

$$\mathcal{G}(\mathcal{V}) = \left\{ \sum_{\lambda \in \Lambda} \alpha_\lambda \gamma_\lambda : \alpha_\lambda \in \mathbb{R}, \forall \lambda \in \Lambda \right\}. \tag{3}$$

Note that, because $\text{Card}(\Gamma) = \text{Card}(\Lambda) = 2^d$, we also have $\dim\big(\mathcal{G}(\mathcal{V})\big) = 2^d$.

Finally, a Clifford algebra is obtained by endowing $\mathcal{G}(\mathcal{V})$ with an associative binary operation called Clifford or geometric product. The geometric product is defined as follows on the basis elements $\gamma_1, \ldots, \gamma_d$ of $\mathcal{V}$:

$$\gamma_i \gamma_j = \begin{cases} -\gamma_j \gamma_i, & i \neq j, \\ +1, & i = j \text{ and } i = 0, \ldots, p, \\ -1, & i = j \text{ and } i = p+1, \ldots, d, \end{cases} \tag{4}$$

where $p \in \{0, \ldots, d\}$. We note that (4) can be computed on multivectors by using the associativity and anti-commutativity properties. For example, assuming $\gamma_1^2 \equiv \gamma_1 \gamma_1 = -1$, the product of γ_{12} and γ_1 yields

$$\gamma_{12}\gamma_1 = (\gamma_1\gamma_2)\gamma_1 = -(\gamma_2\gamma_1)\gamma_1 = -\gamma_2(\gamma_1\gamma_1) = -\gamma_2(-1) = \gamma_2.$$

Table 1. Product of vectors and multivectors in four-dimensional Clifford algebras.

$\mathcal{C}\ell(2,0)$	γ_1	γ_2	γ_{12}	$\mathcal{C}\ell(1,1)$	γ_1	γ_2	γ_{12}	$\mathcal{C}\ell(0,2)$	γ_1	γ_2	γ_{12}
γ_1	1	γ_{12}	γ_2	γ_1	1	γ_{12}	γ_2	γ_1	-1	γ_{12}	$-\gamma_2$
γ_2	$-\gamma_{12}$	1	$-\gamma_1$	γ_2	$-\gamma_{12}$	-1	γ_1	γ_2	$-\gamma_{12}$	-1	γ_1
γ_{12}	$-\gamma_2$	γ_1	-1	γ_{12}	$-\gamma_2$	$-\gamma_1$	1	γ_{12}	γ_2	$-\gamma_1$	-1

The pair (p, q), with $p+q=d$, identifies the Clifford algebra $\mathcal{C}\ell(p,q)$. In this paper, we focus on four-dimensional Clifford algebras. This choice is motivated by the successfull applications of four-dimensional hypercomplex algebras, mostly quaternions, for image processing tasks [9,17,21]. Table 1 shows the product of vectors and multivectors in all four-dimensional Clifford algebras, i.e., the algebras derived from a two dimensional vector space $\mathcal{V}$ whose orthonormal basis is $\{\gamma_1, \gamma_2\}$. Note that the Clifford algebra $\mathcal{C}\ell(0,2)$ corresponds to the quaternions. The algebras $\mathcal{C}\ell(1,1)$ and $\mathcal{C}\ell(2,0)$ are isomorphic and can be identified with the coquaternions, also called split-quaternions.

3 Clifford Neural Networks

Neural networks (NNs) are powerful machine learning techniques inspired by the human brain processing capabilities. Convolutional neural networks refer to the broad class of neural networks that combine convolutional and pooling layers sequentially, followed by one or more dense layers. This section offers a brief overview of dense (fully-connected), convolutional, and pooling layers for real-valued and Clifford algebras.

3.1 Dense Layers

Dense layers are the building block of several NN architectures, such as the famous multi-layer perceptron (MLP) network. Dense layers are composed of several neurons in parallel, in which each neuron receives all inputs through synaptic connections. They are also commonly known as fully-connected layers.

Dense layers process data by means of a linear combination of its inputs by the synaptic weights (trainable parameters), to which a scalar bias term is added. A non-linear activation function can be applied to yield the neuron's output. Formally, let $x_1, \ldots, x_N$ denote the inputs, the output of the ith neuron in a dense layer is given by

$$y_i = \varphi(s_j), \quad \text{with} \quad s_i = \left(\sum_{j=1}^{N} w_{ij} x_j \right) + b_i \tag{5}$$

where w_{ij} denotes the weight associated with the jth input variable, b_i is the bias term of the ith neuron, and φ represents the activation function.

Despite being computationally expensive due to the numerous parameters, dense layers are widely used since they support the universal approximation theorem. In a few words, the universal approximation theorem asserts that the family of neural networks with at least two dense layers is dense in the set of continuous functions on a compact subset. The universal approximation theorem ensures that simple dense feedforward networks can approximate any continuous function within any desired precision. We would like to remark that, although Cybenko proved the universal approximation theorem in the late 1980s for real-valued dense neural networks, the universal approximation theorem also holds for several hypercomplex-valued neural networks. Indeed, the universal approximation property was proven for complex- and quaternion-valued dense networks by Arena and collaborators [2,3]. The universal approximation theorem has been further extended for Clifford-valued dense neural networks by Buchholz and Sommer in the early 2000s [6].

Clifford dense layers are analogous to the real-valued case but the trainable parameters as well as the inputs and the outputs are all Clifford numbers. They are given by (5) but the products and sums are carried out in a Clifford algebra. Additionally, in hypercomplex-valued networks it is common to use split activation functions. A split activation function $\varphi : \mathcal{G}(\mathcal{V}) \rightarrow \mathcal{G}(\mathcal{V})$ in a Clifford algebra $\mathcal{G}(\mathcal{V})$ is defined using a real-valued function $\varphi_{\mathbb{R}} : \mathbb{R} \rightarrow \mathbb{R}$ as follows

$$\varphi\left(\sum_{\lambda \in \Lambda} \alpha_\lambda \gamma_\lambda\right) = \sum_{\lambda \in \Lambda} \varphi_{\mathbb{R}}(\alpha_\lambda)\gamma_\lambda, \tag{6}$$

where Λ is the index set given by (2). In other words, the split-activation function is merely the application of the associated real-valued function to each component's scalar part individually. It is important to remark that the universal approximation theorem holds for hypercomplex-valued neural networks with split activation functions [2,6]. This paper only considers this kind of activation functions.

3.2 Convolutional Layers

Convolutional layers are particular types of layers in which the trainable parameters are arranged in spatial structures called filters [12]. The filter structures allow the network to process data in a locally cohesive manner, learning local patterns. Convolutional neural networks are named so because the filters act as the kernels in convolutions, and the image being processed acts as the input. These networks have been widely applied to image processing tasks, taking full advantage of the spatial nature of its learning mechanism and the translation invariance of filters.

Let $\mathcal{G}(\mathcal{V})$ be a Clifford algebra and let us take an image $\mathbf{I}$ with C channels, where $\mathbf{I}(p, c) \in \mathcal{G}(\mathcal{V})$ denotes the Clifford number of the cth channel at the pth pixel. A filter in a convolutional layer that receives $\mathbf{I}$ as input is a spatial structure, in general a rectangular grid G, with the same number C of channels. We express the synaptic weight associated with the qth pixel in the grid G of the cth channel of the kth filter as $\mathbf{F}(q, c, k) \in \mathcal{G}(\mathcal{V})$, with $k \in \{1, 2, \ldots, K\}$, where K is the number of filters in the layer. In other words, a Clifford convolutional layer is represented by a three-dimensional array $\mathbf{F}$ with entries in $\mathcal{G}(\mathcal{V})$. The output of a convolutional layer with K filters is an image $\mathbf{J}$ with K channels, each of which is produced by applying the convolution operation of one of the K filters to the image $\mathbf{I}$. Formally, the convolution of the image by the filter k at pixel p is denoted by $(\mathbf{I} * \mathbf{F})(p, k)$ and defined as the linear combination of the filter weights by the pixel values in a window defined by the filter domain. Intuitively this can be seen as superposing the filter grid over the image centered at the pixel p and multiplying the corresponding weights by the underlying intensities. In mathematical terms, let $S(q)$ denote the translation relative to a pixel p, for every $q \in G$. Then, we can represent the convolution by the equation

$$(\mathbf{I} * \mathbf{F})(p, k) = \sum_{c=1}^{C} \sum_{q \in G} \mathbf{I}(p + S(q), c)\mathbf{F}(q, c, k) \tag{7}$$

where $c = 1, \ldots, C$ are the channels of $\mathbf{I}$. Finally, the intensity of the kth channel at pixel p of the output is given by

$$\mathbf{J}(p, k) = \varphi\big((\mathbf{I} * \mathbf{F})(p, k) + b(k)\big), \tag{8}$$

where $\varphi : \mathcal{G}(\mathcal{V}) \to \mathcal{G}(\mathcal{V})$ is the activation function and $b(k)$ is the bias term.

Remark 1. We note that the sum and product operations in equations (7) and (8) are carried out in the underlying Clifford algebra. In fact, the definitions above are the same for real-valued convolutional layers, except in that case the sum and product operations are simpler since they are real sums and products.

3.3 Pooling Layer

A pooling layer operates a downsampling effect in the input. Moreover, this layer structure contains no trainable parameters. The most common pooling layers are the max and average pooling layers. In this work in particular we use the max pooling layer, exclusively. Roughly speaking, a max pooling layer has a kernel shape, usually a rectangular grid G, and it operates by collapsing each set of pixels contained in the grid into the single maximum value present. This operation reduces the dimensionality of the input while also highlighting the "stronger" signal in each window. The max pooling operation is conducted on each filter separately, i.e., it acts as a split maximum function for elements of a Clifford algebra.

4 Lymphoblast Image Classification Task

In this section we describe the experiment conducted to showcase the proposed convolutional Clifford neural network. It consists in a classification task in a medical-image dataset containing blood smear images.

Acute Lymphoblastic Leukemia (ALL) is a rare type of blood cancer that occurs more frequently in children of ages 2–5 and can be lethal in under a

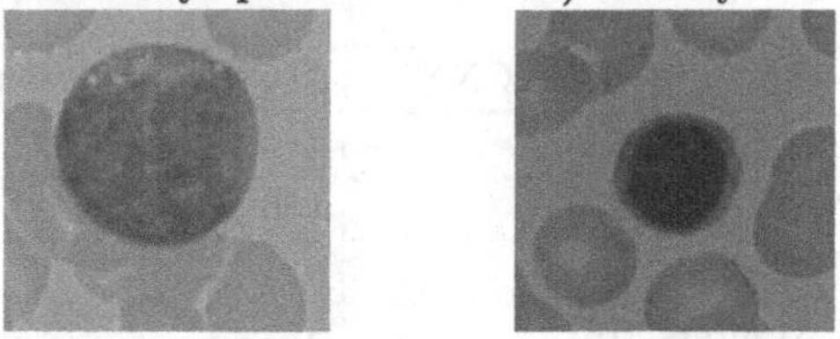

Fig. 1. Example of images from the ALL-IDB dataset used for the classification task.

few weeks if left undiagnosed. The main indicator of ALL is the presence of lymphoblasts, a type of malformed lymphocyte, in the blood. The most common diagnosis method is the inspection of microscopic blood smear images. The ALL-IDB [16] is a public benchmark aimed at computer assisted ALL diagnosis and consists of 2 datasets: one directed at a segmentation and classification, and the other directly aimed at the classification task itself. In this work we use the latter dataset which contains 260 images, each containing a single blood element, and perform a binary classification task in which the model decides whether or not the presented image is a lymphoblast. Figure 1 shows examples of a probable lymphoblast and a healthy cell, respectively.

The baseline Clifford CNN model ($\mathcal{C}\ell$CNN) used in this work is composed by a convolutional layer with 4 filters, followed by two consecutive convolutional layers with 8 filters each and a convolutional layer with 16 filters. All layers use the split-rectified linear unit (split-ReLU) activation, i.e., the real ReLU applied separately to each channel of a Clifford number, and filters of size 3×3. Each of these layers is followed immediately by a max pooling layer with 2×2 kernels. The output of the final max pooling layer is then flattened and fed to a real-valued dense layer containing a single unit whose output is the label, 1 for lymphoblast, 0 otherwise. This defines a total of 3 $\mathcal{C}\ell$CNNs, one based on each algebra with multiplication table presented in Table 1.

For comparison, we propose a real-valued network with similar number of free trainable parameters and, hence, we shall refer to the $\mathcal{C}\ell$CNNs defined above as "**equivalent**". Since each hypercomplex-valued channel is roughly equivalent to four real-valued channels, we take the real-valued architecture with a larger number of filters per layer. Precisely, the real-valued CNN (RvCNN) is composed of the same four convolutional layers with 3×3 filters, each followed by a max pooling operator with 2×2 kernel. The number of filters per layer is 8, 16, 16 and 32, respectively, i.e., twice the number of filters in the corresponding equivalent hypercomplex-valued layer. The activation function used is the ReLU. The output of the fourth max pooling operation is then fed to a real-valued dense layer with a single neuron which outputs the calculated label. Lastly, to illustrate the vast learning capabilities of the $\mathcal{C}\ell$CNNs we take much smaller versions of the equivalent $\mathcal{C}\ell$CNNs and use these to perform the same task. These henceforth called "**small**" models use the same architecture of four convolutional layers followed by max pooling layers and a dense layer with a single neuron for labeling, yet each convolutional layer is taken with half the number of filters in the equivalent

Table 2. Sequential architecture outline and the number of trainable parameters.

		RvCNN	$C\ell$CNN (equivalent)	$C\ell$CNN (small)
Conv Layer 1	(3,3) filters	8	4	2
	Parameters	224	160	80
Max Pooling	2×2	–	–	–
Conv Layer 2	(3,3) filters	16	8	4
	Parameters	1,168	1,184	304
Max Pooling	2×2	–	–	–
Conv Layer 3	(3,3) filters	16	8	4
	Parameters	2,320	2,336	592
Max Pooling	2×2	–	–	–
Conv Layer 4	(3,3) filters	32	16	8
	Parameters	4,640	4,672	1,184
Max Pooling	2×2	–	–	–
Dense Layer	Neurons	1	1	1
	Parameters	1,153	2,305	1,153
Total		9,505	10,657	3,313

model. This leads to 3 small models with considerably less parameters than the real-valued and equivalent Clifford models. Thus, we end up with a total of 7 networks, namely, a real-valued model, 3 **equivalent** Clifford models with similar size to that of the real-valued model, and 3 **small** Clifford networks. All the architectures include a dropout layer before the dense layer with rate 0.5. This layer acts on a random behavior of setting inputs of the layer to the value 0 with a 0.5 rate. This layer helps avoiding overfitting the network to the training examples. Table 2 outlines the architectures and shows a comparison of the total number of parameters. Despite the architectural similarity, the equivalent and small networks proposed in this paper have respectively 29% and 9% of the trainable parameters of the hypercomplex-valued CNNs considered in [22].

The dataset contains 260 images evenly divided between the two classes. We resize images to 126×126 upon loading. Next, we perform 100 experiments with each of the 7 networks, a total of 700 experiments. We adopted the same 50% train-test split used in [10] and performed vertical/horizontal flips to augment the training set. To showcase the proposed model's ability to learn on scarce datasets, we prioritized the use of compact (i.e. lower total number of parameters) networks, which help reduce the risk of overfitting on small training sets, a frequent issue with deep-learning applications in the medical field due to the inherent data scarcity. The proposed neural networks were implemented using `Tensorflow v2.9` and `Keras`. We trained for 300 epochs, using the Adam optimizer, with learning rate of 0.001, batch size of 32, and binary cross-entropy loss function. Performance is gauged using the accuracy in the test set.[1]

[1] The complete code is available at https://github.com/mevalle/Hypercomplex-valued-Convolutional-Neural-Networks.

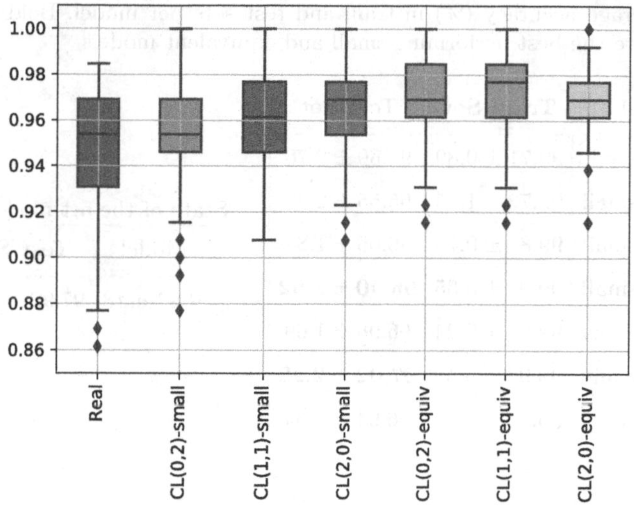

Fig. 2. Boxplot of test set accuracy performance by model.

Remark 2. The original images are encoded in RGB channels. This means each pixel contains 3 values representing the red, green and blue values respectively. An alternate encoding for images is the HSV, which stands for hue, saturation, and value. This color scheme displays colors in a radial slice, and better represents the human eye perception of color elements. In mathematical terms, a HSV encoded color pixel is represented as follows in a four-dimensional Clifford algebra derived from an orthonormal basis $\{\gamma_1, \gamma_2\}$:

$$\mathbf{I}(p) = \big(S(p) + V(p)\gamma_1\big)\big(\cos(H(p)) + \sin\big(H(p)\big)\gamma_2\big), \tag{9}$$

where $H(p) \in [0, 2\pi)$ and $S(p), V(p) \in [0, 1]$ denote respectively the hue, saturation, and value, of pixel p. We tested the 7 networks on both RGB- and HSV-encoded images and the results reported here are the best for each network. Namely, the RvCNN uses RGB-encoded images while all $\mathcal{C}\ell$CNNs use the HSV-encoded images.

Results for the 100 experiments of each model are depicted in Fig. 2. The real-valued model shows a larger range and a wider interquartile range (IQR) when compared to the Clifford models. Furthermore, the maximum and minimum values attained by the RvCNN are lower than the maximum and minimum for the remaining models respectively. This clearly indicates that the Clifford models outperformed the RvCNN.

When comparing the 6 $\mathcal{C}\ell$CNNs we observe that the equivalent models show smaller IQRs and ranges than the small models. Also, the equivalent models achieve a superior mean accuracy in the test set than their respective small counterparts. Thus, the equivalent networks perform better and are statistically more reliably than the respective small models. Nonetheless, the accuracy showcased by the small models is impressive in the light of their reduced number

Table 3. Average accuracy (%) in train and test sets per model. Bold numbers are used to indicate the best performing small and equivalent models.

Model	**Train Set**	**Test Set**
Real-valued	99.71 ± 0.89	94.60 ± 2.76
$\mathcal{C}\ell(0,2)$-small	99.72 ± 1.45	95.55 ± 2.30
$\mathcal{C}\ell(1,1)$-small	99.81 ± 0.75	96.05 ± 1.88
$\mathcal{C}\ell(2,0)$-small	99.91 ± 0.35	$\mathbf{96.40 \pm 1.92}$
$\mathcal{C}\ell(0,2)$-equiv	99.97 ± 0.24	96.96 ± 1.69
$\mathcal{C}\ell(1,1)$-equiv	99.92 ± 0.52	$\mathbf{97.02 \pm 2.22}$
$\mathcal{C}\ell(2,0)$-equiv	99.96 ± 0.25	96.94 ± 1.54

State of the art model [11]:

Model	**Test Set**
ResNet18	97.92 ± 1.62

of parameters. Table 3 shows the detailed metrics for all models, including the accuracy achieved by the ResNet18 combined with histopathological transfer learning [11].

Finally, Fig. 3 presents a Hasse diagram of the 7 models used. This diagram represents a hypothesis test with 95% significance level. Models higher up in the hierarchy perform better than the ones to which they are linked and also better than the ones below those models. As expected, at the top are located the equivalent $\mathcal{C}\ell$CNN models. The RvCNN model is the poorest performer, being at the bottom of the diagram. The small $\mathcal{C}\ell$CNNs lie in the middle, with a

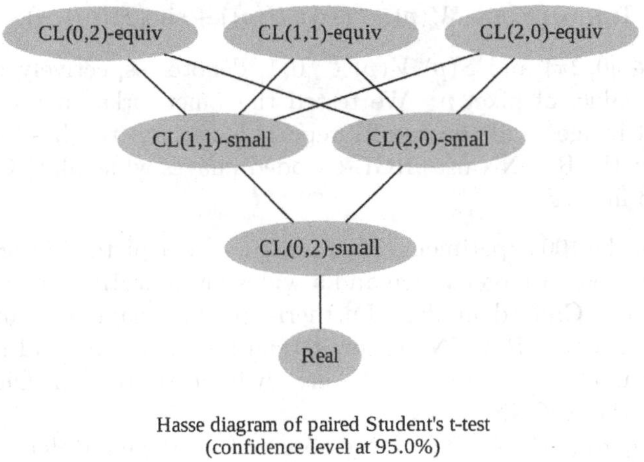

Fig. 3. Hasse diagram of the seven models present. A solid line linking two models indicates that the one on top performs better than the one below on a hypothesis test of 95% significance.

special mention to the $\mathcal{C}\ell(0,2)$ (quaternion-valued) model showcasing the worst performance of the three small $\mathcal{C}\ell$CNNs. Notably, the small $\mathcal{C}\ell$CNNs outperform the RvCNN despite having significantly less parameters.

5 Concluding Remarks

In this work we propose an implementation of a convolutional neural network on Clifford algebras, $\mathcal{C}\ell$CNN, and outline the operations involved in the model. We present an application of the proposed model to a medical image classification task used in a computer-aided diagnosis task. We compare the results attained by the $\mathcal{C}\ell$CNNs to a real-valued CNN of equivalent size to find that the proposed model outperformed the real counterpart by a significant margin. Then, we introduce a more compact version of the $\mathcal{C}\ell$CNN which features significantly less trainable parameters, and show that this model's performance sits between the initial $\mathcal{C}\ell$CNN's and the real-valued CNN's performance. On the one hand, this shows that despite having around 34.8% of the trainable parameters of the real model the small $\mathcal{C}\ell$CNN model performs noticeably better. On the other hand, the $\mathcal{C}\ell$CNN with a number of trainable parameters similar to the real-valued one performs vastly better, and close to state-of-the-art models in the literature. Indeed, the best result attained, namely the CNN based on the algebra $\mathcal{C}\ell(1,1)$, corresponds to 99.08% of the average accuracy reported in [11] but uses only 0.0093% of the approximately 11.4 million trainable parameters of the ResNet18.

In sum, we implement the proposed $\mathcal{C}\ell$CNN architecture and apply it to a real world dataset of a medical image classification task. The results attained show that $\mathcal{C}\ell$CNNs are extremely more compact than real-valued networks, all while presenting a gain in generalization capability. Moreover, the proposed models show performance levels close to state-of-the-art models up to 6 orders of magnitude larger in terms of parameters, hence posing $\mathcal{C}\ell$CNNs as more adequate solutions for small portable devices.

References

1. Aizenberg, I.N.: Complex-Valued Neural Networks with Multi-Valued Neurons. Studies in Computational Intelligence, vol. 353. Springer, Heidelberg (2011). https://doi.org/10.1007/978-3-642-20353-4
2. Arena, P., Fortuna, L., Muscato, G., Xibilia, M.G.: Multilayer perceptrons to approximate quaternion valued functions. Neural Networks **10**(2), 335–342 (1997)
3. Arena, P., Fortuna, L., Re, R., Xibilia, M.G.: On the capability of neural networks with complex neurons in complex valued functions approximation. In: 1993 IEEE International Symposium on Circuits and Systems, pp. 2168–2171. IEEE (1993)
4. Bibi, N., Sikandar, M., Din, I.U., Almogren, A., Ali, S.: IOMT-based automated detection and classification of leukemia using deep learning. J. Healthcare Eng. 2020 (2020)

5. Breuils, S., Tachibana, K., Hitzer, E.: New applications of Clifford's geometric algebra. Adv. Appl. Clifford Algebras **32**(2), 1–39 (2022)
6. Buchholz, S., Sommer, G.: Clifford algebra multilayer perceptrons. In: Sommer, G. (ed.) Geometric Computing with Clifford Algebras, pp. 315–334. Springer, Heidelberg (2001). https://doi.org/10.1007/978-3-662-04621-0_13
7. Claro, M., et al.: Convolution neural network models for acute leukemia diagnosis. In: International Conference on Systems, Signals, and Image Processing 2020-July, pp. 63–68 (2020)
8. Comminiello, D., Lella, M., Scardapane, S., Uncini, A.: Quaternion convolutional neural networks for detection and localization of 3D sound events. In: ICASSP 2019–2019 IEEE International Conference on Acoustics, Speech and Signal Processing (ICASSP), pp. 8533–8537. IEEE (2019)
9. Gaudet, C.J., Maida, A.S.: Deep quaternion networks. In: Proceedings of the International Joint Conference on Neural Networks 2018-July (2018)
10. Genovese, A., Hosseini, M.S., Piuri, V., Plataniotis, K.N., Scotti, F.: Acute lymphoblastic leukemia detection based on adaptive unsharpening and deep learning. In: ICASSP, IEEE International Conference on Acoustics, Speech and Signal Processing - Proceedings 2021-June, pp. 1205–1209 (2021)
11. Genovese, A., Hosseini, M.S., Piuri, V., Plataniotis, K.N., Scotti, F.: Histopathological transfer learning for acute lymphoblastic leukemia detection. In: CIVEMSA 2021 - IEEE International Conference on Computational Intelligence and Virtual Environments for Measurement Systems and Applications, Proceedings (2021)
12. Goodfellow, I., Bengio, Y., Courville, A.: Deep Learning. MIT Press, Cambridge (2016)
13. Granero, M.A., Hernández, C.X., Valle, M.E.: Quaternion-valued convolutional neural network applied for acute lymphoblastic leukemia diagnosis. In: Britto, A., Valdivia Delgado, K. (eds.) BRACIS 2021. LNCS (LNAI), vol. 13074, pp. 280–293. Springer, Cham (2021). https://doi.org/10.1007/978-3-030-91699-2_20
14. Hestenes, D., Sobczyk, G.: Clifford Algebra to Geometric Calculus: A Unified Language for Mathematics and Physics, vol. 5. Springer, Dordrecht (2012). https://doi.org/10.1007/978-94-009-6292-7
15. Hirose, A.: Complex-Valued Neural Networks. Studies in Computational Intelligence, 2nd edn. Springer, Heidelberg (2012). https://doi.org/10.1007/978-3-642-27632-3
16. Labati, R.D., Piuri, V., Scotti, F.: All-idb: the acute lymphoblastic leukemia image database for image processing. In: 2011 18th IEEE International Conference on Image Processing, pp. 2045–2048. IEEE (2011)
17. Parcollet, T., Morchid, M., Linarès, G.: A survey of quaternion neural networks. Artif. Intell. Rev. **53**(4), 2957–2982 (2020)
18. Prellberg, J., Kramer, O.: Acute lymphoblastic leukemia classification from microscopic images using convolutional neural networks. In: Gupta, A., Gupta, R. (eds.) ISBI 2019 C-NMC Challenge: Classification in Cancer Cell Imaging. LNB, pp. 53–61. Springer, Singapore (2019). https://doi.org/10.1007/978-981-15-0798-4_6
19. Shafique, S., Tehsin, S.: Acute lymphoblastic leukemia detection and classification of its subtypes using pretrained deep convolutional neural networks. Technol. Can. Res. Treat. **17**, 1–7 (2018)
20. Vaz, J., da Rocha, R.: An Introduction to Clifford Algebras and Spinors. Oxford University Press, Oxford (2016)
21. Vieira, G., Valle, M.E.: A general framework for hypercomplex-valued extreme learning machines. J. Comput. Math. Data Sci. **3**, 100032 (2022)

22. Vieira, G., Valle, M.E.: Acute lymphoblastic leukemia detection using hypercomplex-valued convolutional neural networks (2022). https://doi.org/10.48550/arxiv.2205.13273
23. Zolfaghari, M., Sajedi, H.: A survey on automated detection and classification of acute leukemia and WBCs in microscopic blood cells. Multimedia Tools Appl. 1–31 (2022)

Geometric Algebra and Distance Matrices

Vinicius Riter[1], Rafael Alves[2(✉)], and Carlile Lavor[1]

[1] University of Campinas, IMECC, Campinas, SP, Brazil
viniriter@gmail.com, clavor@unicamp.br
[2] Federal University of ABC, CMCC, Sao Bernardo, SP, Brazil
alves.rafael@ufabc.edu.br

Abstract. We present a new approach to the problem of recognizing an Euclidean distance matrix, based on Conformal Geometric Algebra. Such matrices are symmetric and hollow with non negative entries that are equal to the squared distances among the set of points. In addition to find these points, the method presented here also provides the minimal dimension of the related space. A comparison with a linear algebra approach is also provided.

Keywords: Geometric Algebra · Euclidean Distance Matrices · Sphere Intersection

1 Introduction

In Distance Geometry (DG) the fundamental object of study is the concept of distance [9], being established as a field in mathematics after the works of Blumenthal [2]. In recent years, DG has been applied to model problems in several areas of computer science, engineering and mathematics, such as sensor network localization, molecular geometry, GPS modelling among others [10].

An $n \times n$ matrix with real entries is called a distance matrix if there exists an ordered set $\{x_1, \dots, x_n\}$ of points in $\mathbb{R}^m$ such that each entry a_{ij} is the squared distance between x_i and x_j. When the Euclidean metric is used, we refer to a matrix of this type as an Euclidean Distance Matrix (EDM). In this case, the set $\{x_1, \dots, x_n\}$ is called a realization of the EDM. It is clear that an EDM is symmetric, has zeros in its main diagonal and all other entries are non-negative real numbers.

We present a geometric algebra (GA) based method to recognize an EDM, which provides also a realization in a space with the minimum possible dimension. The motivation for the use of GA was the geometric description based on sphere intersections of the approach presented in [1].

In the next section, we provide some important theoretical results on EDM's. In Sect. 3, we describe a linear algebra approach for recognizing an EDM, based on the method presented in [1]. Finally, in the Sect. 4, we present the main contribution of this work, which is a Conformal GA (CGA) approach developed to simplify the notation and the understanding of the problem.

Supported by organization CAPES, FAPESP, CNPq.

D. W. Silva et al. (Eds.): ICACGA 2022, LNCS 13771, pp. 88–98, 2024.
https://doi.org/10.1007/978-3-031-34031-4_8

2 EDM Recognition Problem

We start giving a formal definition of an EDM to make it clear that this does not depend on a specific set of points.

Definition 1. *Let D be an $n \times n$ matrix with real entries given by $D(i,j)$. If there exists a sequence $\{x_i\}_{i=1}^n \subset \mathbb{R}^K$, for some positive integer K, such that*

$$D(i,j) = \|x_i - x_j\|^2, i,j \in \{1,\ldots,n\}, \tag{1}$$

we call D an EDM.

The EDM recognition problem consists in finding a sequence of points that satisfies (1), called a realization. If there is a solution, there are infinitely many realizations for a given EDM, since any isometric transformation preserves the distances among the points of the realization. Also, if we add null coordinates at the right of each point, we obtain realizations in spaces with dimensions greater than K. We synthesize these results in the next proposition, given in [1].

Proposition 1. *Let an $n \times n$ matrix D be an EDM. If D has a realization $\{x_i\}_{i=1}^n$, $x_i \in \mathbb{R}^K$, then there are infinitely many realizations of D in $\mathbb{R}^p$, for any $p \geq K$.*

The idea of the method we discuss here is to find the minimum K such that there is a realization for an EDM. This number is called the embedding dimension of the EDM [1].

Definition 2. *Let an $n \times n$ matrix D be an EDM and let us suppose that there is a realization for D in $\mathbb{R}^K$. If for any other realization of D in $\mathbb{R}^m$, $m \geq K$, then K is called the embedding dimension of D, denoted by $dim(D)$.*

There is an upper bound for the embedding dimension related to the dimension of the matrix. In [1], the authors prove that

$$dim(D) \leq n - 1,$$

for an $n \times n$ EDM D with $n \geq 2$.

3 A Linear Algebra Approach for the EDM Recognition Problem

The method presented in [1] is based on the proof of Theorem 1, given below, which depends on the two following lemmas.

Lemma 1. *Let an $(n+1) \times (n+1)$ matrix D be an EDM and let D_n be the submatrix of D given by its first n rows and columns. If $dim(D_n) = K$, then $dim(D)$ is either K or $K+1$.*

Lemma 2. *Let D as in Lemma 1 and let m be such that $dim(D) \leq m$. If $\{x_i\}_{i=1}^n$ in $\mathbb{R}^m$ is a set of points that realizes D_n, then there exists a point $x_{n+1} \in \mathbb{R}^m$ such that $\{x_i\}_{i=1}^{n+1}$ realizes D.*

The proof of the next result, given in [1], yields an algorithm to recognize an EDM, which also finds a realization for that.

Theorem 1. *Let K be a positive integer and D a symmetric matrix $n \times n$, $n \geq 2$, with null diagonal and no negative entries. D is an EDM with embedding dimension K if, and only if, there exists a set of points $\{x_i\}_{i=1}^n$ in $\mathbb{R}^K$ and an index set $I = \{i_1, \ldots, i_{K+1}\} \subset \{1, 2, \ldots, n\}$, such that*

$$\begin{cases} x_{i_1} & = 0 \\ x_{i_j}(j-1) & \neq 0,\ j \in I_{2,K+1} \\ x_{i_j}(i) & = 0, j \in I_{2,K},\ i \in I_{j,K} \end{cases}$$

where $\{x_i\}_{i=1}^n$ realizes D and $I_{a,b} = \{a, a+1, \ldots, b\}$ ($x_h(p)$ is the p-th component of the h-th vector).

The idea of the algorithm is to build the given matrix from its submatrices checking whether each one is an EDM and finding a solution for them. In the positive case, the results above are used to ensure that the initial matrix is an EDM and to construct a solution.

Given a hollow $n \times n$ symmetric matrix $A = (a_{ij})$ with non-negative entries, let A_k, $k = 1, \ldots, n$, be the principal submatrices of A. Let us consider the submatrix A_2, which is an EDM with a realization in $\mathbb{R}$ given by $x_1 = 0$ and $x_2 = \sqrt{a_{12}}$, where $dim(A_2) = 1$. From Lemma 1, if A_3 is an EDM, then $dim(A_3)$ is 1 or 2. From Lemma 2, there is $x_3 \in \mathbb{R}^2$ such that the set of points $x_1 = (0, 0)$, $x_2 = (\sqrt{a_{12}}, 0)$, and x_3 realize A_3. Therefore, if we find a solution for x_3, we guarantee that A_3 is an EDM, we give a realization for it, and also determine its embedding dimension. To find x_3, it is necessary to solve the following nonlinear system:

$$\begin{cases} \|x_1 - x_3\|^2 & = a_{13} \\ \|x_2 - x_3\|^2 & = a_{23}. \end{cases}$$

Geometrically, it means that x_3 lies on the intersection of spheres centered at x_1 and x_2, with radius $\sqrt{a_{13}}$ and $\sqrt{a_{23}}$, respectively. Since $x_1 = (0, 0)$, the first equation is simply $\|x_3\|^2 = a_{13}$, and subtracting it from the second one, we have

$$x_2^\top x_3 = \frac{1}{2}(\|x_2\|^2 - a_{23} + a_{13}).$$

This equation returns a unique solution for the first coordinate of x_3, say x_{31}, but not the second since $x_2 = (\sqrt{a_{12}}, 0)$. To find x_{32}, we use the first equation to get

$$x_{32}^2 = a_{13} - x_{31}^2.$$

If x_{32}^2 is non-negative, we ensure that A_3 is an EDM, otherwise it is not. If $x_{32} > 0$, we have two solutions for x_3, say x_3^+ and x_3^-, and we increase the embedding dimension to 2, i.e., A_3 is an EDM with $dim(A_3) = 2$ and $\{x_1, x_2, x_3^+\}$ is a realization for A_3 (x_3^- could be used instead of x_3^+). However, if $x_{32} = 0$, we have only one solution for x_3 and we can get rid of the second coordinate of the three points to have a realization for A_3, which means that the embedding dimension was kept in 1, and the realization would be simply given by $\{0, \sqrt{a_{12}}, x_{31}\}$. For both cases, the realizations satisfy the conditions of Theorem 1. The procedure above is repeated until we reach the whole matrix A, increasing the size of the system to be solved if all submatrices are indeed EDM's. As the realizations satisfy the conditions of Theorem 1, supposing that $dim(A_{n-1}) = K$, we have at the end the following configuration for a realization $\{x_i\}_{i=1}^{n-1} \in \mathbb{R}^K$ of A_{n-1}:

$$\begin{aligned} x_1 &= (0, \ldots, 0) \\ x_2 &= (x_{21}, 0, \ldots, 0) \\ &\vdots \\ x_{n-1} &= (x_{n-1,1}, \ldots, x_{n-1,K}). \end{aligned}$$

We do the same we did before and insert zeros in the last coordinate of each point. Again, from Lemmas 1 and 2, if A is an EDM, $dim(A)$ is either K or $K+1$ and there exists $x_n \in \mathbb{R}^{K+1}$ such that $\{x_1, \ldots, x_n\}$ is a realization for A. The system to be solved is given by

$$\begin{cases} \|x_1 - x_n\|^2 &= a_{1n} \\ &\vdots \\ \|x_{n-1} - x_n\|^2 &= a_{n-1,n} \end{cases},$$

and the procedure is a generalization of what was made for A_3. That is, we subtract the first equation from all the others, recalling that $x_1 = (0, \ldots, 0)$, to obtain:

$$\begin{cases} \|x_n\|^2 &= a_{1n} \\ x_2^\top x_n &= b_2 \\ &\vdots \\ x_{n-1}^\top x_n &= b_{n-1} \end{cases},$$

where $b_j = \dfrac{\|x_j\|^2 - a_{j,n} + a_{1,n}}{2}$, for $2 \le j \le n-1$. Now, from the structure of the points x_i, $i = 2, \ldots, n-1$, there is a triangular linear system that has either a unique solution or no solution for the first K coordinates of x_n. If no solution is found, A is not an EDM. Otherwise, we use the solution found, say x_n^*, to find the last coordinate $x_{n,K+1}$. As we did before, we get

$$x_{n,K+1}^2 = a_{1n} - \|x_n^*\|^2.$$

If $x_{n,K+1}^2$ is negative, A is not an EDM. Otherwise, if $x_{n,K+1}^2 > 0$ and $dim(A) = K+1$, we have two solutions for x_n, say x_n^+ and x_n^-, and choose one to give a realization for A. If $x_{n,K+1}^2 = 0$, there will be only one solution for x_n, we get rid of the last 0 coordinate of each point, and the embedding dimension remains unchanged, implying that $dim(A) = K$.

4 A Conformal Geometric Algebra Approach

Now, using CGA, we present a new formulation for the problem taking advantage of the geometric interpretation of the approach described in the previous section.

We recall that in CGA over $\mathbb{R}^n$ we use two extra basis vectors $\{e_\infty, e_0\}$ together with the canonical Euclidean basis to work in a space with dimension $n+2$. In this space, we can easily represent geometric objects such as points, planes, and spheres by vectors. A powerful tool of CGA is that operators and operands are entities of the same algebra, which means that transformations like reflections, rotations or translations are performed by elements of the algebra. It is important to notice that the increase in the dimension of the space along with the metric used make these transformations to be orthogonal. Another highlight of CGA is the intuitiveness of intersecting objects. A circle, for instance, can be constructed by the intersection of two spheres, the same for a line in the intersection of two planes. These intersections are achieved with the exterior product. The geometric objects we mentioned have also another representation, given by points that lie on them, and there is also another way to intersect these objects, using the inner product. Here, we mainly focus on the first representation and in the intersections with the exterior product. For more details about CGA, we recommend [4,5,11].

It can be proved (see, for instance, [11]) that a sphere in $\mathbb{R}^n$, with center $c \in \mathbb{R}^n$ and radius $r \in \mathbb{R}$, can be represented in CGA by the vector

$$S = C - \frac{1}{2}r^2 e_\infty, \tag{2}$$

where C is the representation of c in the conformal space $\mathbb{R}^{n+1,1}$. This result can be achieved developing the inner product $S \cdot C$, regarding S as the conformal representation of $s \in \mathbb{R}^n$ and using one of the most important relations between vectors in $\mathbb{R}^n$ and its conformal representations, which says that the inner product $S \cdot C$ is proportional to the square distance between s and c:

$$S \cdot C = -\frac{1}{2}\|s - c\|^2. \tag{3}$$

The next key definition is the intersection of spheres through the exterior product. Given two spheres S_1 and S_2 as in (2), the bivector $S_1 \wedge S_2$ represents their intersection. This result can be directly extended to any number of spheres and we also check a prior if, in fact, there is any intersection. For more details, see [7]. The next result, given in [7,8], will be important for the new approach.

Proposition 2. *The intersection of k spheres with affine independent centers in $\mathbb{R}^n$ is either an empty set, a single point or a $(n-k+1)$-sphere*[1].

Also from [7,8], it is possible to check the nature of the intersection, computing the parameter

$$t = \sigma \cdot \tilde{\sigma},$$

[1] An i-sphere is the intersection of a sphere with an affine subspace of dimension i.

where $\sigma = \bigwedge_{i=1}^{k} S_i$ is the intersection of spheres S_i, $i = 1, \ldots, k$, and $\tilde{\sigma}$ is the *reverse*[2] of σ. If $t < 0$, there is no intersection; if $t = 0$, it occurs in a single point; and if $t > 0$, the intersection is a $(n-k+1)$-sphere. The parameter t can be computed as a determinant of a matrix with ij-th entries given by $S_i \cdot S_j$. It is also possible to compute explicitly the radius and the center of the intersection by the following formulas. If σ is the intersection of k spheres, then

$$C_\sigma = -\frac{1}{2}\frac{\sigma e_\infty \sigma}{(e_\infty \cdot \sigma)^2} \tag{4}$$

$$r_\sigma^2 = \frac{(-1)^{(k+1)}\sigma^2}{(e_\infty \cdot \sigma)^2} \tag{5}$$

are respectively the conformal center and squared radius of σ. Note that r_σ also returns the nature of the intersection analogously to what we did for t.

The idea of the method we are developing is to have always at most two points in the intersection. So, from Proposition 2, to satisfy this requirement in $\mathbb{R}^n$ it is necessary to have n spheres. Another important remark is that in the case the intersection is exactly a point pair, these points cannot be in the hyperplane generated by all centers. In fact, they must be symmetric (relatively to this hyperplane) in order to satisfy all distance restraints. Note also that the center of this point pair[3] lies on this hyperplane.

Let $n+1$ spheres in $\mathbb{R}^{n+1}$ with different centers in $\mathbb{R}^n$, i.e., with the $n+1$-th entry equal to zero. The space generated by these $n+1$ points is normal to the vector e_{n+1}. In fact, if c_i is the center of each sphere and C_i is its conformal representation, $i = 1, \ldots n+1$, then

$$C_i = \alpha_1 e_1 + \alpha_n e_n + \alpha_\infty e_\infty + e_0,$$

for $i = 1, \ldots, n+1$, and $\alpha_j \in \mathbb{R}$, $j = 1, \ldots, n$. Since the $n+1$-th coordinate of each point is null, we have that $C_1 \wedge \cdots \wedge C_{n+1}$ is a linear combination of $(n+1)$-blades that does not contain e_{n+1}. Moreover, there is only one $(n+1)$-blade in this combination that does not have the vector e_∞, given by $e_1 \wedge \cdots \wedge e_n \wedge e_0$. We are interested in this one because the plane given by all of those centers is

$$\Pi = C_1 \wedge \cdots \wedge C_{n+1} \wedge e_\infty,$$

and since $e_\infty \wedge e_\infty = 0$, the plane Π is a scalar multiple of $e_1 \wedge \cdots \wedge e_n \wedge e_\infty \wedge e_0$, which means that the vector e_{n+1} is normal to the plane Π.

Now, let us suppose that those spheres intersect at a point pair given by $\{p_+, p_-\}$, implying that p_+ and p_- are symmetric relatively to the plane given by the centers. Let m be the center of this point pair, which lies in the plane Π as we commented earlier. It is easy to see that the segment connecting each

[2] We recall that the reverse of a blade is another blade with the reverted order of the factors in the exterior product.

[3] The center of a point pair is regarded as the midpoint of the segment connecting the two points.

center c_i to m is perpendicular to the line given by the point pair. Indeed, for each c_i this is the height of an isosceles triangle whose equal sides meet at c_i and are the radius of the sphere S_i (the base is exactly the segment connecting the point pair). So, given m, it is possible to obtain p_+ and p_- walking from m through the directions $\pm e_{n+1}$:

$$p_+ = m + re_{n+1}, \quad p_- = m - re_{n+1}, \tag{6}$$

where $r \in \mathbb{R}$ is the radius of the point pair. This is an alternative manner to extract the points of a point pair that takes advantage of the knowledge of the direction of the point pair.

We can now state the main result of this section, which suggests a CGA method to solve the EDM recognition problem. Let us first define the application $\mathcal{P}$ that maps a conformal point into its corresponding point in the Euclidean space:

$$\begin{aligned} \mathcal{P} : \mathbb{R}^{n+1,1} &\to \mathbb{R}^n \\ X &\mapsto x. \end{aligned}$$

Theorem 2. *Let K be a positive integer and A an $n \times n$ hollow and symmetric matrix with non-negative entries, for $n \geq 2$. A is an EDM with $dim(A) = K$ if, and only if, there exists a realization $\{x_i\}_{i=1}^n \subset \mathbb{R}^K$ for A and a set of indexes $I = \{i_1, \dots, i_{K+1}\} \subset \{1, \dots, n\}$, such that*

$$\begin{cases} x_{i_1} &= 0, \\ x_{i_j} &= \mathcal{P}(\bigwedge_{p=1}^{j-1} S_{j_p})^+, \quad j \in I_{2,K+1}, \end{cases} \tag{7}$$

where $S_{j_p} = \mathcal{C}(x_{i_p}) - \frac{1}{2} a_{i_p,i_j} e_\infty$, for each j_p, are the conformal representations of the spheres in $\mathbb{R}^{j-1}$.

Remark 1. Note that $\mathcal{P}(\bigwedge_{p=1}^{j-1} S_{j_p})^+$ refers to the corresponding point in $\mathbb{R}^n$ of one of the conformal points in the point pair computed by the exterior product.

Proof. The proof is by induction on the dimension of the matrix A. Beginning with $n = 2$, the matrix A is given by

$$A = \begin{bmatrix} 0 & a_{12} \\ a_{12} & 0 \end{bmatrix}.$$

Supposing that all the entries outside the diagonal are strictly positive, we have that $a_{12} > 0$, A is an EDM with $dim(A) = 1$, and the points $x_1 = 0$ and $x_2 = \sqrt{a_{12}}$ define a realization for A.

Let us suppose, by induction, that for any EDM with order $n \geq 2$ and embedding dimension K, the theorem is valid, i.e. there exists a realization $\{x_i\}_{i=1}^n \subset \mathbb{R}^K$ for this EDM, with an index set $I = \{i_1, \dots, i_{K+1}\} \subset \{1, \dots, n\}$, satisfying (7). Let us consider A as an EDM with order $(n+1)$ and $dim(A) = K$, and A_n be the n-th principal submatrix of A. By Lemma 1, A_n is an EDM with

$dim(A_n) = k$, where k is either K or $K-1$. Using the induction hypothesis, we have that A_n has a realization $\{x_i\}_{i=1}^n \subset \mathbb{R}^k$ and that there is an index set $I = \{i_1, \ldots, i_{k+1}\} \subset \{1, \ldots, n\}$, such that

$$\begin{cases} x_{i_1} &= 0, \\ x_{i_j} &= \mathcal{P}(\bigwedge_{p=1}^{j-1} S_{j_p})^+, \quad j \in I_{2,k+1}. \end{cases}$$

Define $y = \mathcal{P}(P)$, where

$$P = \bigwedge_{j=1}^{k+1} S_{(n+1)_j}$$

is the intersection of the $k+1$ spheres $S_{(n+1)_j} = \mathcal{C}(x_{i_j}) - \frac{1}{2}a_{i_j,n+1}e_\infty$ in $\mathbb{R}^{k+1}$. The points x_{i_j}, which are the centers of the spheres, lie in $\mathbb{R}^k$. Then, by (6) and the discussion that led to it, if the intersection P is a point pair, its direction is given by e_{k+1}. Moreover, once we know the center of each sphere and their (squared) radius given by the entries of A, the solution set for y cannot be empty, otherwise A would not be an EDM.

Using formula (5), we check the nature of the intersection looking to the sign of r^2. If $r^2 = 0$, we take

$$x_{n+1} = c,$$

where c is the unique point in the intersection, obtained by (4). On the other hand, if $r^2 > 0$, then the intersection y results in a point pair, where we can still compute c and choose

$$x_{n+1} = p_+, \quad \text{(or } p_- \text{ equivalently)},$$

obtained by (6). For both cases, the sequence $\{x_i\}_{i=1}^{n+1}$ realizes the matrix A and satisfies the theorem conditions for $n+1$. Therefore, the theorem is proved for every $n \geq 2$. □

This proof induces an algorithm (see Algorithm 1) to check if a given matrix is an EDM, to find a realization for it, in the positive case, and also to provide its embedding dimension.

The algorithm starts with the submatrix A_2. From A_3, it computes the exterior product among the spheres to obtain P (step 5) and the value r^2 to find the nature of P (step 6). In the next steps, the algorithm proceeds accordingly. It is very important to note that the increment on the embedding dimension only happens when $r^2 > 0$ (steps 11 to 14). In fact, when $r^2 = 0$, the point to be included in the solution lies on the plane generated by the centers, i.e. the dimension of the space containing the realization does not change. The embedding dimension is incremented by one when the new point is out of this plane, which implies that the realization will be in $\mathbb{R}^{K+1}$, with x_i (in step 12) and all the previous points in $\mathbb{R}^K$ gaining a new null $K+1$-th coordinate.

Algorithm 1

Input: $A = (a_{ij})$, with $a_{ii} = 0$ and $a_{ij} = a_{ji} \geq 0$, $i, j = 1, \ldots, n$

1: $I = \{1, 2\}$
2: $K = 1$
3: $(x_1, x_2) = (0, \sqrt{a_{12}})$
4: **for** $i \in \{3, \ldots, n\}$ **do**
5:     $P = \bigwedge_{j \in I} (\mathcal{C}(x_j) - \frac{1}{2} a_{ij} e_\infty)$
6:     $r^2 = \frac{(-1)^{n+1} P^2}{(e_\infty \cdot P)^2}$
7:     **If** $r^2 < 0$ **then**
8:         **return** "failure"
9:     **else if** $r^2 = 0$ **then**
10:         $x_i = \mathcal{P}\left(-\frac{1}{2} \frac{P e_\infty P}{(e_\infty \cdot P)^2}\right)$
11:     **else** $r^2 > 0$ **then**
12:         $x_i = \mathcal{P}\left(-\frac{1}{2} \frac{P e_\infty P}{(e_\infty \cdot P)^2}\right) + r e_{K+1}$
13:         $I \leftarrow I \cup \{i\}$
14:         $K \leftarrow K + 1$
15:     **end if**
16: **end for**
17: **return** K, x

---

5   An Illustrative Example

In this section, we illustrate the proposed approach with an example. Let us consider

$$A = \begin{bmatrix} 0 & 1 & 1 & 2 \\ 1 & 0 & 2 & 1 \\ 1 & 2 & 0 & 1 \\ 2 & 1 & 1 & 0 \end{bmatrix}.$$

The first submatrix to be used is

$$A_2 = \begin{bmatrix} 0 & 1 \\ 1 & 0 \end{bmatrix},$$

where a realization is given by $x_1 = 0$ and $x_2 = \sqrt{a_{12}} = 1$. Following the procedure, we insert zeros in a second coordinate of x_1 and x_2. The conformal representation of these points are respectively $X_1 = e_0$ and $X_2 = e_1 + 0.5 e_\infty + e_0$, and the related spheres we need to intersect are given by

$$S_1 = X_1 - 0.5 a_{31} e_\infty = e_0 - 0.5 e_\infty,$$
$$S_2 = X_2 - 0.5 a_{32} e_\infty = (e_1 + 0.5 e_\infty + e_0) - e_\infty = e_1 - 0.5 e_\infty + e_0.$$

We can use several tools to compute r^2 and the center C of $P = S_1 \wedge S_2$ (for example, [3,12]). Using Gaalop [6], we obtain:

$$
\begin{aligned}
P &= S_1 \wedge S_2 = 0.5e_1 \wedge e_\infty - e_1 \wedge e_0, \\
r^2 &= \frac{(-1)^{2+1} P^2}{(e_\infty \cdot P)^2} = 1, \\
C &= -\frac{1}{2} \frac{P e_\infty P}{(e_\infty \cdot P)^2} = e_0.
\end{aligned}
$$

Since $r^2 > 0$, we have that $x_3 = C + e_2 = e_2$. Now, we have $I = \{1, 2, 3\}$ and $K = 2$. As we need the conformal points, let us see the current solution given by

$$
X_1 = e_0, \quad X_2 = e_1 + 0.5e_\infty + e_0, \quad X_3 = e_2 + 0.5e_\infty + e_0.
$$

To find X_4, we insert a null coordinate in the previous solution and compute $S = S_1 \wedge S_2 \wedge S_3$. From

$$
\begin{aligned}
S_1 &= X_1 - 0.5a_{41}e_\infty = e_0 - e_\infty, \\
S_2 &= X_2 - 0.5a_{42}e_\infty = (e_1 + 0.5e_\infty + e_0) - 0.5e_\infty = e_1 + e_0, \\
S_3 &= X_3 - 0.5a_{43}e_\infty = (e_2 + 0.5e_\infty + e_0) - 0.5e_\infty = e_2 + e_0,
\end{aligned}
$$

we obtain

$$
\begin{aligned}
P &= \bigwedge_{i=1}^{3} S_i = -e_1 \wedge e_2 \wedge e_\infty + e_1 \wedge e_2 \wedge e_0 + e_1 \wedge e_\infty \wedge e_0 - e_2 \wedge e_\infty \wedge e_0, \\
r^2 &= \frac{(-1)^{3+1} P^2}{(e_\infty \cdot P)^2} = 0, \\
C &= -\frac{1}{2} \frac{P e_\infty P}{(e_\infty \cdot P)^2} = e_1 + e_2 + e_\infty + e_0.
\end{aligned}
$$

Since $r^2 = 0$, $X_4 = C$, $dim(A) = 2$, and the solution is kept in $\mathbb{R}^2$, given by

$$
\{(0,0), (1,0), (0,1), (1,1)\}.
$$

6 Conclusion

The application of CGA to the EDM recognition problem provided a much simpler description of the linear algebra approach used to solve this problem. In fact, the result given by Theorem 2 makes clear how the sequence of points in the realization is constructed and gives a geometric meaning for each of those points as intersections of spheres, which cannot be seen in Theorem 1. Another important remark is that, in Algorithm 1, one does not need to actually change the dimension of the space. The description and the computation are similar for each dimension, implying that it is possible to set the maximum dimension since the beginning of the algorithm and update the embedding dimension according to the value of r^2. The geometric intuition given by the CGA approach will be useful for instances of the problem involving uncertainties in the matrix entries, since the idea of sphere intersections is preserved.

Acknowledgements. We would like to thank the Brazilian research agencies FAPESP, CAPES and CNPq, and the reviewers for the useful comments.

References

1. Alencar, J., Lavor, C., Liberti, L.: Realizing Euclidean distance matrices by sphere intersection. Discret. Appl. Math. **256**, 5–10 (2019)
2. Blumenthal, L.: Theory and Applications of Distance Geometry. Oxford University Press, Oxford (1953)
3. Breuils, S., Nozick, V., Fuchs, L.: Garamon: a geometric algebra library generator. Adv. Appl. Clifford Algebras **29**(4), 1–41 (2019). https://doi.org/10.1007/s00006-019-0987-7
4. Dorst, L.: Geometric Algebra for Computer Science. An Object-Oriented Approach to Geometry. Morgan Kauffmann, San Francisco (2007)
5. Hildenbrand, D.: Foundations of Geometric Algebra Computing. Springer, Berlin, Heidelberg (2013). https://doi.org/10.1007/978-3-642-31794-1
6. Hildenbrand, D., Charrier, P., Steinmetz, C., Pitt, J.: Gaalop Homepage. http://www.gaalop.de. Accessed 24 Jan 2022
7. Lavor, C., Alves, R., Fernandes, L.A.F.: Linear and geometric algebra approaches for sphere and spherical shell intersections in $\mathbb{R}^n$. Expert Syst. Appl. **187**(115993), (2022)
8. Maioli, D., Lavor, C., Gonçalves, D.: A note on computing the intersection of spheres in $\mathbb{R}^n$. ANZIAM J. **59**, 271–279 (2017)
9. Menger, K.: Untersuchungen uber allgemeine Metrik. Math. Ann. **100**, 75–163 (1928)
10. Mucherino, A., Lavor, C., Liberti, L., Maculan, N. (eds.): Distance Geometry: Theory, Methods, and Applications. Springer, New York (2013). https://doi.org/10.1007/978-1-4614-5128-0
11. Perwaß, C.: Geometric Algebra with Applications in Engineering. Springer, Berlin, Heidelberg (2009). https://doi.org/10.1007/978-3-540-89068-3
12. Sousa, E.V., Fernandes, L.A.: TbGAL: a tensor-based library for geometric algebra. Adv. Appl. Clifford Algebras **30**, 1–33 (2020)

GAAlign: Robust Sampling-Based Point Cloud Registration Using Geometric Algebra

Kai A. Neumann[1,2](✉), Dietmar Hildenbrand[1], Florian Stock[1], Christian Steinmetz[3], and Maximilian Michel[1]

[1] Technical University of Darmstadt, Darmstadt, Germany
kai@neumann.to
[2] Fraunhofer IGD, Darmstadt, Germany
[3] :em engineering methods AG, Darmstadt, Germany

Abstract. Geometrical 3D data is often represented in form of point clouds. A common problem is the registration of point clouds with shared underlying geometry, for example to align two 3D scans. This work presents *GAAlign*, a new formulation of a geometric algebra (GA) based algorithm that aims to solve this problem. While the algorithm itself is a gradient descent based approach, the implementation takes advantage of *GAALOP*, which had to be extended with a specific, so far unsupported GA, namely projective GA.

The proposed new robust registration algorithm uses a geometric algebra based motor estimation algorithm in the context of a stochastic gradient descent inspired algorithmic structure and achieves state-of-the-art results. When using synthetically disturbed input data the results show, that *GAAlign* either outperforms other used algorithms (outliers) or is comparable to the best (Gaussian noise) while having a significantly better runtime as soon as the number of correspondences increases. When used in a real world pipeline, *GAAlign* also performs on the same level or above compared to state-of-the-art algorithms.

Keywords: 3D registration · Geometric algebra · Point cloud alignment

1 Introduction

Point clouds are a popular way of representing 3D objects and scenes used in many applications ranging from 3D reconstruction to interactive visualization. A central problem in computer vision is the registration of point clouds with (partially) shared underlying geometry. This can for example be used to tightly align two different 3D scans of the same object.

One of the earliest and most commonly used methods for point cloud registration is called *Iterative Closest Point* (ICP) presented by Besl and McKay [2]. Similar to this, Chen and Medioni [4] introduced a *point-to-plane* distance metric

D. W. Silva et al. (Eds.): ICACGA 2022, LNCS 13771, pp. 99–111, 2024.
https://doi.org/10.1007/978-3-031-34031-4_9

which is optimized as a non-linear least squares problem to achieve a registration. The solution of this problem can be approximated using a linear equation system as shown by Low [15]. Different improved objective functions for ICP were proposed by Yang et al. [24] and Rusinkiewicz [19]. State-of-the-art registration algorithms like *Fast Global Registration* by Zhou et al. [25] combine least squares optimization with a feature-based correspondence search such as FPFH by Rusu et al. [20]. Furthermore, many recent publications like NgeNet by Zhu et al. [26] or GeDi by Poiesi and Boscaini [17] apply deep learning based methods to point cloud registration. A comprehensive survey of recent approaches is presented by Huang et al. [9].

In addition to traditional linear algebra based methods, several methods for point cloud registration using geometric algebra have been introduced. Kleppe et al. introduced a non-linear least squares based registration algorithm [11] as well as two different descriptors that can be used for registration [10,12]. Geometric algebra enables the registration of heterogeneous sets of objects in a unified framework as shown by Valkenburg and Dorst [23] as well as Tingelstad and Egeland [22]. Other approaches for geometric algebra based point cloud registration include the use of a least-mean-squares adaptive filter called *GA-LMS* by Al-Nuaimi et al. [1,13,14] and the formulation of ICP in conformal geometric algebra and subsequent application to airborne laser scanning data by Hitzer et al. [8].

2 Preliminaries

The term geometric algebra (GA) denotes a category of algebras that contain an intuitive representation of geometric objects (e.g. points or spheres) and operations (e.g. intersections). For *GAAlign* we decided to use the 3D projective geometric algebra $\mathbb{R}^*_{3,0,1}$ (PGA), as it enables an elegant representation of points, lines and planes. While many good introductions to geometric algebra exist (see [5,6]), we want to offer a quick overview over the relevant concepts that are used in this paper.

The 3D projective geometric algebra consists of 4 basis vectors, also called *blades of grade 1*, denoted as $\mathbf{e}_0, \mathbf{e}_1, \mathbf{e}_2, \mathbf{e}_3$. Hereby, the algebra defines $\mathbf{e}_0^2 = 0$ and $\mathbf{e}_1^2 = \mathbf{e}_2^2 = \mathbf{e}_3^2 = 1$. The basis vectors can be combined using the *outer product* (denoted as $a \wedge b$) to form blades of higher grade. Any object in GA can be defined by a combination of blades. In 3D PGA there are 16 unique blades, hence any object can be expressed as a 16-dimensional vector. For example, a point can be defined using the equation $\mathbf{p} := x\mathbf{e}_{032} + y\mathbf{e}_{013} + z\mathbf{e}_{021} + \mathbf{e}_{123}$. Furthermore, a line through two points $\mathbf{p}_1, \mathbf{p}_2$ can be defined using the *join product*: $\mathbf{l} = \mathbf{p}_1 \vee \mathbf{p}_2$. Analogous, this can also be used to form a plane from three points.

In addition to geometric objects, it is also possible to define rigid transformations as multivectors. These are called *motors* (see [5]) and are defined on the basis $\{1, \mathbf{e}_{12}, \mathbf{e}_{31}, \mathbf{e}_{23}, \mathbf{e}_{01}, \mathbf{e}_{02}, \mathbf{e}_{03}, \mathbf{e}_{0123}\}$. Hereby, $\mathbf{e}_{12}$, $\mathbf{e}_{31}$, and $\mathbf{e}_{23}$ square to -1 and $\mathbf{e}_{01}, \mathbf{e}_{02}$, and $\mathbf{e}_{03}$ square to 0. This basis is equivalent to dual quaternions as shown by Dorst and De Keninck [5]. Furthermore, every motor is equivalent to the exponential of a bivector, which can be retrieved by using the logarithm. As the bivector space is linear, it can be used for linear interpolation. The resulting interpolated motor can subsequently be calculated by exponentiating the bivector. As a fast approximation, motors can also be interpolated directly. Hereby, to stay on the motor manifold, the scalar of both motors need to have the same sign and the resulting motor needs to be normalized. Notably, this interpolation method is not in equal steps, which can only be guaranteed in the bivector space.

3 Sampling-Based Point Cloud Registration

3.1 GAALOP

The proposed algorithm is based on two major contributions to the geometric algebra optimizer *GAALOP* [7] which were key to its implementation. Those are the introduction of 3D projective geometric algebra (PGA) as well as a more advanced optimization strategy that identifies common subexpressions resulting in an improved runtime performance.

PGA in *GAALOP*. *GAALOP* turns geometric algebra code written in its custom script language into symbolically optimized code for multiple common programming languages. New algebras can be added by defining the base blades and calculating a multiplication table, which is subsequently used to calculate the geometric product. In addition to this, macros for common operations like the regressive product can be defined. This was done to introduce 3D PGA to *GAALOP*.

Common Subexpression Elimination. When generating C code from GA expressions, it often happens, that the coefficients are very similar. These coefficients often contain long terms, that are not the same, but have subexpressions in common. In compiler development, finding subexpressions and eliminating these redundant computations is an often used approach, and several different techniques for it are available. We added such a method for common subexpression elimination [16] to *GAALOP* and were able to improve the speed of the generated code by $\approx 3.5\%$.

3.2 GAAlign

The proposed point cloud registration algorithm consists of two parts. First, a method to calculate a motor from two corresponding triangles (Sect. 3.2) which is then subsequently used inside the algorithmic structure of a stochastic gradient descent (Sect. 3.2) to robustly perform a point cloud registration.

Motor Estimation from Corresponding Triangles. The algorithm used for estimating a motor from two corresponding triangles in *GAALOP* is shown in Listing A.1 in the Appendix. This algorithm defines the points `A_src`, `B_src` and `C_src` of the source triangle as well as the points `A_tar`, `B_tar` and `C_tar` of the target triangle and computes the motor for the transformation between them. The implementation of the algorithm follows the procedure detailed in Sect. 6.8 of the document *A Guided Tour to the Plane-Based Geometric Algebra* by Dorst et al. [5]. The exclamation mark at the beginning of a line indicates that this multivector should explicitly be computed. Hereby, *GAALOP* only computes the coefficients which are needed for further computations. This is done for a better runtime performance of the generated code.

Gradient Descent. While the previously described motor estimation algorithm can accurately estimate a motor based on three corresponding points, any noise or error in these correspondences will directly propagate to the resulting motor and therefore also to the resulting registration. This is especially true for outliers. To mitigate this, the motor estimation is used inside the algorithmic structure of a gradient descent, or more specifically a *stochastic* or *mini batch* gradient descent, which is commonly used in machine learning to enable a tradeoff between robustness and performance for optimization on noisy input data [18].

Algorithm 1. Robust Sampling-based Motor Estimation

Require: Point correspondences $C[i], i = 0, ..., N$
Require: Step size $\alpha \in (0, 1]$
Require: Maximum number of iterations $N_{\text{max}} > 0$
Require: Sampled triangles per iteration $N_{\text{triangle}} > 0$

```
procedure RobustMotorEstimation(C[])
    m_result ← 1                          ▷ Initialize the output with an identity motor
    for i ← 1, N_max do
        m_sum ← 0                         ▷ Initialize the temporary motor sum as zero
        for j ← 1, N_triangle do
            [id1, id2, id3] ← SampleRandomIndices()
            triangle_correspondence ← [C[id1], C[id2], C[id3]]
            m_sum ← m_sum + EstimateMotor(triangle_correspondence)
        end for
        m_sum ← m_sum / abs(m_sum)        ▷ Normalize the motor
        if i > 0 then                     ▷ Do not scale the motor for the first step
            m_sum ← (1 − α) + α m_sum
        end if
        m_result ← m_result m_sum         ▷ Join the new motor with the current result
    end for
end procedure
```

The outline of the proposed algorithm is shown in Algorithm 1. Geometric algebra enables an intuitive and efficient formulation, as the representation of rigid transformations as a motor makes it possible to easily average and interpolate transformations. While interpolating and averaging motors is theoretically only possible in bivector space, this can also be done directly on motor multivectors as a fast and sufficiently accurate approximation (as discussed in Sect. 2). The most common version of gradient descent estimates a number of gradients based on sampled input data, averages them, and takes a step into the direction of the negative averaged gradient. Instead of estimating gradients, our proposed algorithm estimates motors based on three randomly sampled correspondences at a time, averages them and takes a step into the direction of the averaged motor.

As input, the algorithm requires point correspondences, which can be generated using a nearest neighbor or feature based approach. Additionally the algorithm can be configured with three hyperparameters that can be set depending on the use case: N_{max} limits the number of iterations, N_{triangle} specifies how many motors are estimated and subsequently averaged per iteration, and the steps size α, also known as learning rate, controls the strength at which each averaged motor affects the resulting transformation. The influence of the hyperparameters on the runtime and robustness are examined in Sect. 4.2.

Note that the step size does not influence the first iteration, leading to an effective step size of 1 for that iteration. Because of this, the first iteration can be seen as a *coarse* registration, roughly transforming the point clouds onto each other, while the following iterations try to optimize that registration leading to what is called a *fine* registration.

4 Results

To evaluate the properties of the presented algorithm, a number of experiments were conducted. Section 4.1 explains the procedure for generating synthetic test data that was subsequently used to evaluate the influence of hyperparameters (Sect. 4.2), the robustness (Sect. 4.3), runtime performance (Sect. 4.4) as well as the accuracy in a real world pipeline (Sect. 4.5). The experiments were performed on a system with an *Intel Core i5-9600K* processor.

4.1 Synthetic Data Generation

The different properties of the proposed algorithm were evaluated using a same-source approach (see taxonomy in [9]). For this, a single point cloud is used as both the source and target for the reconstruction. Both point clouds are optionally perturbed with Gaussian noise and then the source is transformed with a random rotation and translation. After this, we try to register the source point cloud to the target, effectively undoing the previous transformation. This approach enables us to directly compare the registration result to a ground truth. As source data five different point clouds with varying complexity were used, including three models from the *Stanford 3D Scanning Repository.* Optionally, instead of using the same point cloud as source and target, the original point cloud can be cut into two parts with a partial overlap. This method mimics real world datasets more closely and is used in Sect. 4.5

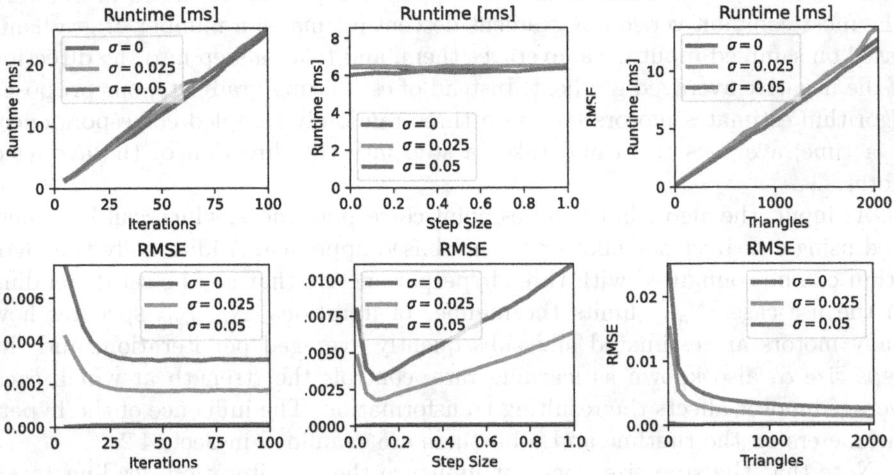

Fig. 1. Influence of the number of iterations, step size, and number of sampled triangles on the runtime and RMSE for Gaussian noise with strength σ. While an increased noise strength leads to an increased RMSE, the runtime appears to be invariant. Furthermore, the runtime increases linearly with the number of iterations, while the error reaches a constant value after approximately 30 iterations. The step size value does not influence the runtime, but the RMSE is minimal for a step size $\alpha \approx 0.1$. Finally, the runtime scales linearly with the number of triangles while the RMSE steadily decreases.

4.2 Influence of Hyperparameters

The proposed algorithm can be configured using its three hyperparameters. These can be set depending on the use case and enable a trade-off between registration accuracy and runtime performance. To test the influence of each of the three parameters, the runtime and root-mean-square error (RMSE) were measured over varying values of each parameter. For each experiment, the two remaining parameters were kept constant. Additionally, multiple noise levels were used in each experiment, to investigate the influence of noise on the choice of hyperparameters.

As a first experiment, the maximum number of iterations was varied, resulting in the runtime and RMSE values depicted in Fig. 1. The runtime linearly increases with the number of iterations and only varies slightly for different noise values. The RMSE reaches a nearly constant value after approximately 30 iterations. While this is true for all three noise levels, the resulting constant value depends on the noise strength. For our combination of hyperparameters, using more than 30 iterations only increases the runtime, but does not result in an improved accuracy and is therefore discouraged.

For the second experiment, the influence of the step size α on the runtime performance and RMSE under different noise strengths was tested. The results are visualized in Fig. 1. The runtime appears constant and independent of the step size, while the RMSE varies substantially. For all three noise levels, a minimum RMSE is achieved for a step size $\alpha \approx 0.1$. For step sizes $\alpha \ll 0.1$ the increased RMSE is likely caused by the algorithm not being able to converge in the set number of iterations. In contrast, step sizes $\alpha > 0.1$ can lead to an increased RMSE, because the influence of each iteration on the final motor is too large.

The third experiment tests the influence of the number of triangles N_{triangle} that are calculated and subsequently averaged in each iteration. As shown in Fig. 1, the runtime increases linearly with the number of triangles and is independent of the noise strength. For small values of N_{triangle}, increasing the number of triangles can lead to a large improvement of the RMSE, while for larger values of N_{triangle} an increase only results in a small gain in accuracy.

4.3 Robustness Against Noise and Outliers

A common property of gradient descent is its robustness against noise and outliers. Two experiments were carried out to evaluate how well this property translates to our proposed algorithm, especially in comparison to other existing algorithms. Here, we used the hyperparameters $N_{\text{max}} = 25$, $\alpha = 0.1$, and $N_{\text{triangle}} = 1024$. We chose to include the three variants *Point2Point* [2], *Point2Plane* [15] and *Symmetric Point2Plane* [19] that are part of the PCL [21] library in the tests.

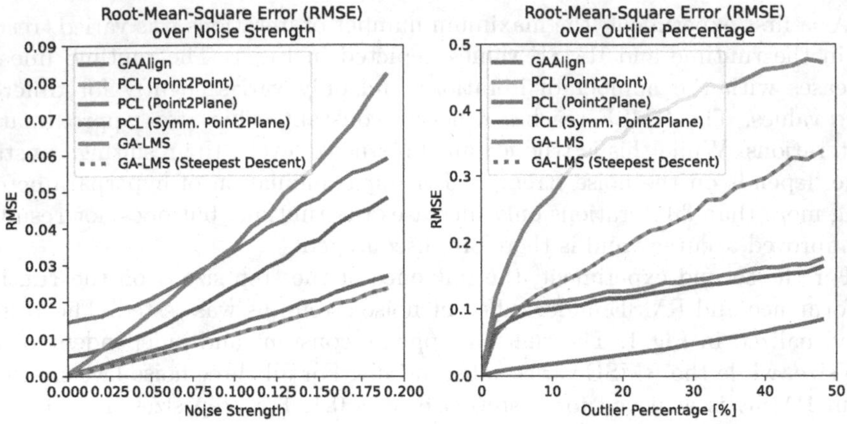

(a) Varying noise strength (b) Varying outlier percentage

Fig. 2. Comparison of the different algorithms to noise and outlier. While the robustness of our algorithm against Gaussian noise is comparable to other state-of-the-art algorithms, it is significantly more robust against outliers than any of the other algorithms.

Additionally, we included the two variants of the state-of the-art geometric algebra-based algorithm *GA-LMS*. First the original algorithm presented in [14], and second the steepest descent based extension presented in [1] that calculates an optimal step size in each iteration. While the authors of *GA-LMS* also proposed several further improvements to the robustness of their algorithm in [1], these were not published as part of their open source project OpenGA [13] and could therefore not be included in the comparison.

For both experiments, the procedure described in Sect. 4.1 was used to generate synthetic test data with perfect correspondences. The correspondences were subsequently perturbed with Gaussian noise (Fig. 2a) or outliers (Fig. 2b). For each noise level/outlier percentage, the tests were repeated 200 times per model and the resulting RMSE values were averaged.

Given perfect correspondences perturbed by Gaussian noise, *Point2Point* ICP and the *GA-LMS* variant using steepest descent consistently lead to the smallest RMSE, with our algorithm performing only slightly worse. The three other tested algorithms appear to be notably less robust against noise.

The second experiment (Fig. 2b) shows that our proposed algorithm is substantially more robust against outliers, than any of the algorithms that are part of the comparison. This is especially true for both *GA-LMS* variants, which is consistent with the description of Al-Nuaimi et al. [1].

4.4 Runtime Performance Evaluation

In addition to the previous experiments focusing on the influence of hyperparameters and the robustness, we also decided to separately test the runtime performance of our algorithm depending on the number of correspondences (Fig. 3).

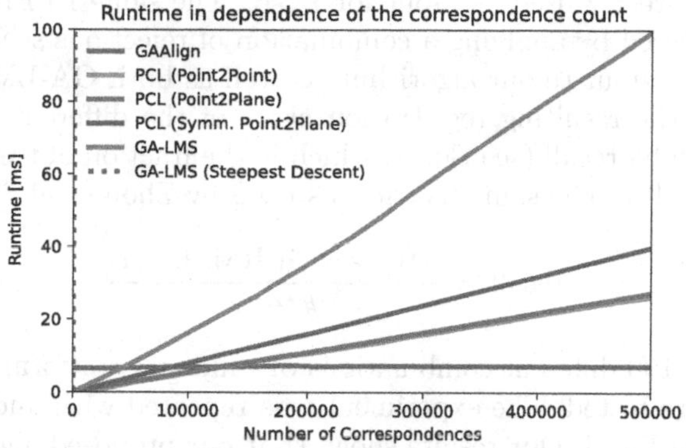

Fig. 3. Runtime depending on the number of correspondences. While the runtime of our proposed algorithm is longer for small numbers of correspondences, it offers a significantly better scalability.

For comparison, we used the same algorithms as in Sect. 4.3. While the runtime of all other used algorithms approximately increases linearly, the runtime of *GAAlign* is approximately constant. This constant runtime is due to the use of the structure of a mini batch gradient descent with a fixed number of iterations. Even though, *GAAlign* has a longer runtime than the other state-of-the-art algorithms for small numbers of correspondences, it is significantly more scalable, making it suitable for applications with large amounts of data. Because of this, high resolution point clouds can directly be used as input, without requiring any further subsampling or keypoint detection.

4.5 Performance in a Real-World Pipeline

As a final comparison, the performance of the proposed algorithm was evaluated in a real world pipeline. In contrast to the previous experiments, no previously known correspondences were used to calculate the registration. Therefore, the correspondence search is now part of the experiment. Furthermore, the test objects were randomly cut into two parts with partial overlap, as described in Sect. 4.1.

In addition to the algorithms in the previous experiments, we also now include the ICP implementation of *OpenCV* [3] as well as the state-of-the-art algorithms *Fast Global Registration* (FGR) presented by Zhou et al. [25], NgeNet by Zhu et al. [26], and GeDi by Poiesi and Boscaini [17]. The latter two algorithms are machine learning based methods, which perform very highly on the 3DMatch Benchmark for point cloud registration. These four algorithms were excluded from the previous experiments, as their implementation prevented the direct input of correspondences.

For the ICP variants from PCL and OpenCV, as well as for NgeNet and GeDi, the respective built-in correspondence search was used. FPFH features [20] were calculated and directly used as input for FGR. The same FPFH features were matched and filtered by applying a combination of rejections schemes built into PCL and used as input to our algorithm, as well as both GA-LMS variants.

To compare the resulting registration error of the different algorithms, we calculated the alpha recall (see Eq. 1), which is the fraction of tests smaller than an error threshold α. The same metric was used by Zhou et al. [25].

$$\alpha\text{-recall} = \frac{\#\text{tests with RMSE} < \alpha}{\#\text{tests}} \tag{1}$$

For each model, 100 different combinations of random transformations and random overlap were tested. The experiment was repeated with and without noise and visualized in Fig. 4. Our results show, that our proposed algorithm outperforms all other tested algorithms in a noise-free environment. Furthermore, in the presence of noise, GAAlign performs on the same level as the best state-of-the-art algorithm that was part of the comparison.

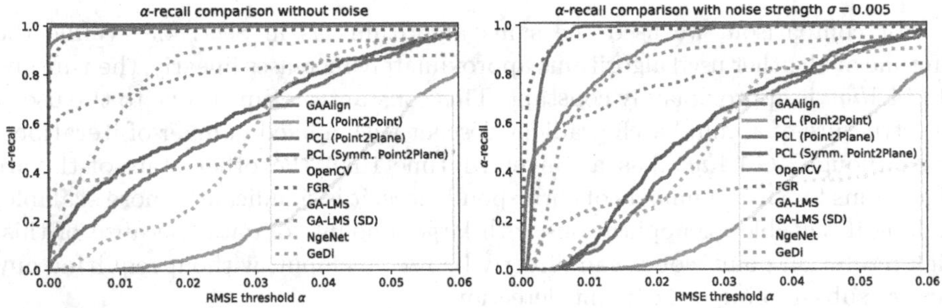

(a) α-recall comparison without noise (b) α-recall comparison with noise

Fig. 4. Visualization of the α-recall for different algorithms with and without noise. Upper-left is better.

5 Conclusion

In our work we presented an original geometric algebra based point cloud registration algorithm, called *GAAlign*, that uses the algorithmic structure of a mini batch gradient descent to achieve an improved robustness against outliers compared to other algorithms. Experiments using a same-source approach on synthetic data showed a high robustness against noise and outliers, as well as a better scalability for large numbers of correspondences. Furthermore, the influence of the hyperparameters on runtime and registration error was tested to suggest an optimal choice of hyperparameters. Finally, the proposed algorithm showed a state-of-the-art accuracy in a real world pipeline when used in conjunction with a feature-based correspondence search.

Acknowledgements. The source code and datasets are publicly available at https://github.com/kai-neumann/GAAlign. We thank the anonymous reviewers whose comments helped to improve this manuscript. Special thanks to Steven De Keninck for his advice on interpolating motors and to Sabrina Falco and Marcel Langer for their critical reading of the manuscript.

Appendix

Listing A.1. GAALOPScript for the Triangle Reconstruction.

```

/* Algorithm based on: Reconstructing a Motor from Exact Point Correspondences
   according to Sect. 6.8 of the tutorial A Guided Tour to the Plane-Based Geometric
   Algebra PGA by Leo Dorst, University of Amsterdam */
// source points
!A_src = createPoint(src1_x, src1_y, src1_z);
!B_src = createPoint(src2_x, src2_y, src2_z);
!C_src = createPoint(src3_x, src3_y, src3_z);

// target points
!A_tar = createPoint(tar1_x, tar1_y, tar1_z);
!B_tar = createPoint(tar2_x, tar2_y, tar2_z);
!C_tar = createPoint(tar3_x, tar3_y, tar3_z);

// Transformation from A_src to A_tar (translation)
!VA_unnormalized = (1+A_tar/A_src);
!VA_norm = abs(VA_unnormalized);
!VA = VA_unnormalized/VA_norm;

!A2 = VA * A_src * ~VA;
!B2 = VA * B_src * ~VA;
!C2 = VA * C_src * ~VA;

// Transformation from B2 to Bt; based on the rotation from the line L2 to L1
!L1 = *(*A_tar ^ *B_tar);
!L2 = *(*A_tar ^ *B2);

!VB_unnormalized = (1+L1/L2);
!VB_norm = abs(VB_unnormalized);
!VB = VB_unnormalized/VB_norm;

!B3 = VB * B2 * ~VB;
!C3 = VB * C2 * ~VB;

// Transformation from C3 to Ct; based on the rotation of two planes
!P1 = *(*L1 ^*C_tar);
!P2 = *(*L1 ^*C3);

!VC_unnormalized = (1+P1/P2);
!VC_norm = abs(VC_unnormalized);
!VC = VC_unnormalized/VC_norm;

// complete transformation
```

```
!combined_motor = VC * VB * VA;

// Get the norm
!motor_norm = abs(combined_motor);

// Make sure the out motor is normalized
?out_motor = combined_motor/motor_norm;
```

References

1. Al-Nuaimi, A., Steinbach, E., Lopes, W.B., Lopes, C.G.: 6DOF point cloud alignment using geometric algebra-based adaptive filtering. In: 2016 IEEE Winter Conference on Applications of Computer Vision (WACV), pp. 1–9. IEEE (2016). https://doi.org/10.1109/WACV.2016.7477642
2. Besl, P.J., McKay, N.D.: Method for registration of 3-D shapes. In: Sensor fusion IV: control paradigms and data structures, vol. 1611, pp. 586–606. Spie (1992). https://doi.org/10.1117/12.57955
3. Bradski, G.: The OpenCV Library. Dr. Dobb's Journal of Software Tools (2000)
4. Chen, Y., Medioni, G.: Object modelling by registration of multiple range images. Image Vis. Comput. **10**(3), 145–155 (1992). https://doi.org/10.1016/0262-8856(92)90066-C
5. Dorst, L., De Keninck, S.: Guided tour to the plane-based geometric algebra PGA, version 2.0 (2022). https://bivector.net/PGA4CS.html
6. Gunn, C.G., De Keninck, S.: Geometric algebra and computer graphics. In: ACM SIGGRAPH 2019 Courses, pp. 1–140 (2019). https://doi.org/10.1145/3305366.3328099
7. Hildenbrand, D., Steinmetz, C.: Gaalop (geometric algebra algorithms optimizer) (2020). http://www.gaalop.de/
8. Hitzer, E., Benger, W., Niederwieser, M., Baran, R., Steinbacher, F.: Foundations for strip adjustment of airborne laserscanning data with conformal geometric algebra. Adv. Appl. Clifford Algebras **32**(1), 1–34 (2022)
9. Huang, X., Mei, G., Zhang, J., Abbas, R.: A comprehensive survey on point cloud registration. arXiv preprint arXiv:2103.02690 (2021). https://doi.org/10.48550/arXiv.2103.02690
10. Kleppe, A.L., Egeland, O.: A curvature-based descriptor for point cloud alignment using conformal geometric algebra. Adv. Appl. Clifford Algebras **28**(2), 1–16 (2018). https://doi.org/10.1007/s00006-018-0864-9
11. Kleppe, A.L., Tingelstad, L., Egeland, O.: Initial alignment of point clouds using motors. In: Proceedings of the Computer Graphics International Conference, pp. 1–5 (2017). https://doi.org/10.1145/3095140.3097282
12. Kleppe, A.L., Tingelstad, L., Egeland, O.: Coarse alignment for model fitting of point clouds using a curvature-based descriptor. IEEE Trans. Autom. Sci. Eng. **16**(2), 811–824 (2018). https://doi.org/10.1109/TASE.2018.2861618
13. Lopes, W.B.: OpenGA. https://openga.org/index.html. Accessed 10 May 2022
14. Lopes, W.B., Al-Nuaimi, A., Lopes, C.G.: Geometric-algebra LMS adaptive filter and its application to rotation estimation. IEEE Signal Process. Lett. **23**(6), 858–862 (2016). https://doi.org/10.1109/LSP.2016.2558461
15. Low, K.L.: Linear least-squares optimization for Point-to-Plane ICP surface registration. Chapel Hill, University of North Carolina **4**(10), 1–3 (2004)
16. Muchnick, S.S.: Advanced compiler design and implementation, pp. 378–396 (1997)

17. Poiesi, F., Boscaini, D.: Learning general and distinctive 3d local deep descriptors for point cloud registration. IEEE Trans. Pattern Anal. Mach. Intell. (2022). https://doi.org/10.1109/TPAMI.2022.3175371
18. Ray, S.: A quick review of machine learning algorithms. In: 2019 International Conference on Machine Learning, Big Data, Cloud and Parallel Computing (COMIT-Con), pp. 35–39 (2019). https://doi.org/10.1109/COMITCon.2019.8862451
19. Rusinkiewicz, S.: A symmetric objective function for ICP. ACM Trans. Graph. (TOG) **38**(4), 1–7 (2019). https://doi.org/10.1145/3306346.3323037
20. Rusu, R.B., Blodow, N., Beetz, M.: Fast point feature histograms (FPFH) for 3D registration. In: 2009 IEEE International Conference on Robotics and Automation, pp. 3212–3217. IEEE (2009). https://doi.org/10.1109/ROBOT.2009.5152473
21. Rusu, R.B., Cousins, S.: 3D is here: Point Cloud Library (PCL). In: IEEE International Conference on Robotics and Automation (ICRA), Shanghai, China, May 9–13 2011
22. Tingelstad, L., Egeland, O.: Motor estimation using heterogeneous sets of objects in conformal geometric algebra. Adv. Appl. Clifford Algebras **27**(3), 2035–2049 (2016). https://doi.org/10.1007/s00006-016-0692-8
23. Valkenburg, R., Dorst, L.: Estimating motors from a variety of geometric data in 3D conformal geometric algebra. In: Guide to Geometric Algebra in Practice, pp. 25–45. Springer (2011). https://doi.org/10.1007/978-0-85729-811-9_2
24. Yang, J., Li, H., Campbell, D., Jia, Y.: Go-ICP: a globally optimal solution to 3d ICP point-set registration. IEEE Trans. Pattern Anal. Mach. Intell. **38**(11), 2241–2254 (2015). https://doi.org/10.1109/TPAMI.2015.2513405
25. Zhou, Q.-Y., Park, J., Koltun, V.: Fast global registration. In: Leibe, B., Matas, J., Sebe, N., Welling, M. (eds.) ECCV 2016. LNCS, vol. 9906, pp. 766–782. Springer, Cham (2016). https://doi.org/10.1007/978-3-319-46475-6_47
26. Zhu, L., Guan, H., Lin, C., Han, R.: Neighborhood-aware geometric encoding network for point cloud registration. arXiv preprint arXiv:2201.12094 (2022). https://doi.org/10.48550/arXiv.2201.12094

Quantum Register Algebra: The Basic Concepts

J. Hrdina[1], D. Hildenbrand[2], A. Návrat[1], C. Steinmetz[4], R. Alves[3(✉)], C. Lavor[5], P. Vašík[1], and I. Eryganov[1]

[1] Brno University of Technology, Brno, Czech Republic
{hrdina,navrat.a,vasik}@fme.vutbr.cz, xperyga00@vutbr.cz
[2] Technische Universität Darmstadt, 64297 Darmstadt, Germany
dietmar.hildenbrand@gmail.com
[3] Federal University of ABC, Sao Bernardo, Brazil
alves.rafael@ufabc.edu.br
[4] :em engineering methods AG, Darmstadt, Germany
christian.steinmetz@e-mail.de
[5] University of Campinas, Campinas, Brazil
clavor@unicamp.br

Abstract. We introduce Quantum Register Algebra (QRA) as an efficient tool for quantum computing. We show the direct link between QRA and Dirac formalism. We present GAALOP (Geometric Algebra Algorithms Optimizer) implementation of our approach. Using the QRA basis vectors definitions given in Sect. 4 and the framework based on the de Witt basis presented in Sect. 5, we are able to fully describe and compute with QRA in GAALOP using the geometric product. We illustrate the intuitiveness of this computation by presenting the QRA form for the well known SWAP operation on a two qubit register.

Keywords: quantum computing · geometric algebra · quantum register algebra

1 Introduction

Geometric Algebra proved in recent years that it is a mathematical system covering many other mathematical systems as conventionally used in engineering and physical science, [4–6,9]. Examples are linear algebra, quaternions, Dirac and Pauli matrices, Plucker coordinates to mention only a few. This means, the big advantage of Geometric Algebra is that you do not have to learn different mathematical systems in order to handle various application areas. You simply have to learn only one mathematical system to handle them. Another simplification based on Geometric Algebra is that both objects and operations are handled in one algebra, means you do not have to distinguish, for instance, between points or vectors as objects and matrices for the description of operations on them.

Supported by FAPESP and CNPq.

D. W. Silva et al. (Eds.): ICACGA 2022, LNCS 13771, pp. 112–122, 2024.
https://doi.org/10.1007/978-3-031-34031-4_10

Now, the question is: can Geometric Algebra also advantageously be used for quantum computing? Quantum computing, currently, is based on working with complex numbers, matrices of complex numbers and tensor computing in order to handle operations with arbitrary numbers of qubits.

Some papers on geometric algebras and quantum computing demonstrate application of geometric algebras $\mathbb{G}_{4,1}$ (relativistic case) and $\mathbb{G}_{3,0}$ (nonrealtivistic case), [3], geometric algebra $\mathbb{G}_{n,n}$, [1] and finally a complex Clifford algebra $\mathbb{C}_n$, [2].

In this paper, we show that quantum computing can simply be based on a Geometric Algebra called QRA (quantum register algebra). In Sect. 2, we show that complex numbers can easily be identified within Geometric Algebra.

The Dirac formalism is a very useful formalism to describe quantum computing. Although it is originally based on complex matrices (see Sect. 3), we show in our definition of QRA according to Sect. 4 and Sect. 5, that it can still be used with QRA.

In order to compute with QRA, we use GAALOP, a stand alone geometric algebra algorithm optimizer. We extended the GAALOP tool presented in [1] for the QRA support based on the definitions presented in Sects. 4 and 5. The full implementation is shown in Sect. 6. In order to simplify the computing as much as possible, we also integrated it in our online tool GAALOPWeb. In Sect. 7, we present how easy it is to compute with QRA using GAALOP, for this we present the QRA form for the SWAP gate. The big advantage for quantum computing beginners is that they only have to know Geometric Algebra in order to describe the objects and operations of quantum computing.

2 From Complex Numbers to Geometric Algebras and Back Again

The algebra of complex numbers $\mathbb{C}$ is an essential tool for quantum computing. Qubits are realised by vectors in complex vector space $\mathbb{C}^n$ and the gates by matrices $n \times n$ over the complex numbers. In our approach, the complex linear algebra is the language for quantum computing. More precisely, we are using the natural concept of complex geometric algebras with the complex numbers as one of their instances. In the sequel, we investigate these concepts in a more detailed way.

Formally, geometric algebra $\mathbb{G}_{p,q}$ is a free, associative, distributive and unitary algebra over the set of abstract elements $\{e_1, \ldots, e_n\}$ endowed with the following identities:

$$\begin{aligned} e_ie_j &= -e_je_i, \text{ where } i \neq j, \\ e_i^2 &= 1, \text{ where } i = 1, \ldots, p, \\ e_i^2 &= -1, \text{ where } i = p+1, \ldots, n. \end{aligned} \tag{1}$$

In computer science notation, we understand $\mathbb{G}_{p,q}$ as a vector space where vectors are built as words over the alphabet $\{e_1, \ldots, e_n\}$, including an empty word, using the following equivalency. Two words are equivalent (they are different representations of the same object) if the first can be rewritten into the second and vice versa with the help of identities (1), distributivity and associativity.

For example, let us consider geometric algebra $\mathbb{G}_{1,1}$, so we have words over the alphabet $\{e_1, e_2\}$ and identities $e_1^2 = 1$, $e_2^2 = -1$ together with anti–commutativity $e_1e_2 = -e_2e_1$. Then the vector space basis is of the form

$$\{1, e_1, e_2, e_1e_2\}. \tag{2}$$

Hence, for example the word $e_2e_1e_2e_1e_1e_2e_2$ can by rewritten to $-e_1$ with the help of identities (1) in the following way:

$$e_2e_1e_2e_1e_1e_2e_2 = e_2e_1e_2(1)(-1) = e_1e_2e_2 = e_1(-1) = -e_1.$$

The element e_1 is in the basis (2). We can see that the elements of geometric algebra $\mathbb{G}_{1,1}$ are the linear combinations over the basis (2), i.e.

$$\mathbb{G}_{1,1} = \{x_1 + x_2e_1 + x_3e_2 + x_4e_1e_2 | x_i \in \mathbb{R}\} \tag{3}$$

together with multiplication given by identities (1). For example

$$\begin{aligned}(e_1 + 2e_1e_2)(1 + e_1 - 3e_2) &= e_1(1 + e_1 - 3e_2) + 2e_1e_2(1 + e_1 - 3e_2)\\ &= e_1 + e_1e_1 - e_13e_2 + 2e_1e_2 + 2e_1e_2e_1 - 2e_1e_23e_2\\ &= e_1 + (1) - 3e_1e_2 + 2e_1e_2 - 2e_2e_1e_1 - 6e_1(-1)\\ &= 1 + e_1 - e_1e_2 - 2e_2(1) + 6e_1\\ &= 1 + 7e_1 - 2e_2 - e_1e_2 \in \mathbb{G}_{1,1}.\end{aligned}$$

In any geometric algebra $\mathbb{G}(p, q)$ such that $p > 1$ or $q > 1$ we can find a subalgebra isomorphically equivalent to $\mathbb{C}$ in the following way: if $p > 1$ we have two elements $e, f \in \mathbb{G}(p, q)$ such that $e^2 = f^2 = 1$ and $ef = -fe$, so

$$(ef)^2 = efef = -eeff = -e^2f^2 = -1$$

and

$$\mathbb{C} \cong \bar{\mathbb{C}} = \{a + b(ef) | a, b \in \mathbb{R}\} \subset \mathbb{G}(p, q).$$

In the same way, if we have $q > 1$ then we have two elements e, f such that $e^2 = f^2 = -1$ and $ef = -fe$, so

$$(ef)^2 = efef = -eeff = -e^2f^2 = -1(-1)(-1) = -1.$$

Note that the element ef commutes with the other basis elements in both cases.

Remark 1. The complex numbers $\mathbb{C}$ are in fact isomorphically equivalent geometric algebra $\mathbb{G}_{0,1}$. Indeed, $\mathbb{G}_{0,1}$ is a space

$$\mathbb{G}_{0,1} = \{x_1 + x_2e_1 | x_i \in \mathbb{R}\} \tag{4}$$

such that $e_1^2 = -1$ which is the vector space isomorphically equivalent to $\mathbb{C}$. The specific concept for the imaginary unit used in QRA is described in Sect. 4.

3 Matrices vs. Dirac Formalism

In this section, we show the direct link between Dirac formalism and geometric algebras. Original quantum mechanics is based on Dirac formalism, but quantum computing can be built up of matrices because the Hilbert space of qubits' states is finite. We briefly recall the link between matrices and Dirac formalism, [10].

– the n-qubit (ket)

$$|i\rangle = |a_1 \cdots a_{n-1}a_n\rangle, \text{ where } a_i \in \{0,1\} \text{ and } i = a_1 2^{n-1} + \cdots + a_n 2^0$$
$$\Longleftrightarrow \begin{pmatrix} 0 \\ \vdots \\ 1 \\ \vdots \\ 0 \end{pmatrix}, \text{ where 1 is on i+1 position, 0 otherwise.}$$

– the dual n-qubit (bra)

$$\langle i| = \langle a_1 \cdots a_{n-1}a_n|, \text{ where } a_i \in \{0,1\} \text{ and } i = a_1 2^{n-1} + \cdots + a_n 2^0$$
$$\Longleftrightarrow \begin{pmatrix} 0 \cdots 1 \cdots 0 \end{pmatrix}, \text{ where 1 is on i+1 position, 0 otherwise.}$$

The n-qubit gates are matrices $A = (a_{ij})$, where $i, j = 1, \ldots, 2^n$. So if a matrix A acts on coordinates in canonical basis, then its element a_{ij} sends the j–th element of canonical basis to i–th element of canonical basis in the same way as Dirac expression $|i-1\rangle\langle j-1|$, so

$$\begin{pmatrix} a_{11} & \ldots & a_{1n} \\ \vdots & & \vdots \\ a_{n1} & \ldots & a_{nn} \end{pmatrix} \Longleftrightarrow \sum_{i,j=1}^{n} a_{ij}|i-1\rangle\langle j-1|.$$

For example, the representations of 1-qubits (ket and bra) are

$$|0\rangle \Longleftrightarrow \begin{pmatrix} 1 \\ 0 \end{pmatrix}, |1\rangle \Longleftrightarrow \begin{pmatrix} 0 \\ 1 \end{pmatrix}, \langle 0| \Longleftrightarrow (1\ 0), \langle 1| \Longleftrightarrow (0\ 1)$$

and the NOT gate is of the form

$$\begin{pmatrix} 0 & 1 \\ 1 & 0 \end{pmatrix} \Longleftrightarrow |0\rangle\langle 1| + |1\rangle\langle 0|.$$

4 Definition of QRA

Recall that a Geometric Algebra $\mathbb{G}_n = \mathbb{G}_{n,0}$ is a free, associative, unitary algebra over the set of anti–commuting generators $\{e_1, \ldots, e_n\}$ such that $e_i^2 = 1$, $i \in \{1, \ldots, n\}$. Now we finally define the Quantum Register Algebra (QRA).

First, let us consider a geometric algebra $\mathbb{G}_{n+2}$ with its basis elements $\{e_1, \ldots, e_n, r_1, r_2\}$ together with the following identities

$$e_1^2 = e_2^2 = \cdots = e_n^2 = r_1^2 = r_2^2 = 1. \tag{5}$$

Then we define a bivector

$$\iota = r_1 r_2$$

and show that the set $\tilde{\mathbb{C}} = \{a + b\iota | a, b \in \mathbb{R}\}$ is isomorphic to an algebra $\mathbb{C}$, so ι plays the role of a complex unit. The set $\tilde{\mathbb{C}}$ is closed with respect to addition and multiplication. The element ι is in square equal to -1. Indeed,

$$\iota^2 = r_1 r_2 r_1 r_2 = -r_1^2 r_2^2 = -1,$$

so $\tilde{\mathbb{C}} \cong \mathbb{C}$. Now we define QRA as a geometric subalgebra $\mathbb{G}_n$ with the coefficients in $\tilde{\mathbb{C}}$, i.e.

$$\text{QRA} = \{a_0 g_0 + \cdots + a_n g_n | a_i \in \tilde{\mathbb{C}}, g_i \in \mathbb{G}_n\} \subset \mathbb{G}_{n+2}.$$

Hence for any element $A \in \mathbb{G}_n$ we have $\iota A = A\iota$ because ι is a bivector and $\iota \notin \mathbb{G}_n$.

5 QC in the QRA Framework

To use QRA to model quantum computing we choose a different basis of QRA based on geometric algebra $\mathbb{G}_{2n}$. This basis is called Witt basis and it is formed by elements $\{f_1, f_1^\dagger, \ldots, f_n, f_n^\dagger\}$ satisfying

$$f_i = \frac{1}{2}(e_i + \iota e_{i+n}), \tag{6}$$

$$f_i^\dagger = \frac{1}{2}(e_i - \iota e_{i+n}), \tag{7}$$

where $\iota = r_1 r_2$. Now, we define an element $I = f_1 f_1^\dagger \cdots f_n f_n^\dagger$ satisfying

$$I^2 = I, \tag{8}$$

$$f_i I = 0, \tag{9}$$

$$f_i f_i^\dagger I = I \tag{10}$$

then we have a straightforward identification of bra and ket vectors with the elements of QRA as follows:

$$\langle a_1 \ldots a_n | \hookrightarrow I(f_n)^{a_n} \ldots (f_1)^{a_1}, \text{ where } a_i \in \{0, 1\}, \tag{11}$$

$$|a_1 \ldots a_n\rangle \hookrightarrow (f_1^\dagger)^{a_1} \ldots (f_n^\dagger)^{a_n} I, \text{ where } a_i \in \{0, 1\}. \tag{12}$$

For example, let us consider the space of 2–qubit states. Then the identification (11) and (12) for the ket vectors reads

$$\begin{aligned}|00\rangle &\mapsto (f_1^\dagger)^0(f_2^\dagger)^0 I = I,\\ |01\rangle &\mapsto (f_1^\dagger)^0(f_2^\dagger)^1 I = f_2^\dagger I,\\ |10\rangle &\mapsto (f_1^\dagger)^1(f_2^\dagger)^0 I = f_1^\dagger I,\\ |11\rangle &\mapsto (f_1^\dagger)^1(f_2^\dagger)^1 I = f_1^\dagger f_2^\dagger I.\end{aligned} \tag{13}$$

So the ket vectors in 2-qubit state space are linear combinations of the basis $\{f_1, f_1^\dagger, \ldots, f_n, f_n^\dagger\}$ elements:

$$|\psi\rangle = (\psi_{00} + \psi_{01} f_2^\dagger + \psi_{10} f_1^\dagger + \psi_{11} f_1^\dagger f_2^\dagger) I. \tag{14}$$

To define the quantum gates we have to identify the bra vectors in the similar way:

$$\begin{aligned}\langle 00| &\mapsto I(f_2)^0(f_1)^0 = I,\\ \langle 01| &\mapsto I(f_2)^1(f_1)^0 = I f_2,\\ \langle 10| &\mapsto I(f_2)^0(f_1)^1 = I f_1,\\ \langle 11| &\mapsto I(f_2)^1(f_1)^1 = -I f_1 f_2.\end{aligned} \tag{15}$$

Thus the bra vectors in 2-qubit state space are combinations of the basis $\{f_1, f_1^\dagger, \ldots, f_n, f_n^\dagger\}$ elements as follows:

$$\langle\psi| = I(\psi_{00} + \psi_{10} f_1 + \psi_{01} f_2 - \psi_{11} f_1 f_2).$$

To demonstrate our approach we show the design of the SWAP gate [8]. Recall that in Dirac notation, the SWAP gate is represented by

$$|00\rangle\langle 00| + |01\rangle\langle 10| + |10\rangle\langle 01| + |11\rangle\langle 11|.$$

Using the identification (13) and (15), the SWAP gate may be rewritten as:

$$\begin{aligned}\mathrm{SWAP} &= |00\rangle\langle 00| + |01\rangle\langle 10| + |10\rangle\langle 01| + |11\rangle\langle 11|\\ &= f_1 f_1^\dagger f_2 f_2^\dagger + f_1^\dagger f_2 - f_1 f_2^\dagger + f_1^\dagger f_1 f_2^\dagger f_2.\end{aligned} \tag{16}$$

Before we present how the gate acts on 2-qubits, let us mention the rules for calculations with the Witt basis $\{f_i, f_i^\dagger\}$ elements in the form of a list of properties which can be verified by straightforward computations:

$$(f_i)^2 = (f_i^\dagger)^2 = 0 \tag{17}$$

$$f_i f_j = -f_j f_i, \quad f_i^\dagger f_j^\dagger = -f_j^\dagger f_i^\dagger \tag{18}$$

$$f_i f_i^\dagger f_i = f_i, \qquad f_i^\dagger f_i f_i^\dagger = f_i^\dagger \tag{19}$$

We present the SWAP gate functionality on 2-qubits step by step. Thus we can calculate:

$$\begin{aligned}
&\mathrm{SWAP}|\psi\rangle = \\
&= (f_1 f_1^\dagger f_2 f_2^\dagger + f_1^\dagger f_2 - f_1 f_2^\dagger + f_1^\dagger f_1 f_2^\dagger f_2)(\psi_{00} + \psi_{01} f_2^\dagger + \psi_{10} f_1^\dagger + \psi_{11} f_1^\dagger f_2^\dagger)I \\
&= f_1 f_1^\dagger f_2 f_2^\dagger (\psi_{00} + \cancel{\psi_{01} f_2^\dagger} + \cancel{\psi_{10} f_1^\dagger} + \cancel{\psi_{11} f_1^\dagger f_2^\dagger})I \text{ by (17)} \\
&\quad + f_1^\dagger f_2 (\cancel{\psi_{00}} + \psi_{01} f_2^\dagger + \cancel{\psi_{10} f_1^\dagger} + \cancel{\psi_{11} f_1^\dagger f_2^\dagger})I \text{ by (9) and (17)} \\
&\quad - f_1 f_2^\dagger (\cancel{\psi_{00}} + \cancel{\psi_{01} f_2^\dagger} + \psi_{10} f_1^\dagger + \cancel{\psi_{11} f_1^\dagger f_2^\dagger})I \text{ by (9) and (17)} \\
&\quad + f_1^\dagger f_1 f_2^\dagger f_2 (\cancel{\psi_{00}} + \cancel{\psi_{01} f_2^\dagger} + \cancel{\psi_{10} f_1^\dagger} + \psi_{11} f_1^\dagger f_2^\dagger)I \text{ by (9) and (17)} \\
&= \underline{(f_1 f_1^\dagger f_2 f_2^\dagger)}(\psi_{00})I + (f_1^\dagger \underline{f_2})(\psi_{01} \underline{f_2^\dagger})I \text{ by (10)} \\
&\quad + f_2^\dagger \underline{f_1} \psi_{10} \underline{f_1}^\dagger I + (f_1^\dagger f_1 f_2^\dagger f_2)(\psi_{11} f_1^\dagger f_2^\dagger)I \\
&= (\psi_{00} + \psi_{01} f_1^\dagger + \psi_{10} f_2^\dagger + \psi_{11}(f_1^\dagger f_1 f_1^\dagger)(f_2^\dagger f_2 f_2^\dagger))I \text{ the last step by (10)} \\
&= (\psi_{00} + \psi_{01} f_1^\dagger + \psi_{10} f_2^\dagger + \psi_{11} f_1^\dagger f_2^\dagger)I,
\end{aligned}$$

where $\cancel{g}$ and $\underline{g}$ stand for $g = 0$ and $g = 1$. Finally, by means of (13), the final 2-qubit can be rewritten in Dirac notation as

$$(\psi_{00}|00\rangle + \psi_{01}|10\rangle + \psi_{10}|01\rangle + \psi_{11}|11\rangle)I$$

which is the expected result.

The other gates may be interpreted in the very same way and thus we showed how quantum computation in QRA is realized. We used the Witt basis axioms (17)–(19) together with additional axioms (9) and (10). The use of the additional axioms may seem redundant and complicated but they only simplify the written form of calculations and thus the functionality is easier for demonstration. Indeed, these rules may be avoided completely for which we point out the following two reasons:

- If we interpret the element I as $f_1 f_1^\dagger \cdots f_n f_n^\dagger$, the rules (9) and (10) do not apply because they are simple consequences of (17)–(19). Only the expressions will be longer.
- We are using the axioms (17)–(19) for calculations in the Witt basis which naturally corresponds to Dirac formalism. But Witt basis is just a different set of generators for QRA elements. Thus the axioms (17)–(19) are derived from geometric algebra axioms (5) which are very simple.

Our approach is based on the fact that QRA and the Witt basis provide nice language for written QC schemes. The expressions are very similar to Dirac formalism, so they are very simple to understand for people not familiar with geometric algebra. But unlike abstract Dirac formalism, our objects are elements of QRA which is a geometric algebra based on very simple axioms and, furthermore, it is very easy for implementation as shown in the following section.

6 GAALOP Implementation

The integration of QRA into GAALOP is done based on the file Definition.csv according to Chapt. 9 of [7]. Listing 1.1 shows the file for the definition of one qubit.

Listing 1.1. Definition.csv for QRA for one qubit

```
1 1,e1,e2,er1,er2
2
3 1,e1,e2,er1,er2
4 e1=1,e2=1,er1=1,er2=1
```

In general, this file consists of 5 lines for the definition of the algebra. In the case of QRA, the lines 2 and 5 are left blank since the used basis in Line 1 and the standard basis in Line 3 are the same and no transformations between the two bases are needed. The basis is defined by the basis vectors e1, e2, er1 and er2 according to e_1, e_2, r_1, r_2 as defined in the previous sections. All their squares are defined to 1 according to line 4.

For each additional qubit we need two additional basis vectors, while er1 and er2 are always the same. The definition of a register with two qubits is shown in Listing 1.2.

Listing 1.2. Definition.csv for QRA based on two qubits

```
1 1,e1,e2,e3,e4,er1,er2
2
3 1,e1,e2,e3,e4,er1,er2
4 e1=1,e2=1,e3=1,e4=1,er1=1,er2=1
```

For the second qubit we need the additional basis vectors e3 and e4. For this paper, we also made a new version of GAALOPWeb[1]. It allows online computations with a number of n qubits without the installation of a specific software.

7 Example in GAALOP

As an example we implemented the SWAP gate on GAALOP for a register with two qubits. Following the Eqs. (6) and (7), the vectors f_i and f_i^+ as

$$f_1 = \tfrac{1}{2}(e_1 + \iota e_3), \quad f_1^\dagger = \tfrac{1}{2}(e_1 - \iota e_3)$$
$$f_2 = \tfrac{1}{2}(e_2 + \iota e_4), \quad f_2^\dagger = \tfrac{1}{2}(e_2 + \iota e_4)\ ,$$

where $\iota = r_1 r_2$. Then, we use the definitions in (13) to implement the basis elements, Eq. (14) for $|\psi\rangle$ and (16) to define the operator of the SWAP gate. We apply the SWAP operator on the element $|\psi\rangle$. But first, let us see the coordinates of each ket vector basis when they are multiplied by I separately:

[1] http://www.gaalop.de/gaalopweb/.

Listing 1.3. SWAP gate in QRA for two qubits.

```
// Imaginary unit
i = er1*er2;
// Witt basis
f1 = 0.5*(e1+i*e3);
f1T = 0.5*(e1-i*e3);
f2 = 0.5*(e2+i*e4);
f2T = 0.5*(e2-i*e4);
// Element "I"
Id = f1*f1T*f2*f2T;
// ket basis vectors multiplied by "Id"
?ket00 = 1*Id;
?ket01 = f2T*Id;
?ket10 = f1T*Id;
?ket11 = f1T*f2T*Id;
```

This is important to show which coordinates correspond to each vector in order to see the amplitudes interchanging when SWAP is applied to some linear combination of these vectors. The output for this code is shown below:

Listing 1.4. Basic elements multiplied by I.

```
ket00[0] = 0.25; // 1.0
ket00[42] = 0.25; // e1 ^ (e2 ^ (e3 ^ e4))
ket00[50] = -0.25; // e1 ^ (e3 ^ (er1 ^ er2))
ket00[55] = -0.25; // e2 ^ (e4 ^ (er1 ^ er2))
ket01[2] = 0.25; // e2
ket01[26] = -0.25; // e1 ^ (e3 ^ e4)
ket01[41] = -0.25; // e4 ^ (er1 ^ er2)
ket01[59] = 0.25; // e1 ^ (e2 ^ (e3 ^ (er1 ^ er2)))
ket10[1] = 0.25; // e1
ket10[32] = 0.25; // e2 ^ (e3 ^ e4)
ket10[40] = -0.25; // e3 ^ (er1 ^ er2)
ket10[60] = -0.25; // e1 ^ (e2 ^ (e4 ^ (er1 ^ er2)))
ket11[7] = 0.25; // e1 ^ e2
ket11[16] = -0.25; // e3 ^ e4
ket11[51] = -0.25; // e1 ^ (e4 ^ (er1 ^ er2))
ket11[54] = 0.25; // e2 ^ (e3 ^ (er1 ^ er2))
```

We remark that all identities given in (17)–(10) can be observed in GAALOP. Now, let us choose a vector $|\psi\rangle$[2] given by:

$$
\begin{aligned}
|\psi\rangle &= (|00\rangle + 2|01\rangle + 3|10\rangle + 4|11\rangle)I \\
&= (1 + 3f_2^\dagger + 3f_1^\dagger + 4f_1^\dagger f_2^\dagger)I,
\end{aligned}
$$

[2] The vector $|\psi\rangle$ is not normalized, so the final result should be divided by $\sqrt{29}$.

and apply the SWAP gate on it. The expected result is given by:

$$|\bar{\psi}\rangle = (1 + 3f_2^\dagger + 2f_1^\dagger + 4f_1^\dagger f_2^\dagger)I$$
$$= (|00\rangle + 3|01\rangle + 2|10\rangle + 4|11\rangle)I.$$

We add 3 lines to the code in 1.3:

Listing 1.5. Definition and application of the SWAP gate on GAALOP.

```
//SWAP
SWAP=(f1*f1T*f2*f2T)+(f1T*f2)-(f1*f2T)+(f1T*f1*f2T*f2);
?psi = ket00 + 2*ket01 + 3*ket10 + 4*ket11;
?SwapPsi = SWAP*psi;
```

The output is shown below and then we analyze the result.

Listing 1.6. Output of the application of the SWAP gate on GAALOP.

```
psi[0] = 0.25; // 1.0
psi[1] = 0.75; // e1
psi[2] = 0.5; // e2
psi[7] = 1.0; // e1 ^ e2
psi[16] = -1.0; // e3 ^ e4
psi[26] = -0.5; // e1 ^ (e3 ^ e4)
psi[32] = 0.75; // e2 ^ (e3 ^ e4)
psi[40] = -0.75; // e3 ^ (er1 ^ er2)
psi[41] = -0.5; // e4 ^ (er1 ^ er2)
psi[42] = 0.25; // e1 ^ (e2 ^ (e3 ^ e4))
psi[50] = -0.25; // e1 ^ (e3 ^ (er1 ^ er2))
psi[51] = -1.0; // e1 ^ (e4 ^ (er1 ^ er2))
psi[54] = 1.0; // e2 ^ (e3 ^ (er1 ^ er2))
psi[55] = -0.25; // e2 ^ (e4 ^ (er1 ^ er2))
psi[59] = 0.5; // e1 ^ (e2 ^ (e3 ^ (er1 ^ er2)))
psi[60] = -0.75; // e1 ^ (e2 ^ (e4 ^ (er1 ^ er2)))
SwapPsi[0] = 0.25; // 1.0
SwapPsi[1] = 0.5; // e1
SwapPsi[2] = 0.75; // e2
SwapPsi[7] = 1.0; // e1 ^ e2
SwapPsi[16] = -1.0; // e3 ^ e4
SwapPsi[26] = -0.75; // e1 ^ (e3 ^ e4)
SwapPsi[32] = 0.5; // e2 ^ (e3 ^ e4)
SwapPsi[40] = -0.5; // e3 ^ (er1 ^ er2)
SwapPsi[41] = -0.75; // e4 ^ (er1 ^ er2)
SwapPsi[42] = 0.25; // e1 ^ (e2 ^ (e3 ^ e4))
SwapPsi[50] = -0.25; // e1 ^ (e3 ^ (er1 ^ er2))
SwapPsi[51] = -1.0; // e1 ^ (e4 ^ (er1 ^ er2))
SwapPsi[54] = 1.0; // e2 ^ (e3 ^ (er1 ^ er2))
SwapPsi[55] = -0.25; // e2 ^ (e4 ^ (er1 ^ er2))
SwapPsi[59] = 0.75; // e1 ^ (e2 ^ (e3 ^ (er1 ^ er2)))
SwapPsi[60] = -0.5; // e1 ^ (e2 ^ (e4 ^ (er1 ^ er2)))
```

Note that the groups of coordinates $\{0, 42, 50, 55\}$ and $\{7, 16, 51, 54\}$, which refer to the coefficients of $|00\rangle$ and $|11\rangle$ respectively, remained unchanged from `psi` to `SwapPsi`, as expected. However, note that the coefficients in the coordinates $\{2, 26, 41, 59\}$ of `psi` that corresponds to $|01\rangle$ was exactly 2 times the coefficients of `ket01` given by 1.4 and, in `SwapPsi`, they became 3 times those same coefficients. On the other hand, the coordinates $\{1, 32, 40, 60\}$ changed from 3 (in `psi`) to 2 (in `SwapPsi`) times the coefficients of `ket10` in 1.4, that corresponds to $|10\rangle$. So, the output vector `SwapPsi` is exactly the vector $|\bar{\psi}\rangle$.

8 Conclusion

In this paper, we presented how naturally Geometric Algebra can be used for quantum computing. Based on the newly developed QRA and its support by the extended GAALOP tool, the handling of quantum computing is strongly simplified, especially for beginners. The big advantage for quantum computing beginners is that they only have to know Geometric Algebra in order to describe the objects and operations of quantum computing. We hope that this can be the reason for Geometric Algebra to become "the" language for quantum computing in the future.

References

1. Alves, R., Hildenbrand, D., Hrdina, J., et al.: An online calculator for quantum computing operations based on geometric algebra adv. Appl. Clifford Algebras **32**, 4 (2022)
2. Hrdina, J., Návrat, A., Vašík, P.: Quantum computing based on complex Clifford algebras. Quant. Inf. Process **21**, 310 (2022). https://doi.org/10.1007/s11128-022-03648-w
3. Cafaro, C., Mancini, S.: A geometric algebra perspective on quantum computational gates and universality in quantum computing. Adv. Appl. Clifford Algebras **21**, 493–519 (2011). https://doi.org/10.1007/s00006-010-0269-x
4. Doran, C., Lasenby, A.: Geometric Algebra for Physicists. Cambridge University Press (2003). https://doi.org/10.1017/CBO9780511807497
5. Dorst, L.: Geometric Algebra for Computer Science. An Object-Oriented Approach to Geometry. Morgan Kauffmann, San Francisco (2007)
6. Hildenbrand, D.: Foundations of Geometric Algebra Computing. Springer Science & Business Media (2013)
7. Hildenbrand, D.: The Power of Geometric Algebra Computing. CRC Press, Taylor & Francis Group (2022)
8. Nielsen, M., Chuang, I.: Quantum Computation and Quantum Information: 10th Anniversary Edition. Cambridge University Press, Cambridge (2010) https://doi.org/10.1017/CBO9780511976667
9. Perwass, Ch.: Geometric Algebra with Applications in Engineering (1st edn). Springer (2009)
10. de Lima Marquezino Franklin, Portugal R., Lavor C.: A Primer on Quantum Computing. Springer Publishing Company (2019)

Technological Applications

A Geometric Procedure for Computing Differential Characteristics of Multi-phase Electrical Signals Using Geometric Algebra

Ahmad H. Eid[1] and Francisco G. Montoya[2(✉)]

[1] Port Said University, Port Said, Egypt
ahmad.eid@eng.psu.edu.eg
[2] University of Almeria, 04120 Almeria, Spain
pagilm@ual.es

Abstract. This paper presents exploratory investigations on the concept of generalized geometrical frequency in electrical systems with an arbitrary number of phases by using Geometric Algebra and Differential Geometry. By using the concept of Darboux bivector it is possible to find a bivector that encodes the invariant geometrical properties of a spatial curve named electrical curve. It is shown how the traditional concept of instantaneous frequency in power networks can be intimately linked to the Darboux bivector. Several examples are used to illustrate the findings of this work.

Keywords: Geometric Algebra · Geometric Electricity · Power systems · Geometric Frequency

1 Introduction

Multi-phase power systems play a crucial role in modern society due to the tremendous increase in energy needs. Moreover, with the proliferation of new generation smart grids based on a decentralized paradigm and focused on renewable energies, it is of utmost importance to investigate new methodologies that can deal with the distortion and unbalance scenarios produced by nonlinear loads. In this regard, voltage and frequency control is essential to achieve adequate stability, so appropriate tools are necessary for a better understanding of transient phenomena that can potentially disturb the grid. Currently, the concept of instantaneous frequency, widely used in electrical power systems, presents some issues in its classical definition as the time derivative of the phase angle of a signal. Apparently, this definition only holds for a sine representation of such a signal but fails for other representations where harmonics or transients are included [9]. Existing techniques based on the Fourier or Hilbert Transform, for example, are unable to deal with these problems and give rise to a number of paradoxes, described in [10].

Geometric Algebra (GA) and Differential Geometry (DG) are among the most promising tools recently proposed, as evidenced by recent works [13,14].

D. W. Silva et al. (Eds.): ICACGA 2022, LNCS 13771, pp. 125–137, 2024.
https://doi.org/10.1007/978-3-031-34031-4_11

For example, in [11,12] a geometrical interpretation of the frequency for three-phase electric circuits is proposed. Through the analysis of the invariants of spatial curves, it is possible to find a direct relation with the generalized concept of frequency that can be of interest in electrical power systems. However, a generalization to an arbitrary number of phases has yet to be presented. In particular, the study of multiphase electrical machines or electrical power systems with a number of phases greater than three can benefit from the investigations presented in this paper.

This approach not only allows a generalization through the use of exterior algebra or geometric algebra but also provides a unifying mathematical framework.

In this paper, GA and DG is applied to multi-phase electrical systems comprising any number of electrical phases by characterizing the voltage as a vector describing a curve in n-dimensional space. We will refer to such curves as "electrical curves" (EC). A new procedure is proposed that allows to obtain a multivector representation of the geometrical angular frequency, known as Darboux Bivector.

2 Electrical Curves and Geometric Properties

The electric curve approach is an effort for the application of concepts related to spatial curves within electrical systems. More specifically, geometric invariants (e.g. curvature) of the curve can accurately describe properties of interest for the power community. For example, the instantaneous or average grid frequency in power systems can be linked easily to curvature properties. This approach lead us to the concept of "geometric frequency" as introduced by Milano [11]. For higher dimensional, i.e., multi-phase power systems, the procedure depends on the derivations of Hestenes [5] (Chap. 6) where arc-length parameterization is assumed. Here we provide a procedure that expresses this idea using time-dependent formulation of the original arc-length parameterization formulation.

2.1 Electrical Curve Definition and Parameterization

In practical power systems applications, we are typically given a uniformly sampled multi-phase and periodic voltage (or current) signals $v_i[k]$ with k the sample index and $i = 1, 2, \ldots, n$, the electrical phase index, and n the total number of electrical phases. We are allowed to create a discrete vector signal $\boldsymbol{v}[k]$ from $v_i[k]$ that describes a curve in an n-dimensional Euclidean space. We call this object an "electrical curve" (EC).

Because the method depends on differential geometric characteristics of curves, the first step in this procedure is to use a suitable interpolation or fitting method to obtain a time-dependent differentiable curve $\boldsymbol{v}(t)$ that closely approximates the sampled signal $\boldsymbol{v}[k]$. However, real-world signals can typically contain a fair amount of noise and artifacts because of the Analog to Digital Converter (ADC) quantization sampling process, transient phenomena, etc., that make this a hard task. Fortunately, many procedures exist in practice [3,8]. We will assume this step is already implemented using any suitable method.

Assume now a time-dependent vector $\boldsymbol{v}(t) = \sum_{i=1}^{n} v_i(t)\,\boldsymbol{\sigma}_i$ that describes a curve in n-dimensional Euclidean space defined on the interval $t \in [t_0, t_1]$. Theoretically, we can express the curve by a reparameterization $\boldsymbol{v}(s) = \boldsymbol{v}(t(s)) = \sum_{i=1}^{n} v_i(t(s))\,\boldsymbol{\sigma}_i$ using the arc-length variable $s(t) = \int_{t_0}^{t} \|\boldsymbol{v}'(\alpha)\|\, d\alpha$, where the relation $s(t)$ and its functional inverse $t(s) = s^{-1}(t)$ between parameters s and t are one-to-one. Note that this reparameterization $s(t), v_i(t(s))$ is not always possible to express in closed form in most cases. Therefore, in-depth new knowledge is required to overcome the exposed challenges and issues.

2.2 Time and Arc-Length Derivatives of Electrical Curves

The computation of derivatives for the EC is of paramount importance. They can be obtained in two different ways: with respect to parameter t or with respect to arc-length s. The relationship between s and t is crucial in this regard.

To investigate this point, we can start by calculating the t-derivatives of the arc-length parameter $s(t)$ using the vector $\boldsymbol{v}$ and basic rules of vector differentiation [7]. The following illustrates the first four derivatives:

$$
\begin{aligned}
s'(t) &= \sqrt{\boldsymbol{v}' \cdot \boldsymbol{v}'} \\
s''(t) &= \frac{1}{s'}\boldsymbol{v}' \cdot \boldsymbol{v}'' \\
s'''(t) &= \frac{1}{s'}\left(\boldsymbol{v}'' \cdot \boldsymbol{v}'' + \boldsymbol{v}' \cdot \boldsymbol{v}''' - (s'')^2\right) \\
s''''(t) &= \frac{1}{s'}\left(3\boldsymbol{v}'' \cdot \boldsymbol{v}''' + \boldsymbol{v}' \cdot \boldsymbol{v}'''' - 3\, s'' s'''\right)
\end{aligned}
\tag{1}
$$

Having a suitable closed form time-dependent curve $\boldsymbol{v}(t) = \sum_{i=1}^{n} v_i(t)\,\boldsymbol{\sigma}_i$ of class C^p (i.e. differentiable up to p times in t), it is simple to find the t-derivatives of arbitrary degree $0 < m \leq p$ using:

$$
\begin{aligned}
\boldsymbol{v}'(t) &= \partial_t \boldsymbol{v}(t) = \sum_{i=1}^{n} v_i'(t)\,\boldsymbol{\sigma}_i \\
\boldsymbol{v}''(t) &= \partial_t^2 \boldsymbol{v}(t) = \sum_{i=1}^{n} v_i''(t)\,\boldsymbol{\sigma}_i \\
&\;\;\vdots \\
\boldsymbol{v}^{(m)}(t) &= \partial_t^m \boldsymbol{v}(t) = \sum_{i=1}^{n} \left[\partial_t^m v_i(t)\right]\boldsymbol{\sigma}_i
\end{aligned}
\tag{2}
$$

On the other hand, the derivatives of the curve with respect to arc-length parameter[1] s can be obtained in a similar fashion:

$$\begin{aligned}
\dot{\boldsymbol{v}}(s) &= \partial_s \boldsymbol{v}(s) = \sum_{i=1}^{n} \dot{v}_i(s)\,\boldsymbol{\sigma}_i \\
\ddot{\boldsymbol{v}}(s) &= \partial_s^2 \boldsymbol{v}(s) = \sum_{i=1}^{n} \ddot{v}_i(s)\,\boldsymbol{\sigma}_i \\
\dddot{\boldsymbol{v}}(s) &= \partial_s^2 \boldsymbol{v}(s) = \sum_{i=1}^{n} \dddot{v}_i(s)\,\boldsymbol{\sigma}_i \\
&\vdots \\
\boldsymbol{v}^{(m)}(s) &= \partial_s^m \boldsymbol{v}(s) = \sum_{i=1}^{n} \left[\partial_s^m v_i(s)\right]\boldsymbol{\sigma}_i
\end{aligned} \tag{3}$$

Interestingly, if one wants to express the above set of equations (3) in terms of the parameter t, it is found that they are much more intricate. For example, using the chain rule, the first s-derivative in terms of t can be obtained for every $v_i(s)$:

$$\dot{v}_i(s(t)) = \partial_s v_i(s(t)) = \frac{dv_i(s(t))}{dt}\frac{dt}{ds} = \frac{1}{s'(t)} v_i'(t) \tag{4}$$

From now on, we remove the t symbol to indicate dependency on time in s and v for simplicity. The second and third derivatives follow:

$$\begin{aligned}
\ddot{v}_i(s) &= \partial_s^2 v_i(s) = \partial_s \dot{v}_i(s) = \frac{1}{s'}\frac{d}{dt}\left(\frac{v_i'}{s'}\right) = \frac{1}{(s')^3}\left(s' v_i'' - s'' v_i'\right) \\
\dddot{v}_i(s) &= \partial_s^3 v_i(s) = \partial_s\left[\partial_s^2 v_i(s)\right] = \frac{1}{s'}\frac{d}{dt}\left(\frac{1}{s'}\frac{d}{dt}\left(\frac{1}{s'}\frac{dv_i}{dt}\right)\right) \\
&= \frac{1}{(s')^5}\left(\left[3(s'')^2 - s's'''\right]v_i' - 3s's''v_i'' + (s')^2 v_i'''\right)
\end{aligned} \tag{5}$$

The relation between t-derivatives and s-derivatives of degree $k \geq 2$ in v_i is algebraically complicated, not at all as simple as the first derivative where $\dot{\boldsymbol{v}}(s) = \frac{1}{s'}\boldsymbol{v}'$. To illustrate this complex relation between t-derivatives and s-derivatives, the first 4 s-derivatives $\dot{\boldsymbol{v}}(s), \ddot{\boldsymbol{v}}(s), \dddot{\boldsymbol{v}}(s), \ddddot{\boldsymbol{v}}(s)$ in terms of t-derivatives are presented:

[1] We use a dot for s-derivatives instead of a prime used in t-derivatives.

$$
\begin{aligned}
\dot{\boldsymbol{v}}(s) &= \frac{1}{s'}\boldsymbol{v}' \\
\ddot{\boldsymbol{v}}(s) &= \frac{1}{(s')^3}\left[s'\boldsymbol{v}'' - s''\boldsymbol{v}'\right] \\
\dddot{\boldsymbol{v}}(s) &= \frac{1}{(s')^5}\left[(s')^2\boldsymbol{v}''' - 3\,s's''\boldsymbol{v}'' - \left(s's''' - 3\,(s'')^2\right)\boldsymbol{v}'\right] \\
\ddddot{\boldsymbol{v}}(s) &= \frac{1}{(s')^7}\left[(s')^3\boldsymbol{v}'''' - 6\,(s')^2 s''\boldsymbol{v}''' - \left(4\,(s')^2 s''' - 15\,s'\,(s'')^2\right)\boldsymbol{v}'' \right.\\
&\quad \left. + \left(10\,s's''s''' - 15\,(s'')^3 - (s')^2 s''''\right)\boldsymbol{v}'\right]
\end{aligned}
\tag{6}
$$

Additionally, the following relations hold:

$$\|\dot{\boldsymbol{v}}\| = 1 \tag{7}$$

$$\|\ddot{\boldsymbol{v}}\| = \frac{1}{(s')^2}\sqrt{\|\boldsymbol{v}''\|^2 - 2\frac{s''}{s'}\left(\boldsymbol{v}'\cdot\boldsymbol{v}''\right) + (s'')^2} \tag{8}$$

$$\dot{\boldsymbol{v}}\cdot\ddot{\boldsymbol{v}} = 0 \tag{9}$$

The quantity $\|\ddot{\boldsymbol{v}}\|$ is important as it's the base for computing the first curvature coefficient κ_1 of the curve $\boldsymbol{v}$ which has important implications for the geometrical frequency as illustrated later on.

2.3 Local Orthogonal Frames in Electrical Curves

The next step in the proposed procedure involves the application of an orthogonalization method, such as the Gram-Schmidt process, to the s-derivative vectors $\dot{\boldsymbol{v}}, \ddot{\boldsymbol{v}}$ and so on. For a symbolic expression of the s-derivative vectors, this is simple to compute using either the Classical Gram-Schmidt (CGS) [2], or the Geometric Algebra-based Gram-Schmidt (GAGS) [6]. For practical numerical computations, however, care must be taken when applying the CGS\GAGS as they are highly unstable numerically. A much more numerically stable alternative is the Modified Gram-Schmidt (MGS) procedure [1,2], which we found giving much better results when orthogonalizing higher (i.e. degree 3 or higher) s-derivatives. In any case, we essentially compute at each instant of time a local orthogonal frame $\{\boldsymbol{u}_1, \boldsymbol{u}_2, \ldots, \boldsymbol{u}_m\}, m \leq p$ from the set of p local s-derivative vectors $\{\dot{\boldsymbol{v}}, \ddot{\boldsymbol{v}}, \ldots, \partial_s^p \boldsymbol{v}\}$. We can also normalize the frame $\{\boldsymbol{u}_1, \boldsymbol{u}_2, \ldots, \boldsymbol{u}_m\}$ to get a fully orthonormal local frame $\{\boldsymbol{e}_1, \boldsymbol{e}_2, \ldots, \boldsymbol{e}_m\}$ where $\boldsymbol{e}_m = \frac{1}{\|\boldsymbol{u}_m\|}\boldsymbol{u}_m$.

According to the chain rule presented in (4), the arc-length derivatives of the arc-length frame can be readily computed:

$$\dot{\boldsymbol{e}}_i(s) = \frac{1}{s'}\boldsymbol{e}_i' \tag{10}$$

It is also interesting to find the explicit relation between vectors $\boldsymbol{e}_1, \boldsymbol{e}_2$ and vectors $\dot{\boldsymbol{v}}, \ddot{\boldsymbol{v}}, \boldsymbol{v}', \boldsymbol{v}''$. This will be useful later for expressing the "grid angular velocity" blade, on which the concept of geometric frequency is based. Applying the GAGS process to $\dot{\boldsymbol{v}}, \ddot{\boldsymbol{v}}$, we can write:

$$\begin{aligned} \boldsymbol{u}_1 &= \dot{\boldsymbol{v}} = \frac{1}{s'}\boldsymbol{v}' \\ \boldsymbol{e}_1 &= \frac{\boldsymbol{u}_1}{\|\boldsymbol{u}_1\|} = \dot{\boldsymbol{v}} = \frac{\boldsymbol{v}'}{s'} \Longrightarrow \boldsymbol{v}' = s'\boldsymbol{e}_1 \end{aligned} \tag{11}$$

And for the second vector we have:

$$\begin{aligned} \boldsymbol{u}_2 &= \ddot{\boldsymbol{v}} = \frac{1}{(s')^2}\left(\boldsymbol{v}'' - s''\boldsymbol{e}_1\right) \Longrightarrow \|\boldsymbol{u}_2\| = u_2 = \frac{1}{(s')^2}\sqrt{(v'')^2 - 2\frac{s''}{s'}\left(\boldsymbol{v}'\cdot\boldsymbol{v}''\right) + (s'')^2} \\ \boldsymbol{e}_2 &= \frac{\boldsymbol{u}_2}{\|\boldsymbol{u}_2\|} = \frac{\boldsymbol{v}'' - s''\boldsymbol{e}_1}{(s')^2\|\ddot{\boldsymbol{v}}\|} \Longrightarrow \boldsymbol{v}'' = s''\boldsymbol{e}_1 + (s')^2\|\ddot{\boldsymbol{v}}\| \\ \boldsymbol{e}_2 &= s''\boldsymbol{e}_1 + \sqrt{(v'')^2 - 2\frac{s''}{s'}\left(\boldsymbol{v}'\cdot\boldsymbol{v}''\right) + (s'')^2}\boldsymbol{e}_2 \end{aligned} \tag{12}$$

2.4 Frénet-Serret Curvature Coefficients of an Electrical Curve

The Frénet-Serret equations were formulated for three dimensions by Jean Frédéric Frénet and Joseph Alfred Serret and generalized to higher dimensions by Camile Jordan in the XIX century. They describe some dynamic properties of moving objects along curves in space by establishing a relationship between an orthonormal frame and its derivatives that also moves along the curve. The coefficients of these equations are known as curvature coefficients $\kappa_i, i = 1, 2, \ldots, n-1$ with n the number of dimensions. They satisfy the Frénet equations [5]:

$$\begin{aligned} \dot{\boldsymbol{e}}_1 &= \kappa_1\boldsymbol{e}_2 \\ \dot{\boldsymbol{e}}_2 &= -\kappa_1\boldsymbol{e}_1 + \kappa_2\boldsymbol{e}_3 \\ \dot{\boldsymbol{e}}_i &= -\kappa_{i-1}\boldsymbol{e}_{i-1} + \kappa_i\boldsymbol{e}_{i+1} \\ \dot{\boldsymbol{e}}_n &= -\kappa_{n-1}\boldsymbol{e}_{n-1} \end{aligned} \tag{13}$$

We can re-write these equations as:

$$\begin{aligned} \dot{\boldsymbol{e}}_1 &= \kappa_1\boldsymbol{e}_2 \\ \dot{\boldsymbol{e}}_2 + \kappa_1\boldsymbol{e}_1 &= \kappa_2\boldsymbol{e}_3 \\ \dot{\boldsymbol{e}}_i + \kappa_{i-1}\boldsymbol{e}_{i-1} &= \kappa_i\boldsymbol{e}_{i+1} \\ \dot{\boldsymbol{e}}_n &= -\kappa_{n-1}\boldsymbol{e}_{n-1} \end{aligned} \tag{14}$$

This directly leads to the solutions:

$$\begin{aligned} \kappa_1 &= \dot{\boldsymbol{e}}_1\cdot\boldsymbol{e}_2 \\ \kappa_i &= \left(\dot{\boldsymbol{e}}_i + \kappa_{i-1}\boldsymbol{e}_{i-1}\right)\cdot\boldsymbol{e}_{i+1} = \dot{\boldsymbol{e}}_i\cdot\boldsymbol{e}_{i+1} \end{aligned} \tag{15}$$

Another simpler alternative for computing κ_i is to use the relation from [4]:

$$\kappa_i = \frac{\|\boldsymbol{u}_{i+1}\|}{\|\boldsymbol{u}_i\|} \tag{16}$$

Here, the vectors $\boldsymbol{u}_i$ are defined as

$$\boldsymbol{u}_i = \partial_s^i \boldsymbol{v} - \sum_{j=1}^{i-1} \frac{(\partial_s^i \boldsymbol{v}) \cdot \boldsymbol{u}_j}{\boldsymbol{u}_j \cdot \boldsymbol{u}_j} \boldsymbol{u}_j \tag{17}$$

and are computed using the CGS or MGS process, not the GAGS as before. However, in this work, a slightly different expression will be used for practical applications in power systems where the frequency is ultimately dependent on the time variable t (instead of the arc-length variable s) through the voltage $v(t)$. A scaled version of κ_i will be used known as "scaled curvature coefficient", k_i:

$$k_i\,(t) = s' \frac{\|\boldsymbol{u}_{i+1}\|}{\|\boldsymbol{u}_i\|} = s' \kappa_i \tag{18}$$

These scaled curvature coefficients depend explicitly on the time variable t.

3 The Darboux Blades

The original Darboux bivector $\boldsymbol{\Omega}_H$ is described in [5]. Note that we use the subscript H to highlight the Hestenes definition. It contains a summary of the local differential geometric information of the electrical curve. In this work, a slight modification of this original definition is used, where we flip the order of multiplication. The rationale behind this decision is to fit the practical definitions of frequency in power systems and the recently proposed geometric frequency for particular cases, such as sinusoidal and balanced conditions. We can use either of the following relations as the definition for the new Darboux bivector $\boldsymbol{\Omega}$:

$$\boldsymbol{\Omega} = \frac{1}{2} s' \sum_{i=1}^{n} \boldsymbol{e}_i \wedge \dot{\boldsymbol{e}}_i \tag{19}$$

$$= \sum_{i=1}^{n-1} k_i \boldsymbol{e}_i \wedge \boldsymbol{e}_{i+1} \tag{20}$$

$$= s' \sum_{i=1}^{n-1} \frac{\boldsymbol{u}_i \wedge \boldsymbol{u}_{i+1}}{\|\boldsymbol{u}_i\|^2} \tag{21}$$

$$= s' \sum_{i=1}^{n-1} \boldsymbol{u}_i^{-1} \wedge \boldsymbol{u}_{i+1} \tag{22}$$

For the special case of $s' = 1$, then $\boldsymbol{\Omega} = -\boldsymbol{\Omega}_H$. Using this definition for the Darboux bivector, the following important relation holds true:

$$\dot{\boldsymbol{e}}_i = \boldsymbol{e}_i \rfloor \boldsymbol{\Omega} = -\boldsymbol{\Omega} \lfloor \boldsymbol{e}_i \tag{23}$$

For practical usage we found that it is advisable to separate the Darboux bivector into several 2-blades $\boldsymbol{\Omega}_i$, which we will call the Darboux Blades (DBs), as follows:

$$\begin{aligned}
&\boldsymbol{\Omega} = \frac{1}{2}\sum_{i=1}^{n-1} \boldsymbol{\Omega}_i \\
&\boldsymbol{\Omega}_1 = k_1 \boldsymbol{e}_1 \wedge \boldsymbol{e}_2 \qquad \rightarrow \qquad \|\boldsymbol{\Omega}_1\| = |k_1| \\
&\boldsymbol{\Omega}_i = k_{i-1}\boldsymbol{e}_{i-1} \wedge \boldsymbol{e}_i + k_i \boldsymbol{e}_i \wedge \boldsymbol{e}_{i+1} = s' \boldsymbol{u}_i^{-1} \wedge \boldsymbol{u}_{i+1} + s' \boldsymbol{u}_{i-1}^{-1} \wedge \boldsymbol{u}_i
\end{aligned}$$

Note that $\boldsymbol{\Omega}_i$ are always 2-blades (i.e. represent planes in n-dimensions), while the Darboux bivector $\boldsymbol{\Omega}$ is generally a bivector, not a 2-blade (except in 3-dimensions where all bivectors are 2-blades).

The first DB $\boldsymbol{\Omega}_1$, called the grid angular velocity blade, has special relevance for our analysis, satisfying the following relations:

$$\begin{aligned}
\boldsymbol{\Omega}_1 &= \dot{\boldsymbol{v}} \wedge \ddot{\boldsymbol{v}} \\
&= \frac{1}{(s')^2} \boldsymbol{v}' \wedge \boldsymbol{v}'' \\
&= k_1 \boldsymbol{e}_1 \wedge \boldsymbol{e}_2 \\
&= s' \|\ddot{\boldsymbol{v}}\| \, \boldsymbol{e}_1 \wedge \boldsymbol{e}_2 \\
&= \frac{1}{s'} \sqrt{\|\boldsymbol{v}''\|^2 - 2\frac{s''}{s'}(\boldsymbol{v}' \cdot \boldsymbol{v}'') + (s'')^2} \boldsymbol{e}_1 \wedge \boldsymbol{e}_2 \\
\dot{\boldsymbol{v}} \rfloor (\boldsymbol{\Omega} - \boldsymbol{\Omega}_1) &= \sum_{i=2}^{n-1} \boldsymbol{\Omega}_i = 0
\end{aligned}$$

The difference $\boldsymbol{\Omega} - \boldsymbol{\Omega}_1$ is proportional to the bivector $\boldsymbol{B}$ of relation (3.8) in [5], which always satisfies $\dot{\boldsymbol{v}} \rfloor \boldsymbol{B} = 0$. When the curve $\boldsymbol{v}(t)$ is a planar one (such as in the case for 3-phase line-to-line voltages signal), all curvature coefficients are zero except κ_1. In this specific case, we have $\boldsymbol{\Omega} - \boldsymbol{\Omega}_1 = 0$, and the two quantities are equivalent $\boldsymbol{\Omega} = \boldsymbol{\Omega}_1$.

4 Example Signals

A number of theoretical examples are now presented to validate the proposed method. Sinusoidal and non-sinusoidal multi-phase systems with an arbitrary number of phases are studied. The goal is to obtain a geometric representation of the generalized frequency grid by using the concept of DB.

4.1 Multi-phase Balanced Sinusoidal Signal

Assume we have a balanced multi-phase sinusoidal electrical signal. It means that the amplitude is the same for all phases and the phase angle among them is $\frac{2\pi m}{n}$ with $m = 0, 1, \ldots, n-1$ and n the number of phases.

$$\begin{aligned}
\boldsymbol{v}(t) &= V\sum_{m=0}^{n-1}\cos\left(\omega t - 2\pi\frac{m}{n}\right)\boldsymbol{\sigma}_{m+1} = V\sum_{m=0}^{n-1}\left[\cos\left(2\pi\frac{m}{n}\right)\cos(\omega t) + \sin\left(2\pi\frac{m}{n}\right)\sin(\omega t)\right]\boldsymbol{\sigma}_{m+1} \\
&= \cos(\omega t)\, V\left[\sum_{m=0}^{n-1}\cos\left(2\pi\frac{m}{n}\right)\boldsymbol{\sigma}_{m+1}\right] + \sin(\omega t)\, V\left[\sum_{m=0}^{n-1}\sin\left(2\pi\frac{m}{n}\right)\boldsymbol{\sigma}_{m+1}\right] \\
&= \cos(\omega t)\,\boldsymbol{a} + \sin(\omega t)\,\boldsymbol{b}
\end{aligned} \tag{24}$$

with

$$\boldsymbol{a} = V\sum_{m=0}^{n-1}\cos\left(2\pi\frac{m}{n}\right)\boldsymbol{\sigma}_{m+1}$$
$$\boldsymbol{b} = V\sum_{m=0}^{n-1}\sin\left(2\pi\frac{m}{n}\right)\boldsymbol{\sigma}_{m+1}$$

In this case the two n-dimensional vectors $\boldsymbol{a}, \boldsymbol{b}$ are orthogonal with $\|\boldsymbol{a}\|^2 = \|\boldsymbol{b}\|^2 = \frac{n}{2}V^2$. This signal describes a perfect circular curve in the plane spanned by the n-dimensional orthogonal vectors $\boldsymbol{a}$, $\boldsymbol{b}$ inside the larger signal space defined by orthonormal basis vectors $\{\boldsymbol{\sigma}_i\}_{i=1}^{n}$. We can express all relevant quantities using vectors $\boldsymbol{a}$ and $\boldsymbol{b}$ as follows:

$$\begin{aligned}
\boldsymbol{v}' &= -\omega\left(\sin(\omega t)\,\boldsymbol{a} - \cos(\omega t)\,\boldsymbol{b}\right) \\
\boldsymbol{v}'' &= -\omega^2\left(\cos(\omega t)\,\boldsymbol{a} + \sin(\omega t)\,\boldsymbol{b}\right) \\
\|\boldsymbol{v}'\|^2 &= \frac{\omega^2}{2}nV^2 \\
s' = \|\boldsymbol{v}'\| &= \frac{\omega}{\sqrt{2}}\sqrt{n}V \\
\boldsymbol{u}_1 = \dot{\boldsymbol{v}} &= -\frac{\sqrt{2}}{\sqrt{n}V}\left[\sin(\omega t)\,\boldsymbol{a} - \cos(\omega t)\,\boldsymbol{b}\right] \\
\boldsymbol{u}_2 = \ddot{\boldsymbol{v}} &= -\frac{2}{nV^2}\left[\cos(\omega t)\,\boldsymbol{a} + \sin(\omega t)\,\boldsymbol{b}\right] \\
\boldsymbol{e}_1 = \frac{\boldsymbol{u}_1}{\|\boldsymbol{u}_1\|} &= -\frac{\sqrt{2}}{\sqrt{n}V}\left[\sin(\omega t)\,\boldsymbol{a} - \cos(\omega t)\,\boldsymbol{b}\right] \\
\boldsymbol{e}_2 = \frac{\boldsymbol{u}_2}{\|\boldsymbol{u}_2\|} &= -\frac{\sqrt{2}}{\sqrt{n}V}\left[\cos(\omega t)\,\boldsymbol{a} + \sin(\omega t)\,\boldsymbol{b}\right] \\
k_1 &= s'\frac{\|\boldsymbol{u}_2\|}{\|\boldsymbol{u}_1\|} = \omega \\
\boldsymbol{\Omega}_1 &= k_1\boldsymbol{e}_1 \wedge \boldsymbol{e}_2 = \omega\boldsymbol{e}_1 \wedge \boldsymbol{e}_2 = 2\omega\left[\frac{1}{nV^2}\right]\boldsymbol{a} \wedge \boldsymbol{b}
\end{aligned}$$

Note the constant nature of $\|\boldsymbol{\Omega}_1\| = \omega$ for this signal, which confirms the curve being a perfect circle.

4.2 Multi-phase Unbalanced Sinusoidal Signal

Assume we have a general (possibly unbalanced) multi-phase sinusoidal electrical signal. This now means that the amplitude can be different among phases and phase angle is not regularly spaced by $\frac{2\pi m}{n}$. In this case the voltage vector is:

$$\begin{aligned}
\boldsymbol{v}(t) &= \sum_{m=1}^{n} V_m \cos(\omega t - \varphi_m)\,\boldsymbol{\sigma}_m = \sum_{m=1}^{n} \left[V_m \cos(\varphi_m)\cos(\omega t) + V_m \sin(\varphi_m)\sin(\omega t)\right]\boldsymbol{\sigma}_m \\
&= \cos(\omega t)\left[\sum_{m=1}^{n} V_m \cos(\varphi_m)\,\boldsymbol{\sigma}_m\right] + \sin(\omega t)\left[\sum_{m=1}^{n} V_m \sin(\varphi_m)\,\boldsymbol{\sigma}_m\right] \\
&= \cos(\omega t)\,\boldsymbol{a} + \sin(\omega t)\,\boldsymbol{b}
\end{aligned}$$

with

$$\boldsymbol{a} = \sum_{m=1}^{n} V_m \cos(\varphi_m)\,\boldsymbol{\sigma}_m$$

$$\boldsymbol{b} = \sum_{m=1}^{n} V_m \sin(\varphi_m)\,\boldsymbol{\sigma}_m$$

This signal describes an ellipse in the plane spanned by the n-dimensional vectors $\boldsymbol{a}$, $\boldsymbol{b}$ inside the signal space defined by orthonormal basis vectors $\{\boldsymbol{\sigma}_i\}_{i=1}^{n}$. We can express all relevant quantities using vectors $\boldsymbol{a}$ and $\boldsymbol{b}$ as follows:

$$\begin{aligned}
\boldsymbol{v}' &= -\omega\left(\sin(\omega t)\,\boldsymbol{a} - \cos(\omega t)\,\boldsymbol{b}\right) \\
\boldsymbol{v}'' &= -\omega^2\left(\cos(\omega t)\,\boldsymbol{a} + \sin(\omega t)\,\boldsymbol{b}\right) \\
\|\boldsymbol{v}'\|^2 &= \frac{\omega^2}{2} g^2 \\
s' = \|\boldsymbol{v}'\| &= \frac{\omega}{\sqrt{2}} g \\
g &= \sqrt{(\boldsymbol{b}^2 - \boldsymbol{a}^2)\cos(2\omega t) - 2(\boldsymbol{a}\cdot\boldsymbol{b})\sin(2\omega t) + (\boldsymbol{b}^2 + \boldsymbol{a}^2)} \\
\boldsymbol{u}_1 = \dot{\boldsymbol{v}} &= -\frac{\sqrt{2}}{g}\left[\sin(\omega t)\,\boldsymbol{a} - \cos(\omega t)\,\boldsymbol{b}\right] \\
\boldsymbol{u}_2 = \ddot{\boldsymbol{v}} &= -\frac{4}{g^4}\left[\left(\boldsymbol{b}^2\cos(\omega t) - (\boldsymbol{a}\cdot\boldsymbol{b})\sin(\omega t)\right)\boldsymbol{a} + \left(\boldsymbol{a}^2\sin(\omega t) - (\boldsymbol{a}\cdot\boldsymbol{b})\cos(\omega t)\right)\boldsymbol{b}\right] \\
\boldsymbol{e}_1 = \frac{\boldsymbol{u}_1}{\|\boldsymbol{u}_1\|} &= -\frac{\sqrt{2}}{g}\left[\sin(\omega t)\,\boldsymbol{a} - \cos(\omega t)\,\boldsymbol{b}\right] \\
\boldsymbol{e}_2 = \frac{\boldsymbol{u}_2}{\|\boldsymbol{u}_2\|} &= -\frac{\sqrt{2}}{g}\left[\frac{\boldsymbol{b}^2\cos(\omega t) - (\boldsymbol{a}\cdot\boldsymbol{b})\sin(\omega t)}{\sqrt{\boldsymbol{a}^2\boldsymbol{b}^2 - (\boldsymbol{a}\cdot\boldsymbol{b})^2}}\boldsymbol{a} + \frac{\boldsymbol{a}^2\sin(\omega t) - (\boldsymbol{a}\cdot\boldsymbol{b})\cos(\omega t)}{\sqrt{\boldsymbol{a}^2\boldsymbol{b}^2 - (\boldsymbol{a}\cdot\boldsymbol{b})^2}}\boldsymbol{b}\right]
\end{aligned}$$

$$k_1 = s' \frac{\|\boldsymbol{u}_2\|}{\|\boldsymbol{u}_1\|} = \left(\frac{2}{g^2} \sqrt{\boldsymbol{a}^2 \boldsymbol{b}^2 - (\boldsymbol{a} \cdot \boldsymbol{b})^2} \right) \omega$$

$$\boldsymbol{\Omega}_1 = k_1 \boldsymbol{e}_1 \wedge \boldsymbol{e}_2 = \left(\frac{2}{g^2} \sqrt{\boldsymbol{a}^2 \boldsymbol{b}^2 - (\boldsymbol{a} \cdot \boldsymbol{b})^2} \right) \omega \boldsymbol{e}_1 \wedge \boldsymbol{e}_2$$

$$= 2\omega \left[\frac{1}{g^2} - (\boldsymbol{a} \cdot \boldsymbol{b}) \frac{\sin(2\omega t)}{g^4} \right] \boldsymbol{a} \wedge \boldsymbol{b}$$

In this signal we have $\|\boldsymbol{\Omega}_1\| = h(t)\,\omega$, where the constant angular frequency of the grid ω is scaled by the periodic time dependent factor:

$$h(t) = \frac{2\sqrt{\boldsymbol{a}^2 \boldsymbol{b}^2 - (\boldsymbol{a} \cdot \boldsymbol{b})^2}}{(\boldsymbol{b}^2 - \boldsymbol{a}^2)\cos(2\omega t) - 2(\boldsymbol{a} \cdot \boldsymbol{b})\sin(2\omega t) + (\boldsymbol{b}^2 + \boldsymbol{a}^2)}$$

Which has unit average value $\overline{h} = \frac{1}{T} \int_0^T h(t)\,dt = 1$ where $T = \frac{\pi}{\omega}$ is the time of a single cycle of $h(t)$, which is half the time of a single cycle of the signal $\boldsymbol{v}(t)$. This clearly indicates that by averaging $\boldsymbol{\Omega} = \boldsymbol{\Omega}_1$ on one half cycle of this signal, followed by taking the norm of the resulting bivector, we again get the grid frequency ω as intuitively expected.

4.3 Multi-phase Balanced Harmonic Signal

Finally, assume we have the following harmonic electrical signal:

$$\begin{aligned}
\boldsymbol{v}(t) &= \sqrt{2}\left[200\sin(\omega t) + 20\sin(2\omega t) - 30\sin(7\omega t)\right]\boldsymbol{\sigma}_1 \\
&+ \sqrt{2}\left[200\sin\left(\omega t - \frac{2\pi}{3}\right) + 20\sin\left(2\left(\omega t - \frac{2\pi}{3}\right)\right) - 30\sin\left(7\left(\omega t - \frac{2\pi}{3}\right)\right)\right]\boldsymbol{\sigma}_2 \\
&+ \sqrt{2}\left[200\sin\left(\omega t + \frac{2\pi}{3}\right) + 20\sin\left(2\left(\omega t + \frac{2\pi}{3}\right)\right) - 30\sin\left(7\left(\omega t + \frac{2\pi}{3}\right)\right)\right]\boldsymbol{\sigma}_3
\end{aligned}$$

This electrical signal traces a planar symmetric curve in the plane orthogonal to vector $(1,1,1)$. The expression for the first DB is:

$$\boldsymbol{\Omega}_1(t) = \frac{5\omega}{\sqrt{3}} \left(\frac{16\cos(3\omega t) + 672\cos(6\omega t) + 84\cos(9\omega t) - 691}{-160\cos(3\omega t) + 840\cos(6\omega t) + 168\cos(9\omega t) - 857} \right) (\boldsymbol{\sigma}_{1,2} - \boldsymbol{\sigma}_{1,3} + \boldsymbol{\sigma}_{2,3})$$

The average DB and its norm are given by:

$$\overline{\boldsymbol{\Omega}_1} = \frac{1}{T} \int_0^T \boldsymbol{\Omega}_1(t)\,dt = \sqrt{3}\omega\,(\boldsymbol{\sigma}_{1,2} - \boldsymbol{\sigma}_{1,3} + \boldsymbol{\sigma}_{2,3}) \qquad T = \frac{2\pi}{\omega}$$

$$\left\|\overline{\boldsymbol{\Omega}_1}\right\| = 3\omega$$

The norm of the average angular velocity blade $\left\|\overline{\Omega}_1\right\|$ is proportional to the grid nominal angular frequency ω.

5 Conclusion

This paper has presented the generalized concept of geometrical angular frequency applied to multi-phase systems with arbitrary number of phases, extending previous works where linear algebra were used to define the concept of geometric frequency. Geometric Algebra and Differential Geometry have been used to represent vectors in n-dimensional spaces and to compute the geometric invariants associated to the generated spatial curves. Voltage signals have been used to create such vectors and to compute the Darboux Bivector, which encodes the specific differential geometric properties in the curves known as electrical curves. The method can be conveniently employed for a variety of engineering problems such as voltage stability, frequency control, to mention a few.

References

1. Beilina, L., Karchevskii, E., Karchevskii, M.: Numerical Linear Algebra: Theory and Applications. Springer, Cham (2017). https://doi.org/10.1007/978-3-319-57304-5
2. Bjërck, Å.: Numerics of gram-schmidt orthogonalization. Linear Algebra and its Applications 197–198, 297–316 (1994). https://doi.org/10.1016/0024-3795(94)90493-6
3. Buzzi-Ferraris, G., Manenti, F.: Interpolation and regression models for the chemical engineer: solving numerical problems. John Wiley & Sons (2010)
4. Gluck, H.: Higher curvatures of curves in Euclidean space. The American Mathematical Monthly **73**(7), 699–704 (1966). https://doi.org/10.1080/00029890.1966.11970818
5. Hestenes, D., Sobczyk, G.: Clifford Algebra to Geometric Calculus. Springer, Netherlands (1987). https://doi.org/10.1007/978-94-009-6292-7
6. Hitzer, E.M.: Gram-schmidt orthogonalization in geometric algebra (2003)
7. Hubbard, J.H., Hubbard, B.B.: Vector calculus, linear algebra, and differential forms: a unified approach, 5th edn. Matrix Editions, Upper Saddle River, N.J (2015)
8. Khoury, R., Harder, D.W.: Numerical Methods and Modelling for Engineering. Springer, Cham (2016). https://doi.org/10.1007/978-3-319-21176-3
9. Kirkham, H., Dickerson, W., Phadke, A.: Defining power system frequency. In: 2018 IEEE Power & Energy Society General Meeting (PESGM), pp. 1–5. IEEE (2018)
10. Leon, C.: Time-frequency analysis: theory and applications. Pnentice Hall, USA (1995)
11. Milano, F.: A geometrical interpretation of frequency. IEEE Trans. Power Syst. **37**(1), 816–819 (2021)
12. Milano, F., Tzounas, G., Dassios, I., Kerci, T.: Applications of the Frenet frame to electric circuits. Regular Papers, IEEE Transactions on Circuits and Systems I (2021)

13. Montoya, F.G., Baños, R., Alcayde, A., Arrabal-Campos, F.M., Roldán-Pérez, J.: Geometric algebra applied to multiphase electrical circuits in mixed time-frequency domain by means of hypercomplex Hilbert transform. Mathematics **10**(9), 1419 (2022)
14. Montoya, F.G., De Leon, F., Arrabal-Campos, F.M., Alcayde, A.: Determination of instantaneous powers from a novel time-domain parameter identification method of non-linear single-phase circuits. IEEE Transactions on Power Delivery (2021)

Notes on Forward Kinematics of Generalised Robotic Snakes Based on Compass Ruler Algebra

Roman Byrtus(✉) and Stanislav Frolík

Institute of Mathematics, Faculty of Mechanical Engineering, Brno University of Technology, Technická 2896/2, 616 69 Brno, Czech Republic
171155@vutbr.cz

Abstract. We use conformal transformations in the kinematic chain of a specific planar mechanism. We present two possible ways to describe the configuration of a three link generalised snake robot using compass ruler algebra, which allows us to point out the nature of the generalised snake as an extension of the classic snake robot.

Keywords: Compass ruler algebra · Clifford algebra · Mathematical · robotics · Non-holonomic mechanisms · Snake robots · Forward kinematics

1 Introduction

In this paper we investigate a generalised planar robotic snake, which is inferred from a standard robotic snake by adding one parameter. The scope of this paper is to construct its forward kinematics using so-called compass ruler algebra. The topic of the snake-like robots goes back to early 1970's when Hirose formulated the essential model design and developed limbless locomotors, for the complex review of his work see [3]. He started the first bio-mechanical study using the real snakes and designed the first snake-like robot based on so-called serpentine locomotion. Original Hirose work has been followed by Downling [1], Chirikjian and Burdick [8], and Ostrowski [6] in the next years. Recently, the theory of robotic snakes have been developed in papers [4,5,7] in the framework of so-called geometric algebras [2].

To investigate the mechanism, let us firstly recall the robotic snake-like mechanism. A robotic snake is a planar mechanism consisting of n rigid links connected by $n-1$ revolute joints. To each link, a pair of wheels is attached, such that the ground friction in the direction perpendicular to the link is considerably higher than the friction of a simple forward move. It means that the pair of supportive wheels can not slip aside. Usually such a condition is referred to as a non-slip condition. The configuration space of an n–link robotic snake is

$$(x, y, \theta, \Phi_1, \ldots, \Phi_{n-1}) \in E(2) \times \mathbb{S}_1^{n-1},$$

D. W. Silva et al. (Eds.): ICACGA 2022, LNCS 13771, pp. 138–146, 2024.
https://doi.org/10.1007/978-3-031-34031-4_12

where $p_0 = (x, y)$ is the position of the beginning of the robotic snake, θ is the absolute rotation w.r.t. the x axis and Φ_i are the relative rotations of the respective revolute joints, see Fig. 1.

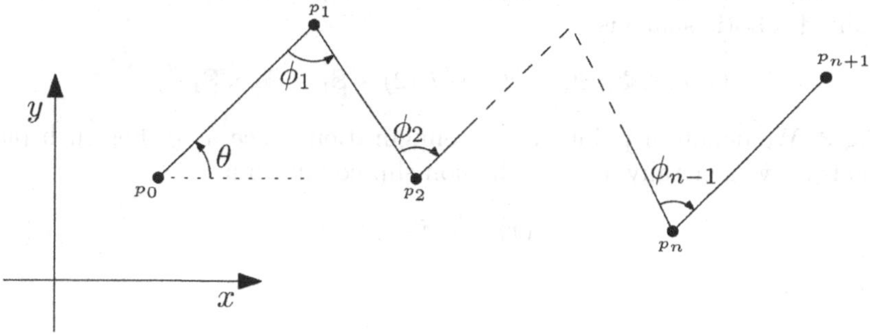

Fig. 1. The description of an n–link robotic snake

In a similar way, other planar mechanisms have been studied. As an example let us mention the so-called trident snake robot. Originally the general trident snake robot has been introduced in [9]. It is a planar robot with a body in the shape of a triangle with three legs, where each leg can be understood as an n–link robotic snake.

In our paper we introduce a new concept of planar mechanisms. We consider generalised robotic snakes consisting of n prismatic joints connected by $n-1$ links equipped with revolute joints. Similarly to robotic snakes, each link is equipped with a pair of wheels with the same physical nature. The configuration space of an n–link generalised robotic snake can be described as

$$(x, y, \theta, \ell_1, \Phi_1, \ell_2 \ldots, \Phi_{n-1}, \ell_n) \in E(2) \times (\mathbb{R} \times \mathbb{S}_1)^{n-1} \times \mathbb{R},$$

where $(x, y, \theta) \in E(2)$ is a position and an orientation in the plane, $(\Phi_1,, \ldots, \Phi_{n-1})$ are parameters of revolute joints and $(\ell_1, \ldots, \ell_n)$ are parameters of prismatic joints, see Fig. 2.

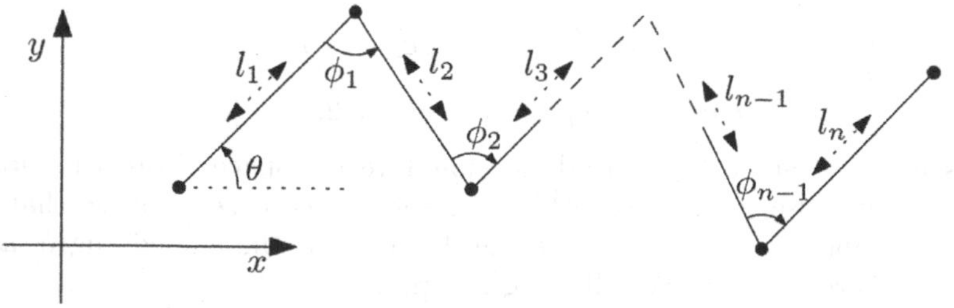

Fig. 2. The description of the generalised n–link robotic snake

From now on we will investigate a special case of such a snake. We consider a 3-link generalised snake where not all links are prismatic. In our case a robotic 3-link generalised snake consists of three links connected by revolute joints, but only the peripheral links are prismatic. Each link is equipped with a pair of wheels with the same physical nature. The configuration space of such a 3–link generalised robotic snake is

$$(x, y, \theta, \Phi_1, \Phi_2, \ell_1, \ell_2) \in E(2) \times \mathbb{S}_1 \times (\mathbb{R} \times \mathbb{S}_1)^2,$$

see Fig. 3. We denote a point in the configuration space as q. For such planar mechanisms we generally assume the non-slip condition as

$$(x_i, y_i) \cdot \vec{n} = 0,$$

where $p_i = (x_i, y_i)$.

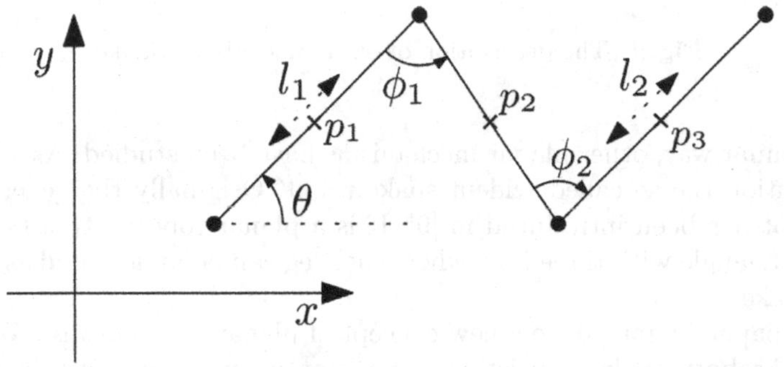

Fig. 3. The description of the studied generalised 3–link robotic snake

2 Compass Ruler Algebra - CRA

Let $\mathbb{R}^{3,1}$ denote the vector space $\mathbb{R}^4$ equipped with the scalar product of the signature $(3, 1)$. Thus we have the corresponding Clifford algebra $\mathcal{C}l(3, 1)$ such that the set $\{e_1, e_2, e_+, e_-\}$ is the basis. To describe elements of $\mathcal{O} := \mathcal{C}l(3, 1)$ we have to determine a free, associative and distributive algebra as a span of the set $\{e_1, e_2, e_+, e_-\}$ such that the following identities are satisfied:

$$e_1^2 = e_2^2 = e_+^2 = 1, e_-^2 = -1,$$

$$e_i e_j = -e_j e_i, i \neq j, i, j \in \{1, 2, +, -\}.$$

In this case, we get a $2^4 = 16$-dimensional vector space. Note that for simplicity we will use a basis of $\mathbb{R}^{3,1}$ as a set $\{e_1, e_2, e_0, e_\infty\}$ such that $e_0 = \frac{1}{2}(e_- + e_+)$ and $e_\infty = e_- - e_+$, see Fig. 4. In CRA, the embedding of a point $(x, y) \in \mathbb{R}^2$ is realised by the following mapping:

$$\text{point } x \mapsto P = x_1 e_1 + x_2 e_2 + \frac{1}{2}(x_1^2 + x_2^2) e_\infty + e_0,$$

Scalars	1
Vectors	e_1, e_2, e_0, e_∞
Bivectors	$e_1 \wedge e_2, e_1 \wedge e_0, e_1 \wedge e_\infty, e_2 \wedge e_0, e_2 \wedge e_\infty, e_\infty \wedge e_0$
Trivectors	$e_1 \wedge e_2 \wedge e_\infty, e_1 \wedge e_2 \wedge e_0, e_1 \wedge e_\infty \wedge e_0, e_2 \wedge e_\infty \wedge e_0$
Pseudoscalar	$e_1 \wedge e_2 \wedge e_\infty \wedge e_0$

Fig. 4. A basis of the compass ruler algebra

and a line is given by

$$L = P_1 \wedge P_2 \wedge e_\infty.$$

Moreover we can represent a point pair as

$$P_p = P_1 \wedge P_2.$$

Note that we have used the IPNS representation for a point construction whereas we used the OPNS representations of a line and a point pair, which is more convenient in the latter case. Notice that we can represent links of the generalised snake-like robot as point pairs. In GA generally any transformation of the object O is realized by the sandwich product

$$O \mapsto MO\tilde{M},$$

where M is an appropriate multivector from $\mathcal{O}$. For instance, the translation in the direction $t = xe_1 + ye_2$ is realized by the multivector

$$T = 1 - \frac{1}{2} te_\infty$$

and the rotation around an axis L by an angle ϕ is realized by the multivector

$$R = \cos \frac{\phi}{2} - L \sin \frac{\phi}{2},$$

where $L = a_1 e_1 e_2$. Moreover we will use another important transformation, which defines our concept. We will introduce the scaling multivector (acting from the origin) as

$$S = e^{-\frac{\ln \gamma}{2} e_0 e_\infty},$$

where γ is the scaling factor.

Remark 1. The multivector $S_{x,y}$ for scaling centered on a point (x, y) can be obtained by a straightforward calculation:

$$S_{x,y} = e^{-\frac{ln \gamma}{2} L_s},$$

where

$$L_s = Te_0 e_\infty \tilde{T} = (1 - \frac{1}{2} te_\infty) e_0 e_\infty (1 + \frac{1}{2} te_\infty) = e_0 e_\infty + xe_1 e_\infty + ye_2 e_\infty.$$

The scalor multivector allows us to elegantly represent the change of the length of a link using the transformation. Note that while the translation and the rotation are euclidean transformations, the scaling is not.

Example 1. Let us demonstrate the application of the scalor centered in a point acting on a point pair. The computation is done using the Python package Clifford (the library imports in the code are omitted for brevity). In short, euclidean points A, B and the origin O (used only for visualisation purposes) are embedded into CRA using the *up* function and a point pair AB is constructed using the wedge product. The midpoint of this point pair is found and used in the construction of the scalor. The scalor is then applied using the sandwich product and visualised. The resulting visualisation can be seen in the Fig. 5.

```
O = up(0)
A = up(2*e1+2*e2)
B = up(-e1+2*e2)
AB = A^B
gamma = log(2)
L = eo^einf

#midpoint of point pair
midpoint = AB*einf*(~AB)

#extract translation
downMidpoint = down(midpoint)
x = downMidpoint.value[1]
y = downMidpoint.value[2]
#translation from origin
T = 1 - 1/2*(x*e1 +y*e2)*einf
L = T*L*~T
S = exp(gamma/2 * L)
SAB = S*AB*~S

sc = GanjaScene()
sc.add_object(O, color=(46, 94, 42), label='O')
sc.add_object(AB, color=(255, 0, 0), label='PP')
sc.add_object(SAB, color=(0, 0, 255), label='SPP')
sc.add_object(midpoint, color=(0,0,0), label='p')

draw(sc, sig=layout.sig, scale=0.5)
```

3 Forward Kinematics of the Generalised 3-Link Snake

As mentioned earlier, the mechanism that we will work with is the 3-link snake robot, where the 1st and the 3rd links include a prismatic joint.

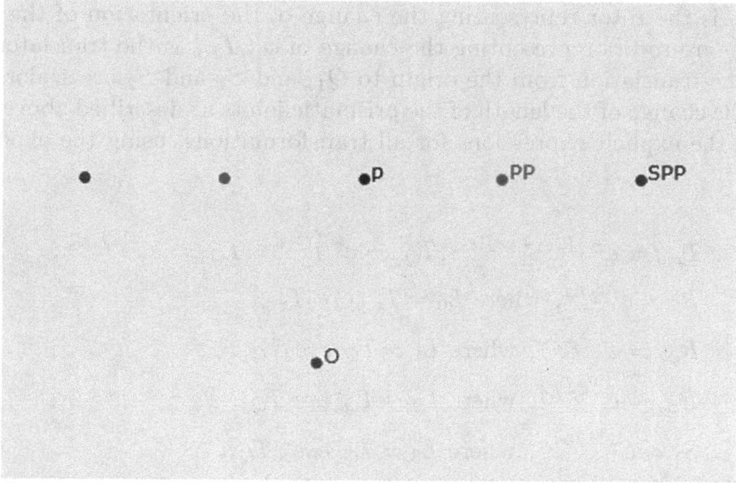

Fig. 5. Point pair PP scaled by a factor of 2 around the point p.

We interpret the prismatic links of the mechanism in the following sense: the change of length of the first link is realised by an extension/retraction in the direction of the vector $Q_1 - Q_2$ and the change of the length of the 3rd link is realised by an extension/retraction in the direction of the vector $Q_4 - Q_3$.

Let us now describe the mechanism using CRA. The i-th link can be described as the point pair $P_i = Q_i \wedge Q_{i+1}$, i.e. the outer product of two points. The mechanism is then described as the triplet of point pairs P_1, P_2, P_3. The point pairs can be represented by their centres p_i, given by

$$p_i = P_i e_\infty \tilde{P}_i.$$

Let us show how the initial position for the point pair looks in terms of elements of CRA. Setting the mechanism's initial configuration as $x = y = \theta = \phi_1 = \phi_2 = 0$, $l_1 = l_3 = 1$, we get

$$\begin{aligned} P_{1,0} &= 2e_0e_1 + e_0e_\infty, \\ P_{2,0} &= 4 + 2e_1e_\infty + 2e_0e_1 + 3e_0e_\infty, \\ P_{3,0} &= 12 + 6e_1e_\infty + 2e_0e_1 + 5e_0e_\infty. \end{aligned}$$

We present two possible ways to describe the configuration of the mechanism. The first approach is as follows. Denote the initial position of the point pair P_i as $P_{i,0}$. Then we can understand P_i as the current configuration of the mechanism's i-th link. The first description of the configuration is taken as

$$\begin{aligned} P_1 &= R_\theta T_{x,y} S_1 P_{1,0} \tilde{S}_1 \tilde{T}_{x,y} \tilde{R}_\theta, \\ P_2 &= R_{\phi_1} R_\theta T_{x,y} P_{2,0} \tilde{T}_{x,y} \tilde{R}_\theta \tilde{R}_{\phi_1}, \\ P_3 &= R_{\phi_2} R_{\phi_1} R_\theta T_{x,y} S_3 P_{1,0} \tilde{S}_3 \tilde{T}_{x,y} \tilde{R}_\theta \tilde{R}_{\phi_1} \tilde{R}_{\phi_2}, \end{aligned}$$

where R_θ is the rotor representing the change of the orientation of the mechanism, R_{ϕ_i} are rotors representing the change of ϕ_i, $T_{x,y}$ is the translator representing the translation from the origin to Q_1, and S_1 and S_2 are scalors representing the change of the length of the prismatic joints as described above. Let us now give the explicit expressions for all transformations, using the exponential notation:

$$T_{x,y} = e^{-\frac{1}{2}(xe_1+ye_2)e_\infty}, T_{Q_2} = e^{-\frac{1}{2}Q_2 e_\infty}, T_{Q_3} = e^{-\frac{1}{2}Q_3 e_\infty}, \tag{1}$$

$$R_\theta = e^{-\frac{\theta}{2}L_0}, \text{ where } L_0 = T_{x,y}e_1e_2\tilde{T}_{x,y}, \tag{2}$$

$$R_{\phi_1} = e^{-\frac{\phi_1}{2}L_1}, \text{ where } L_1 = T_{Q_2}e_1e_2\tilde{T}_{Q_2}, \tag{3}$$

$$R_{\phi_2} = e^{-\frac{\phi_2}{2}L_2}, \text{ where } L_2 = T_{Q_3}e_1e_2\tilde{T}_{Q_3}, \tag{4}$$

$$S_1 = e^{-\frac{\ln \gamma_1}{2}L_3}, \text{ where } L_3 = T_{Q_2}e_0e_\infty\tilde{T}_{Q_2}, \tag{5}$$

$$S_2 = e^{-\frac{\ln \gamma_2}{2}L_4}, \text{ where } L_4 = T_{Q_3}e_0e_\infty\tilde{T}_{Q_3}, \tag{6}$$

where γ_i are scaling factors.

The second, and more interesting approach (shown in Fig. 6) how to describe the configuration is given by

$$\begin{aligned} P_1 &= S_1 R_\theta T_{Q_2} P_{1,0} \tilde{T}_{Q_2} \tilde{R}_\theta \tilde{S}_1, \\ P_2 &= R_{\phi_1} R_\theta T_{Q_2} P_{2,0} \tilde{T}_{Q_2} \tilde{R}_\theta \tilde{R}_{\phi_1}, \\ P_3 &= S_3 \tilde{R}_{\phi_2} R_{\phi_1} R_\theta T_{Q_2} P_{3,0} \tilde{T}_{Q_2} \tilde{R}_\theta \tilde{R}_{\phi_1} R_{\phi_2} \tilde{S}_3. \end{aligned} \tag{7}$$

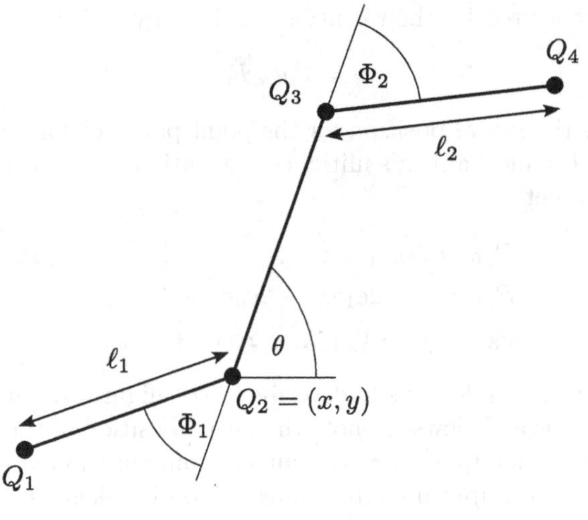

Fig. 6. Second approach to describing the configuration.

The transformations are same as in the first case. The difference lies in the order of transformations, identification of the tracked point (x, y) with the point

Q_2 and the global angle of the orientation θ is also tracked with respect to the second link. The argument for why this approach to formulate the kinematic chain is more interesting is that it better coincides with the realisation of the mechanism with respect to the order of the transformations. Take a look at (7). Notice that if scalors S_1, S_3 were to be removed, the kinematic chain would correspond to a 3-link snake without prismatic joints, thus the fact that the generalised robot snake is an extension of the classic mechanism is clearly visible.

Remark 2. We have applied scaling, a non-euclidean transformation, to describe the configuration of a generalised snake robot. However, a more interesting family of mechanisms for which scaling would lead to a more natural representation would be the generalised trident snake robot. The generalised trident snake mechanism is composed of a regular symmetric polygonal base, with an arm (which can be viewed as an n-link snake robot) attached to every vertex. It would make a sense to consider the last link of such an arm to be extensible, for example to represent end effectors; then our approach of representing prismatic joints by a scalor in CRA becomes much more attractive.

In fact, if we disregard the passive wheel in the middle of the central link for our 3-link snake, we can think of it as a "degenerate" case of the generalised trident snake, for which the base is a line segment.

4 Conclusion

We have presented a representation of the generalised snake robot using point pairs as objects of CRA. The advantage of our approach is that we can easily describe such a non-rigid body motion using transformations represented as bivectors (or their exponentials). This leads to a simple formalism used for the description of the kinematic chain of our mechanism. Applying this approach to other families of robotic mechanisms, such as the generalised trident snake, and using the CRA approach in simulations could lead to further interesting work topics.

References

1. Dowling, K.J.: Limbless locomotion: learning to crawl with a snake robot. Ph.D. thesis, Carnegie Melon University, Pittsburgh, USA (1997)
2. Hildenbrand, D.: Foundations of Geometric Algebra Computing. Springer, Cham (2013)
3. Hirose, S.: Biologically Inspired Robots (Snake-Like Locomotor and Manipulator). Oxford University Press, Oxford (1993)
4. Hildenbrand, D., Hrdina, J., Návrat, A., Vašík, P.: Local controllability of snake robots based on CRA, theory and practice. Adv. Appl. Clifford Algebras **30**(1) (2020)
5. Hrdina, J., Návrat, A., Vašík, P.: Control of 3-link robotic snake based on conformal geometric algebra. Adv. Appl. Clifford Algebras **26**(3), 1069–1080 (2015). https://doi.org/10.1007/s00006-015-0621-2

6. Ostrowski, J.: The mechanics of control of undulatory robotic locomotion. Ph.D. thesis, CIT (1995)
7. Hrdina, J., Návrat, A., Vašík, P., Matoušek, R.: CGA-based robotic snake control. Adv. Appl. Clifford Algebras **27**(1), 621–632 (2017)
8. Chirikjian, G.S., Burdick, J.W.: An obstacle avoidance algorithm for hyper-redundant manipulators. In: Proceedings 1990 IEEE International Conference on Robotics and Automation, pp. 625–631 (1990)
9. Ishikawa, M.: Trident snake robot: locomotion analysis and control. NOLCOS. IFAC Symp. Nonlinear Control Syst. **6**, 1169–1174 (2004). vol. 27 (2017) Trident Snake Robot via CGA 645

Hand-Eye Calibration Using Camera's IMU Sensor in Quadric Geometric Algebra (QGA)

Julio Zamora-Esquivel, Edgar Macias-Garcia, and Leobardo Campos-Macias(✉)

Intel Labs, Human Robot Collaboration, Guadalajara, Jalisco, Mexico
{julio.c.zamora.esquivel,edgar.macias.garcia, leobardo.e.campos.macias}@intel.com

Abstract. In this work, a novel method to solve the hand-eye calibration problem for a camera attached to the end-effector of a robot articulation is proposed using the Geometric Algebra framework. In the proposed method, the linear acceleration and angular velocity acquired from an IMU (Inertial Measurement Unit) sensor attached to the camera is employed with the kinematic model of the robot articulation to find the relative camera position and orientation. Compared with other methods, there is no need to employ a calibration pattern or human intervention, allowing to implement the method easily while saving the computations related to the image processing.

Keywords: Hand-eye calibration · Geometric Algebra · Kinematics

1 Introduction

Cameras are sensors widely used in robotics to acquire information from the environment, like the position of objects, obstacles, and other related objects that may imply some level of interaction with the robot. In this topic, hand-eye calibration is a well-known problem, where the key idea is finding the rigid-body transformation between the camera and the robot reference frame. Currently, there is a rich literature on these methods, which includes the use of calibration patterns and measurements taken from both camera and Inertial Measurement Units (IMU) sensors. Despite the effectiveness of these methods, most of them require several manual manipulations developed by users, which in practice complicate the calibration process.

This paper proposes a real-time method for estimating the camera-grasping point transformation while avoiding the processing of the camera image frames. Instead, our approach is based on the acquired IMU signals: linear acceleration, and angular velocity. First, we impose a sequenced prescribed movement at each joint of the robot arm, and then we use the linear acceleration and angular velocity to compute the transformation. The proposed algorithm is built using the Geometric Algebra (GA) mathematical framework, in which many geometric primitives can be operated naturally and directly.

D. W. Silva et al. (Eds.): ICACGA 2022, LNCS 13771, pp. 147–158, 2024.
https://doi.org/10.1007/978-3-031-34031-4_13

1.1 Related Works

The first challenge encountered during hand-eye calibration is usually the estimation of the pose of the camera relative to the world as the hand pose can be easily acquired by calculating the forward kinematics of the robot articulation. Depending on how the camera pose is estimated, the hand-eye calibration can be regarded as a target-based or target-less approach. In the target-based approach, the calibration employs physical objects with measured distances or dimensions (calibration objects). The basic idea is to estimate the camera pose by observing a set of 3D points provided by a calibration object and their corresponding 2D representations taken from the camera. Once the motions are calculated, the end-effector position is calculated using the forward-kinematic model. Finally, the homogeneous matrix is found through optimization algorithms [1,2].

In the target-less-based approach, the calibration employs just the information acquired from sensors with no measured distances or calibration patterns: In the structure from motion approach, the method employs the same approach as the target approach, but the motions from the camera are estimated by detecting and tracking a set of "features" through different camera frames, avoiding the use of calibration objects [3,4]. On the other hand, there is another technique known as tool motion tracking, which is a variation where an exact CAD model is employed to estimate the forward kinematics of the end-effector grasping an external tool, while the structure from the motion algorithm is employed to calculate the camera motions looking at the end-effector [5].

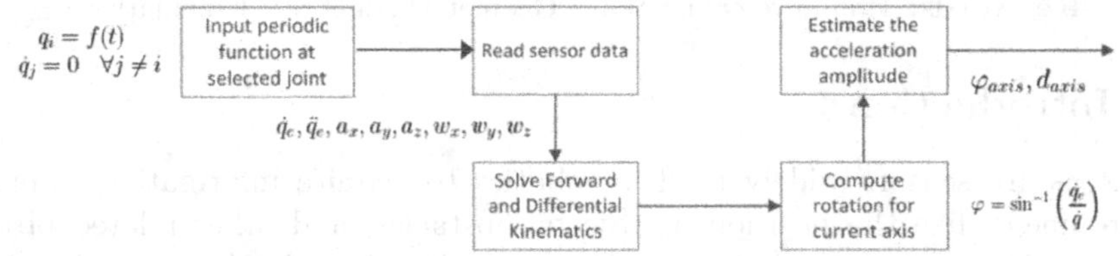

Fig. 1. Algorithm to compute the angle and distance for the selected joint.

1.2 Algorithm Description

The proposed calibration algorithm is summarized in Fig. 1. First, we select a periodic input function for each joint axis and lock all the other robot joints, while we acquire the linear acceleration and angular velocity from the camera attached to the end-effector using the information provided by the IMU. Then, we solve the forward and differential kinematics and compute the rotation transformation corresponding to the axis movement using the angular velocity. Finally, using linear acceleration, we estimate the translation part. Repeating this process for at least six robot joints will return the complete transformation between the grasping point of the end-effector and the camera IMU.

The rest of the paper is organized as follows: In Sect. 2, some preliminaries related to Geometric Algebra and Robots Kinematics are provided. In Sect. 3, the general mathematical framework employed by the proposed algorithm is presented to describe how the calculations are made to determine the camera position. In Sect. 4, a sample applied of the proposed algorithm is presented for a single joint axis. Finally, in Sect. 5 the conclusions and future work of the present paper are covered.

2 Geometric Algebra

A Geometric Algebra (GA) is a n dimensional space $\mathbb{G}_{p,q}$, where p, q stand for the number of basis vectors (e_i) which squares to 1 and -1 respectively, and fulfill $n = p + q$. In this algebra, the Clifford product ($*$) between two basis vectors can be defined as the operation [8,9]

$$e_i e_j = \begin{cases} 1 \text{ for } i = j \in 1, \cdots, p \\ -1 \text{ for } i = j \in p+1, \cdots, p+q \\ e_{ij} = e_i \wedge e_j \text{ for } i \neq j, \end{cases}$$

which leads to a basis for the entire algebra:

$$\{1\}, \{e_i\}, \{e_i \wedge e_j\}, \{e_i \wedge e_j \wedge e_k\}, \ldots, \{e_1 \wedge e_2 \wedge \ldots \wedge e_n\}. \tag{1}$$

2.1 $\mathbb{G}_{6,3}$ Geometry

A stereographic projection can be used to map points from $\mathbb{R}^1$ into points on $\mathbb{R}^{2,1}$, by considering a three-dimensional mapping from e_1, e_2 and e_3 a new $\mathbb{G}_{6,3}$ algebra is obtained:

$$\mathbb{G}_{6,3} = \mathbb{R}^{2,1} \times \mathbb{R}^{2,1} \times \mathbb{R}^{2,1} \tag{2}$$

As described in [7], $\mathbb{G}_{6,3}$ is a 9-dimensional geometry that has six bases squaring to 1, and three squaring to -1, respectively:

$$e_1^2, \cdots, e_6^2 = 1, \qquad e_7^2, \cdots, e_9^2 = -1. \tag{3}$$

Since this geometry allows the manipulation of quadratic entities, the points mapped to this space are denoted by the subindex Q. A point in the Euclidean space $(x, y, z) \in \mathbb{R}^3$ can be mapped to $\mathbb{G}_{6,3}$ by employing the transformation:

$$x_Q = \frac{2xe_1}{x^2+1} + \frac{2ye_2}{y^2+1} + \frac{2ze_3}{z^2+1} + \frac{x^2-1}{x^2+1}e_4 + e_7 + \frac{y^2-1}{y^2+1}e_5 + e_8 + \frac{z^2-1}{z^2+1}e_6 + e_9 \tag{4}$$

where the point at infinity e_∞ is given by $\lim_{x,y,z\to\infty}\{X_Q\}$ as:

$$e_\infty = \frac{1}{3}(e_{\infty x} + e_{\infty y} + e_{\infty z}), \tag{5}$$

where $e_{\infty x} = (e_4 + e_7), e_{\infty y} = (e_5 + e_8), e_{\infty z} = (e_6 + e_9)$. The above expressions represent the basis of a new coordinate frame at the infinity or vanishing coordinate frame, being the vanish vectors in x,y, and z direction. Based on [10], Eq. 4 can be rewritten as:

$$x_Q = xe_1 + ye_2 + ze_3 + \frac{1}{2}(x^2 e_{\infty x} + y^2 e_{\infty y} + z^2 e_{\infty z}) + e_o. \tag{6}$$

2.2 Translation

By definition a translator is given by:

$$T_x = 1 - \frac{1}{2}xe_1e_{\infty x},\ T_y = 1 - \frac{1}{2}ye_2e_{\infty y},\ T_z = 1 - \frac{1}{2}ze_3e_{\infty z}, \tag{7}$$

The mapping x_Q of the Eq. 6 can be represented by

$$x_Q = T_zT_yT_xe_o\widetilde{T}_x\widetilde{T}_y\widetilde{T}_z, \tag{8}$$

$$x_Q = Te_o\widetilde{T}, \tag{9}$$

where T represents the translator and can be written in exponential form as:

$$T = e^{-\frac{1}{2}(xe_1e_{\infty x}+ye_2e_{\infty y}+ze_3e_{\infty z})}. \tag{10}$$

2.3 Rotation

The rotation is the product of two reflections between non-parallel planes as:

$$R = \cos\left(\frac{\theta}{2}\right) - \sin\left(\frac{\theta}{2}\right) l = e^{-\frac{\theta}{2}l}, \tag{11}$$

here l denotes the rotation axis. The screw motion called motor $M = TR\tilde{T}$ represents the rotation related to an arbitrary axis L defined on

$$M = e^{-\frac{q}{2}L}, \tag{12}$$

where q represents the rotation angle or the translation in case of L at infinity. Any geometric entity can be rotated doing $x' = Mx\tilde{M}$. Conformal Geometric algebra can also used to perform this transformation. The reason of using QGA is not only to compute the position of the end effector, but also the orientation of it, because every point has an embedded coordinate frame as in [10].

3 Robot Differential Kinematics

3.1 Forward Kinematics

The forward kinematics of articulations is given by the multiplication of successive rotations, where each rotation is associated with every axis of the robot joints (Fig. 2):

$$x'_j = \prod_{i=1}^{j} M_i x_j \prod_{i=1}^{j} \tilde{M}_{j-i+1}. \tag{13}$$

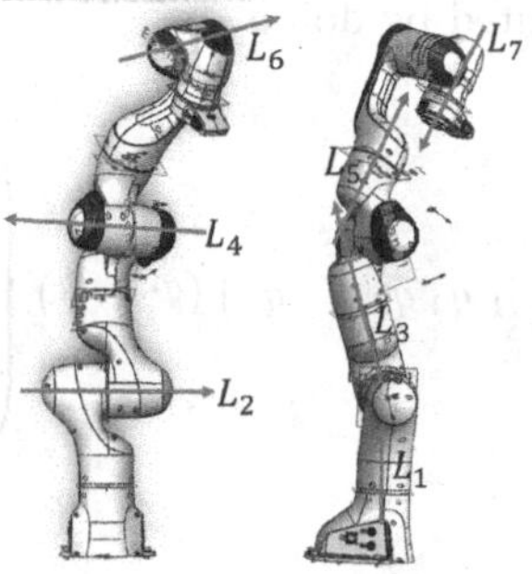

Fig. 2. Robot Axis for a Panda Robot 7 DoF.

3.2 Differential Kinematics

According to [6] the differential kinematics describes the linear velocity v of a point x given by:

$$v = \dot{x}' = \sum_{1}^{n} x' \cdot L'_i \dot{q}_i, \tag{14}$$

$$v = \begin{pmatrix} x' \cdot L'_1 & x' \cdot L'_2 & x' \cdot L'_3 & \cdots & x' \cdot L'_n \end{pmatrix} \begin{pmatrix} \dot{q}_1 \\ \dot{q}_2 \\ \dot{q}_3 \\ \vdots \\ \dot{q}_n \end{pmatrix}, \tag{15}$$

where L_i represents the $i-th$ axis of rotation as show in the Fig. 2, this equation can be written as:

$$v = J\dot{q}, \tag{16}$$

$$J = \left(x' \cdot L'_1 \; x' \cdot L'_2 \; x' \cdot L'_3 \cdots x' \cdot L'_n\right), \tag{17}$$

where:

$$L'_j = \prod_{i=1}^{j-1} M_i L_j \prod_{I=1}^{j-1} \tilde{M}_{j-i}. \tag{18}$$

The computation of the linear acceleration a of the end-effector (camera) as a function of the joint accelerations can be determined using the derivatives of the Jacobian J as:

$$a = \dot{J}\dot{q} + J\ddot{q}, \tag{19}$$

Here $(\dot{J}\dot{q})$ can be computed by doing:

$$\dot{J}\dot{q} = \left(\dot{q}_1 \; \dot{q}_2 \; \dot{q}_3 \cdots \dot{q}_n\right) (\Phi + \Psi) \begin{pmatrix} \dot{q}_1 \\ \dot{q}_2 \\ \dot{q}_3 \\ \vdots \\ \dot{q}_n \end{pmatrix}, \tag{20}$$

where:

$$\Phi = x' \cdot \begin{pmatrix} L'_1 \cdot L'_1 & L'_1 \cdot L'_2 & \cdots & L_1 \cdot L'_n \\ L'_2 \cdot L'_1 & L'_2 \cdot L'_2 & \cdots & L'_2 \cdot L'_n \\ L'_3 \cdot L'_1 & L'_3 \cdot L'_2 & \cdots & L'_3 \cdot L'_n \\ \vdots & \vdots & \vdots & \vdots \\ L'_n \cdot L'_1 & L'_n \cdot L'_2 & \cdots & L'_n \cdot L'_n \end{pmatrix}, \tag{21}$$

and:

$$\Psi = \frac{1}{2} x' \cdot \left[\begin{pmatrix} L'_1 L'_1 & 0 & \cdots & 0 \\ L'_2 L'_1 & L'_2 L'_2 & \cdots & 0 \\ \vdots & \vdots & \ddots & \vdots \\ L'_n L_1 & L_n L'_2 & \cdots & L'_n L'_n \end{pmatrix} - \begin{pmatrix} L'_1 L'_1 & 0 & \cdots & 0 \\ L'_1 L'_2 & L'_2 L'_2 & \cdots & 0 \\ \vdots & \vdots & \ddots & \vdots \\ L_1 L_n & L'_2 L'_n & \cdots & L'_n L'_n \end{pmatrix} \right], \tag{22}$$

then, the acceleration is equal to

$$a = \dot{q}^T (\Phi + \Psi) \dot{q} + J\ddot{q}. \tag{23}$$

This computed position, velocity, and acceleration are compared with the measured position, velocity, and acceleration coming from the encoders, gyroscope, accelerometer, and solving for x, which gives us the position of the camera. In the next section, we will illustrate the method used by estimating the position of the camera using a single joint axis.

4 Algorithm Description

In this section we will describe the steps followed to estimate the position of the camera, we will show step by step using the information provided by a simulator created previously (which has the dynamic model of the isolated robot). To simplify the generated expressions, an example for a single axis is presented using the following procedure: First, the forward kinematics of the end-effector are calculated by manipulating a single joint. Second, the differential kinematics of the end-effector by manipulating the same joint are calculated. Third, the camera orientation is determined using the IMU signals. Finally, the above information is employed to determine the camera position relative to the manipulated joint.

4.1 Compute Forward Kinematics for a Single Joint Movement

First, we produce a periodic motion in one of the joints, without moving the rest of the motors. For the joint q_1, any periodic motion can be selected, for instance:

$$q_1 = \frac{3\pi}{10}(1 - \cos(0.7\pi t + \pi)). \tag{24}$$

Using this equation to move the joint, the end-effector rotates following a circular trajectory around the center of the joint axis, going back and forward over a quarter of the circle. The value of the radius d in the estimated circle is defined by the position of the camera on the robot. Figure 3 shows the x and z coordinate respectively of the end-effector as a result of the forward kinematics simulation, recreated by using the encoder information of every joint. In this case, the position of (x, z) coordinates is given by the equations of the forward kinematics (13).

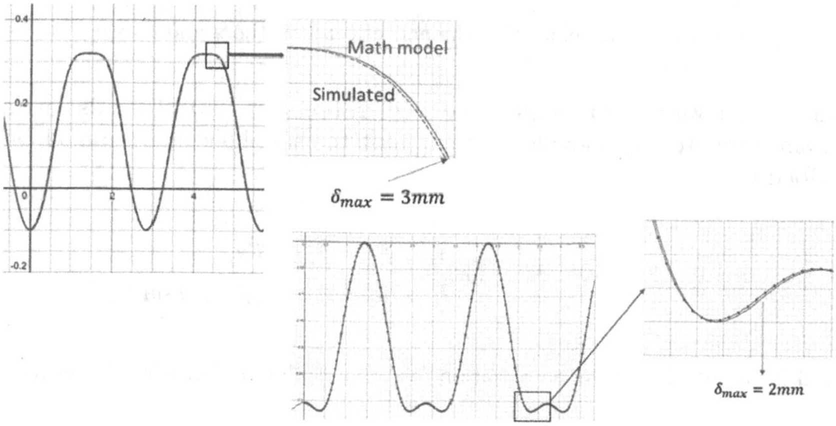

Fig. 3. X and Y - axis movement using the forward kinematics model feed with the encoder measurements of the joints.

4.2 Compute Differential Kinematics

After computing the kinematics, we must compute the differential kinematic to estimate the linear velocity of the end-effector using Eq. (14), in this example, we are moving only the first joint, then in this particular case the Eq. 2, can also be computed as follows:

$$v = \begin{pmatrix} x' \cdot L_1' & x' \cdot L_2' & x' \cdot L_3' & \cdots & x' \cdot L_n' \end{pmatrix} \begin{pmatrix} \dot{q}_1 \\ \dot{q}_2 \\ \dot{q}_3 \\ \vdots \\ \dot{q}_n \end{pmatrix} \quad \xrightarrow{\text{Moving only } q_1} \quad \begin{aligned} v_z &= -d\frac{3(0.7)\pi^2}{10}\cos(q)\left(\sin(0.7\pi t + \pi)\right) \\ v_x &= -d\frac{3(0.7)\pi^2}{10}\sin(q)\left(\sin(0.7\pi t + \pi)\right) \end{aligned}$$

Fig. 4. Linear velocity of the end-effector as a function of the input joint q_1

Where d represents the unknown distance from the camera to the rotation axis. In the following graph the linear velocity for the two end-effector axis is presented $(v_x, 0, v_z)$:

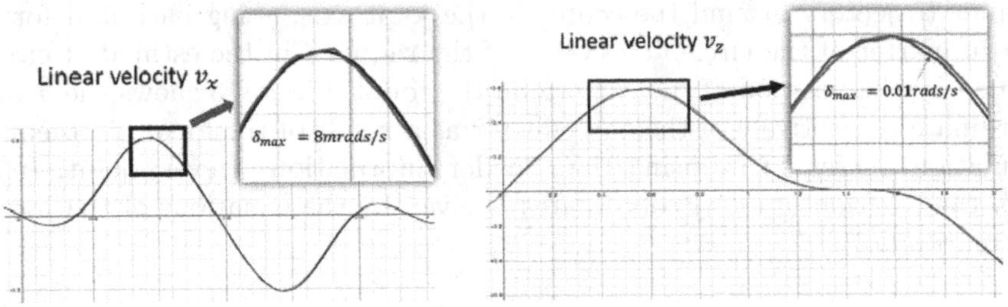

Fig. 5. Linear velocity of the end-effector in the x and z axis.

In this step we have to compute the acceleration (Eq. (23)). For this particular case, since we are only moving a single joint the simplified math model can be described as:

$$a = \dot{q}^T\left(\Phi + \Psi\right)\dot{q} + J\ddot{q} \quad \Longrightarrow \quad \begin{aligned} a_x &= d\sin(q)\dot{q}^2 - d\cos(q)\ddot{q} \\ a_z &= d\cos(q)\dot{q}^2 + d\sin(q)\ddot{q} \end{aligned}$$

Fig. 6. Computation of the acceleration for a particular case of one axis movement.

A comparison between the estimated and modeled acceleration is presented in Fig. 7. The estimated acceleration is given by the red line, while the mathematical model is given by the blue line. As can be seen, the math model can describe the acceleration behavior properly.

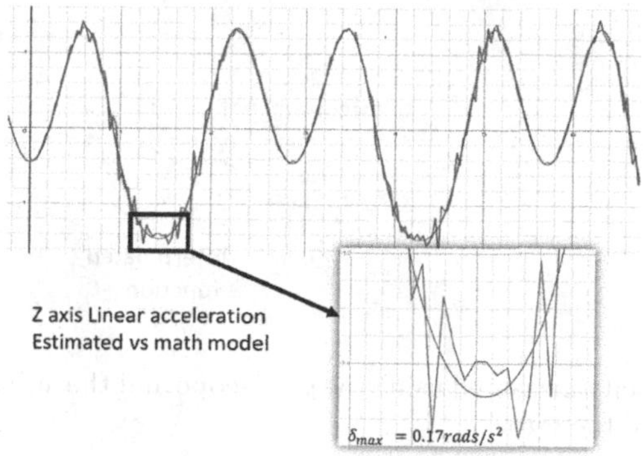

Fig. 7. Linear acceleration in z-axis. Mathematical model vs the differential kinematics estimation of the acceleration based on the encoder of the joint position through time.

4.3 Estimation of the Camera Orientation

Ideally, when the IMU sensor is perfectly aligned with the axis of rotation, every rotation applied on a single axis of the end-effector has no components with other axis. Nevertheless, this is a strange situation in the real life. Whereby it is required finding the orientation of the camera sensor respecting to the end-effector axis. In this step we show one angle computation because we are stimulating the rotation on this axis, to find the other angle we should repeat this step rotating axis by axis, the following equation represents the original rotation speed induced on the robot multiplied by h, where h is a scalar factor that we will adjust to minimize the error interpolating the captured signal.

$$\dot{q}_e = 3h\frac{\pi^2}{10}(0.7)\sin(0.7\pi t + \pi). \tag{25}$$

Figure 8 shows an example of the error minimization for the interpolation of the function, this interpolation helps us to estimate with high precision the measured signal (removing the error of $0.03r/s$ in our experiments). This function $\dot{q}_e$ was generated by the misalignment of the sensor and the angle of miss alignment, which is given by

$$h = \sin(\phi) = \sin^{-1}(\dot{q}_e/\dot{q}). \tag{26}$$

In our example it was h=0.04 with an estimated angle $\phi = 2.29^o$, this orientation in the accelerometer will produce a constant component induced by the gravity in the x-axis this offset according to:

$$a_g x = gh = g\sin(\phi), \tag{27}$$

in this example for instance

$$a_g x = 0.04 * 9.81m/s^2 = 0.39m/s^2, \tag{28}$$

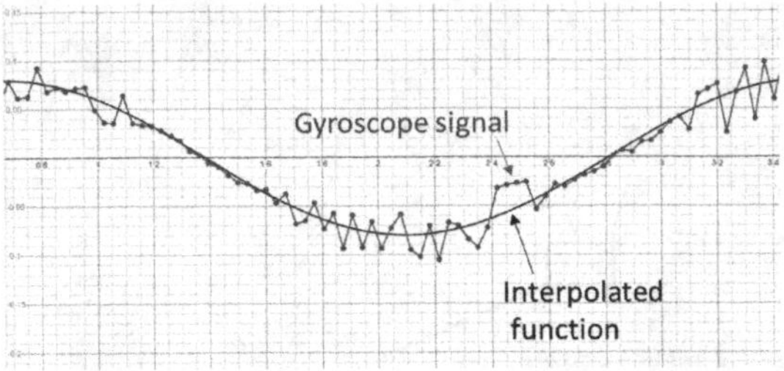

Fig. 8. Angular velocity measured using the gyroscope and the adjusted by interpolated function to minimize the error.

since the IMU is rotating due to the stimulation we introduced to the robot, the accelerations mathematically computed for each axis should be mapped to the IMU new axis and the offset introduced by the gravity should be incorporated as:

$$a_x = d\sin(q)\,\dot{q}^2 - d\cos(q)\,\ddot{q} + a_{gx}, \tag{29}$$

$$a_z = d\cos(q)\,\dot{q}^2 + d\sin(q)\,\ddot{q} + a_{gz}. \tag{30}$$

the estimated acceleration on the rotated x axis is given by

$$a = a_x\cos(q) - a_z\sin(q). \tag{31}$$

graphically for this example is given by:

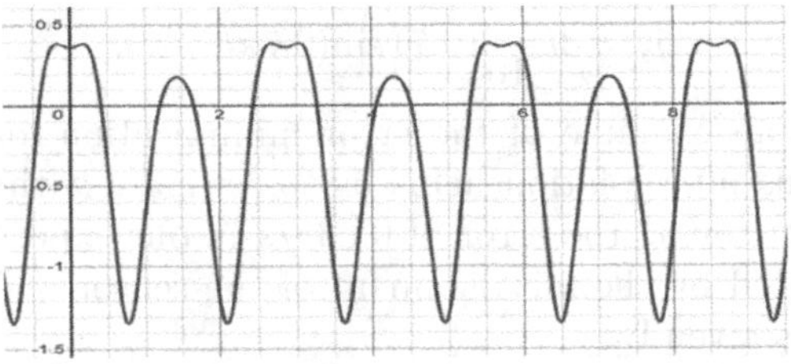

Fig. 9. Estimated Linear acceleration expected to be measured by the accelerometer.

4.4 Determine the Camera Relative Position

Finally, the information from the accelerometer is compared with the estimated previously. This comparison is presented in Fig. 10, where as can be seen, despite the accelerometer is noisy it allows us to minimize the error and find the distance from the sensor to the rotation axis (d). By moving the joint one we have

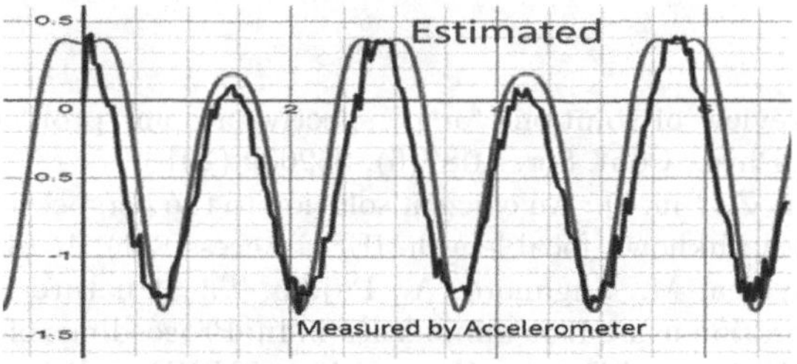

Fig. 10. Estimated Linear acceleration vs real acceleration captured by the accelerometer.

estimated d as the distance to this axis. The process can be repeated to every axis to get a more accurately result for all the axis, by intersecting the circular trajectories generated by every movement.

As the algorithm can determine the position of the camera using periodic movements on a single articulation, some considerations need to be addressed when performing the algorithm for multiple joints: First, every joint is manipulated using periodic inputs while maintaining the other fixed; the algorithm is developed by considering a single joint by a time. And second, to avoid matrix singularities it is important to be sure that there are no overlaps between the robot's axis.

5 Conclusions and Future Work

In this paper, a novel method for the estimation of the position and orientation of a camera attached on the end-effector was proposed, employing the movements measured by the camera's sensors (Accelerometer and Gyroscope) as the reaction of the stimulated motion on the robot joints.

- A computational online method to compute the transformation from the end-effector to a camera attached to it based on Conformal Geometric Algebra was proposed.
- The method only uses IMU information, avoiding the need to know intrinsic calibration parameters specific to each camera.
- The method's accuracy does not depend on the camera resolution, lens attached, update frame, or other camera configuration.
- Calibration can be achieved online, by developing free movements on the robot articulation which has the camera attached.
- Our method does not require a visual pattern and visual processing algorithm, also, this not requires human intervention.

References

1. Lu, X.X.: A review of solutions for perspective-n-point problem in camera pose estimation. J. Phys.: Conf. Ser. **1087**(5), 052009 (2018)
2. Zhou, B., Chen, Z., Liu, Q.: An efficient solution to the perspective n-point problem for camera with unknown focal length. IEEE Access **8**(1), 162838–162846 (2020)
3. Heller, J., Havlena, M., Sugimoto, A., Pajdla, T.: Structure from-motion based hand-eye calibration using L ∞ minimization. In: Proceedings of IEEE Conferences on Computer Vision and Pattern Recognition (CVPR), vol. 1 no.1, pp. 3497–3503 (2011)
4. Andreff, N., Horaud, R., Espiau, B.: Robot hand-eye calibration using structure-from-motion. Int. J. Robot. Res. **20**(3), 228–248 (2001)
5. Pachtrachai, K., Allan, M., et al.: Hand-eye calibration for robotic assisted minimally invasive surgery without a calibration object. In: International Conference on Intelligent Robots and Systems (IROS), pp. 2485–2491 (2016)
6. Bayro-Corrochano, E., Zamora-Esquivel, J.: Differential and inverse kinematics of robot devices using conformal geometric algebra. Robotica **25**(1), 43–61 (2007)
7. Zamora-Esquivel, J.: G 6,3 geometric algebra; description and implementation. Adv. Appl. Clifford Algebras **24**(2), 493–514 (2014). https://doi.org/10.1007/s00006-014-0442-8
8. Li, H., Hestenes, D., Rockwood, A.: Generalized homogeneous coordinates for computational geometry. Geometric Comput. Clifford Algebras, 27–59 (2001). https://doi.org/10.1007/978-3-662-04621-0_2
9. Bayro-Corrochano, E.: Geometric Algebra Applications, Vol. II. Springer, Cham (2020). https://doi.org/10.1007/978-3-030-34978-3
10. Zamora-Esquivel, J.: Vanishing vector rotation in quadric geometric algebra. Adv. Appl. Clifford Algebras **32**(4), 46 (2022). https://doi.org/10.1007/s00006-022-01234-y

Applications to Physics and Mathematics

The Supergeometric Algebra as the Language of Physics

Andrew J. S. Hamilton(✉)

JILA and Department of Astrophysical and Planetary Sciences, U. Colorado Boulder, Box 440, Boulder, CO 80309, USA
Andrew.Hamilton@colorado.edu
http://jila.colorado.edu/ ajsh/

Abstract. It is shown how the fermions and forces of Nature fit elegantly into the Supergeometric Algebra in 11+1 spacetime dimensions.

Keywords: Supergeometric Algebra · Geometric Algebra · Spinors

1 Introduction

The author's interest in spinors is sparked by the fact that spinors seem to be the fundamental objects from which physics is built. All known forms of matter (leptons and quarks) are made from spinors. And all known interactions, namely the three forces of the standard model, plus gravity, emerge from symmetries of spinors. The present paper, which is based on [1], shows how this works.

The present paper is a companion to [2], which presents a pedagogical introduction to the Supergeometric Algebra (SGA), the square root of the Geometric Algebra (GA). A central message of [2] is that a spinor, the fundamental representation of the group Spin(N) of rotations in N spacetime dimensions, is indexed by a bitcode with $[N/2]$ bits.

2 The Electron as a Dirac Spinor

A Dirac spinor is a spinor in 3+1 spacetime dimensions. It has $4/2 = 2$ bits, a boost bit ($\Uparrow$ or $\Downarrow$), and a spin bit ($\uparrow$ or $\downarrow$). A Dirac spinor is said to be right-handed if its boost and spin bits align, left-handed if they anti-align. Altogether, a Dirac spinor has $2^2 = 4$ complex components, or 8 real components. The 4 complex components of a Dirac electron, grouped into right- and left-handed (R and L) are:

$$e_R : \; e_{\Uparrow\uparrow} \, , \; e_{\Downarrow\downarrow} \; , \quad e_L : \; e_{\Downarrow\uparrow} \, , \; e_{\Uparrow\downarrow} \; . \tag{1}$$

The right- and left-handed components e_R and e_L are called the Weyl components of the electron, and they are massless. The massive electrons and positrons observed in Nature are linear combinations of right- and left-handed components.

D. W. Silva et al. (Eds.): ICACGA 2022, LNCS 13771, pp. 161–173, 2024.
https://doi.org/10.1007/978-3-031-34031-4_14

Electrons e and positrons $\bar{e}$ in their rest frames are complex conjugates of each other:

$$e_{\uparrow} = \frac{1}{\sqrt{2}}(e_{\Uparrow\uparrow} - ie_{\Downarrow\uparrow}) , \qquad e_{\downarrow} = \frac{1}{\sqrt{2}}(e_{\Downarrow\downarrow} - ie_{\Uparrow\downarrow}) , \tag{2a}$$

$$i\bar{e}_{\uparrow} = \frac{1}{\sqrt{2}}(e_{\Uparrow\uparrow} + ie_{\Downarrow\uparrow}) , \qquad i\bar{e}_{\downarrow} = \frac{1}{\sqrt{2}}(e_{\Downarrow\downarrow} + ie_{\Uparrow\downarrow}) . \tag{2b}$$

3 The Electron as a Spin(10) Spinor

That the chiral nature, right- or left-handed, of the electron should be taken seriously follows from the fact that only left-handed electrons feel the weak $\mathrm{SU}_L(2)$ force: right-handed electrons feel no weak force.

The standard model of physics is based on $\mathrm{U}_Y(1)\times\mathrm{SU}_L(2)\times\mathrm{SU}(3)$, the product of the hypercharge, left-handed weak, and color groups. At energies less the electroweak scale $\sim 100\,\mathrm{GeV}$, the symmetry of the hypercharge and weak groups breaks to the electromagnetic symmetry, $\mathrm{U}_Y(1)\times\mathrm{SU}_L(2) \to \mathrm{U}_{\mathrm{em}}(1)$. The method of electroweak symmetry breaking proposed by Weinberg (1967) [3], based on the so-called Higgs mechanism [4,5], has received spectacular experimental confirmation, culminating with the detection of the electroweak Higgs boson, with a mass 125 GeV, at the Large Hadron Collider in 2012 [6,7].

The success of the electroweak symmetry-breaking model prompted proposals in the mid-1970s that the three groups of the standard model would themselves become unified in a so-called Grand Unified Theory (GUT) group, at an energy that was estimated from the running of the three coupling parameters to be at $\sim 10^{14}$–10^{16} GeV. Three possible GUT groups fit the observed pattern of charges of fermions, of which the most unifying was Spin(10) (the covering group of SO(10)), first pointed out by [8,9]. The other two possible GUT groups, SU(5) proposed by [10], and the Pati-Salam group Spin(4) × Spin(6) proposed by [11], are subgroups of Spin(10).

As first pointed out by Wilczek in 1998 [12], and reviewed by Baez & Huerta [13], a spinor of Spin(10) is described by a bitcode with $10/2 = 5$ bits, consisting of 2 weak bits and 3 color bits. Wilczek and Baez & Huerta proposed different conventions for naming the bits. My own preference is to label the color bits r, g, b, following [13], and the weak bits y and z, inspired by the fact that y and z are infrared bands to be used by the Vera Rubin Observatory (the LSST) [14], for which first light is expected in 2025. The sequence $yzrgb$ is, in (inverse) order of wavelength,

$$y \sim 1000\,\mathrm{nm} , \quad z \sim 900\,\mathrm{nm} , \quad r \sim 600\,\mathrm{nm} , \quad g \sim 500\,\mathrm{nm} , \quad b \sim 400\,\mathrm{nm} . \tag{3}$$

This is an electron in Spin(10), labeled according to its $yzrgb$ bits (colored silver, bronze, red, green, blue):

$$\bar{e}_R : \downarrow\uparrow\downarrow\downarrow\downarrow , \quad \bar{e}_L : \uparrow\uparrow\downarrow\downarrow\downarrow , \quad e_R : \downarrow\downarrow\uparrow\uparrow\uparrow , \quad e_L : \uparrow\downarrow\uparrow\uparrow\uparrow . \tag{4}$$

Flipping all 5 $yzrgb$ bits flips between electron and positron. Flipping the y-bit flips between right- and left-handed. In the Spin(10) picture, each of the

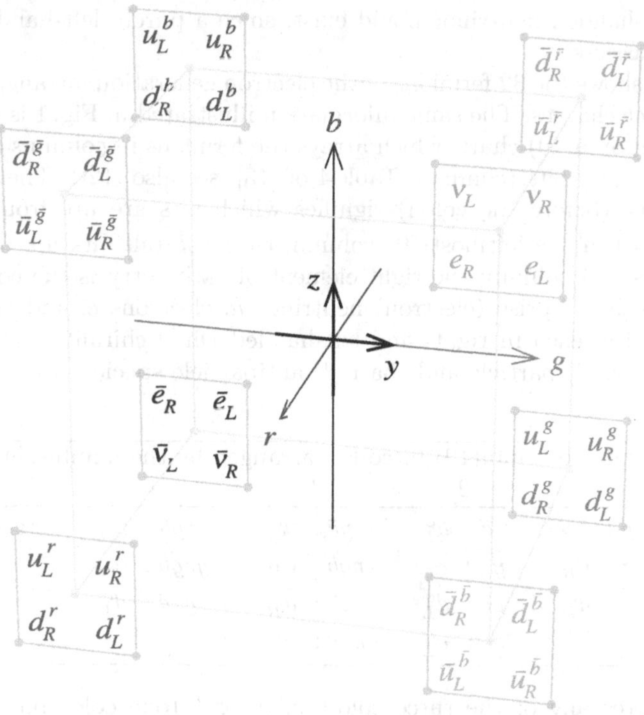

Fig. 1. The electron generation of 32 fermions arranged according to their Spin(10) *yzrgb* charges. A Spin(11, 1) version of this figure is Fig. 2

right- and left-handed components is itself a Weyl spinor, with two complex components.

In the standard model, fundamental fermions come in 3 generations, the electron, muon, and tauon generations. The three generations of fermions differ only in their masses: the standard model charges of each generation replicate each other. Only fermions come in three generations. The gauge bosons that mediate the forces, the interactions between fermions, are the same for all generations: there is only one "boson generation." This suggests that the 3 generations are not just another symmetry to be adjoined to the standard model. What causes the 3 generations remains a deep mystery of physics.

The lightest fermion generation is the electron generation. The fermions of the electron generation comprise 8 species, consisting of electrons and neutrinos, and 3 colors each of down and up quarks. Each of the 8 species comes in right- and left-handed varieties, and in particle and antiparticle versions, for a total of 32 fermion types. Each of those fermion types can be either spin-up or spin-down, for a total of 64 degrees of freedom. The pattern repeats for each of the 3 generations. Although no right-handed neutrino has been observed in Nature, the fact that the left-handed neutrino carries a non-zero mass strongly suggests that a right-handed neutrino should exist, since a purely left-handed neutrino would be massless.

Figure 1 shows the 32 fermions of the electron generation, arranged according to their *yzrgb* charges. The same information illustrated in Fig. 1 is tabulated in the following Spin(10) chart, which arrays the fermions in columns according to the number of up-bits (compare Table 4 of [13]; see also [12]). The left element of each entry (before the colon) signifies which bits are up, from – (no bits up, or ↓↓↓↓↓) in the leftmost (0) column, to *yzrgb* (all bits up, or ↑↑↑↑↑) in the rightmost (5) column; the right element of each entry is the corresponding fermion, which comprise (electron) neutrinos ν, electrons e, and up and down quarks u and d, each in right- and left-handed Dirac chiralities R and L, and each in (unbarred) particle and (barred) antiparticle species, a total of $2^5 = 32$ fermions:

Fermions and their Spin(10) bitcodes, arranged by the number of up-bits						
0	1	2	3	4	5	
$- : \bar{\nu}_L$	$y : \bar{\nu}_R$	$\bar{c} : \bar{u}^{\bar{c}}_L$	$y\bar{c} : \bar{u}^{\bar{c}}_R$	$zrgb : \nu_L$	$yzrgb : \nu_R$	(5)
	$z : \bar{e}_R$	$yz : \bar{e}_L$	$rgb : e_R$	$yrgb : e_L$		
	$c : d^c_R$	$yc : d^c_L$	$z\bar{c} : \bar{d}^{\bar{c}}_R$	$yz\bar{c} : \bar{d}^{\bar{c}}_L$		
		$zc : u^c_L$	$yzc : u^c_R$			

Here c denotes any of the three colors r, g, or b (one color bit up), while $\bar{c}$ denotes any of the three anticolors gb, br, or rg (two color bits up, the bit flip of a one-color-bit-up spinor).

The Spin(10) chart (5) of fundamental fermions is a Christmas puzzle of striking features. The most striking feature is that Dirac chirality (subscripted L or R in the chart) coincides with Spin(10) chirality. Spin(10) chirality counts whether the number of Spin(10) *yzrgb* up-bits is even or odd: the even and odd columns of the chart (5) have respectively left- and right-handed Spin(10) chirality. In any GA, chirality is the eigenvalue, ± 1, of the pseudoscalar (normalized by a phase so the eigenvalues are real). The coincidence of Dirac and Spin(10) chiralities suggests that the pseudoscalars of the Dirac and Spin(10) geometric algebras are somehow the same, in contrast to the usual assumption that the Dirac and GUT algebras are distinct.

The second striking feature of the Spin(10) chart (5) is that standard-model transformations connect fermions vertically, while Lorentz transformations connect fermions (for the most part) horizontally. For example, electrons e and positrons $\bar{e}$ are arrayed along one row of the chart. Every Spin(N) group has a subgroup SU($[N/2]$) that preserves the number of up-bits [15]. The columns of the chart (5) are SU(5) multiplets within Spin(10), with dimensions respectively 1, 5, 10, 10, 5, 1. The standard-model group is a subgroup of SU(5). All standard-model interactions preserve the number of Spin(10) up-bits. With standard-model transformations arrayed vertically and spacetime transformations arrayed horizontally, the chart (5) seems to be signalling that the two are somehow connected.

The third striking feature of the Spin(10) chart (5) is that right- and left-handed versions of the same species (for example electrons e_R and e_L) differ by a flip of the y-bit. In the Spin(10) picture, electroweak symmetry breaking is a loss of y-symmetry. The electroweak Higgs field carries y-charge, and it gives mass to fermions by flipping their y-bit. This is prettier than the somewhat abstruse traditional description $\mathrm{U}_Y(1) \times \mathrm{SU}_L(2) \to \mathrm{U}_{\mathrm{em}}(1)$ of electroweak symmetry breaking.

4 The Electron as a Spin(11,1) Spinor

The two guises of each generation of fermions, on the one hand as spinors of the Spin(3, 1) Dirac algebra under Lorentz transformations, and on the other hand as spinors of the Spin(10) algebra under standard model transformations, cry out for unification in a common algebra. Each of the 2^5 entries in the Spin(10) chart (5) is a Weyl fermion with 2 components, so the unified algebra, if it exists, must have 6 bits and 12 dimensions. And since the Dirac algebra has a time dimension while Spin(10) has none, one of the extra dimensions must be a time dimension, and the extra bit must be a boost bit. The algebra must be that of Spin(11, 1) in 11+1 spacetime dimensions. The extra bit can be labeled the t-bit, or time bit.

The conclusion that the unified algebra should have 11+1 spacetime dimensions conflicts with the usual assumption that the Dirac and Spin(10) algebras combine as a direct product, in which case the 3+1 dimensions of the Dirac algebra and the 10 dimensions of the Spin(10) algebra would yield 13+1 spacetime dimensions.

The standard assumption that Dirac and GUT algebras combine as a direct product is motivated by the Coleman-Mandula no-go theorem [16,17], which says, roughly, that any gauge group that contains the Poincaré group of spacetime symmetries and admits non-trivial analytic elastic scattering is necessarily a direct product of the Poincaré group and a commuting group of internal symmetries. The Coleman-Mandula theorem generalizes to higher dimensions [18].

However, if the grand unified group is Spin(11, 1), then all grand symmetries are spacetime symmetries, and there are no additional internal symmetries, so the higher-dimensional Coleman-Mandula theorem [18] is satisfied trivially. After grand symmetry breaking, the Coleman-Mandula theorem requires only that spacetime and *unbroken* internal symmetries combine as a direct product. In the present context, the Coleman-Mandula theorem requires that the Dirac and standard-model algebras combine as commuting subalgebras of the Spin(11, 1) algebra.

Encouragement that 11+1 dimensions is the right number comes from the period-8 Cartan-Bott periodicity [19–21] of geometric algebras. which guarantees that the discrete symmetries of the Spin(11, 1) algebra are the same as those of the Dirac Spin(3, 1) algebra: the spinor metric is antisymmetric, while the conjugation operator is symmetric.

If indeed the unified algebra is that of Spin(11, 1), then the Spin(10) chart (5) cannot be quite right as it stands. Diagnosing the problem, and then solving it,

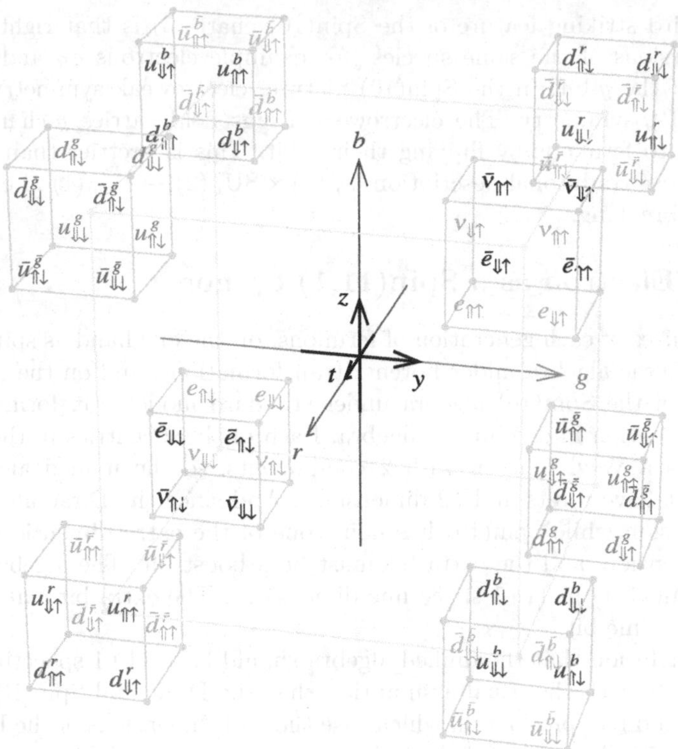

Fig. 2. The electron generation of $2^6 = 64$ fermions arranged according to their Spin(11, 1) *tyzrgb* charges. This is similar to Fig. 1, but with the addition of the *t*-bit

is tricky. It's a Christmas puzzle. The loophole in the chart is that it assigns a definite charge to each fermion based on its Spin(10) charges, whereas the example of Eq. (2) shows that fermions and antifermions, which have opposite charges, are linear combinations of the same chiral components. Fermions and antifermions are distinguished by the fact that they are complex conjugates of each other; more precisely, the antiparticle of a spinor ψ is the anti-spinor $\bar{\psi} \equiv C\psi^*$, where C is the conjugation operator. The conjugation operator in Spin(11, 1) proves to be the same as the conjugation operator in Spin(10): the conjugation operator in Spin(11, 1) flips all bits except the time bit, so flips all 5 *yzrgb* Spin(10) bits, as does the conjugation operator in Spin(10).

The solution to the unification problem is to replace each 2-component Weyl fermion in the Spin(10) chart (5) with a 2-component fermion with t-bit respectively up and down, with opposite Dirac boost but the same Dirac spin, a fermion and an antifermion. The Weyl companion of each fermion is identified as the fermion with all 6 *tyzrgb* bits flipped. This is similar to the Dirac algebra, where the Weyl companion of for example the right-handed electron $e_{\Uparrow\uparrow}$ is its all-bit-flip partner $e_{\Downarrow\downarrow}$.

This is a Dirac electron in Spin(11, 1), labeled according to its $tyzrgb$ bits (colored gold, silver, bronze, red, green, blue):

$$\begin{aligned} &\bar{e}_{\Downarrow\downarrow} : \uparrow\downarrow\uparrow\downarrow\downarrow\downarrow\ , \quad \bar{e}_{\Uparrow\downarrow} : \uparrow\uparrow\uparrow\downarrow\downarrow\downarrow\ , \quad e_{\Uparrow\uparrow} : \uparrow\downarrow\downarrow\uparrow\uparrow\uparrow\ , \quad e_{\Downarrow\uparrow} : \uparrow\uparrow\downarrow\uparrow\uparrow\uparrow\ , \\ &e_{\Uparrow\downarrow} : \downarrow\downarrow\uparrow\downarrow\downarrow\downarrow\ , \quad e_{\Downarrow\downarrow} : \downarrow\uparrow\uparrow\downarrow\downarrow\downarrow\ , \quad \bar{e}_{\Downarrow\uparrow} : \downarrow\downarrow\downarrow\uparrow\uparrow\uparrow\ , \quad \bar{e}_{\Uparrow\uparrow} : \downarrow\uparrow\downarrow\uparrow\uparrow\uparrow\ . \end{aligned} \tag{6}$$

Flipping the time bit t flips between electrons e and positrons $\bar{e}$. Flipping the y bit flips Dirac chirality. Flipping all 6 bits spatially rotates the spin of the electron (or positron) between up and down, preserving chirality.

Figure 2 illustrates one generation (the electron generation) of fermions of the standard model arranged according to their Spin(11, 1) $tyzrgb$ charges. The same information illustrated in Fig. 2 is tabulated in the following Spin(11, 1) chart of spinors, arranged in columns by the number of Spin(10) up-bits as in the earlier Spin(10) chart (5):

0	1	2	3	4	5
$- : \begin{matrix} \bar{\nu}_{\Uparrow\downarrow} \\ \nu_{\Downarrow\downarrow} \end{matrix}$	$y : \begin{matrix} \bar{\nu}_{\Downarrow\downarrow} \\ \nu_{\Uparrow\downarrow} \end{matrix}$	$\bar{c} : \begin{matrix} \bar{u}^{\bar{c}}_{\Uparrow\downarrow} \\ u^{c}_{\Downarrow\downarrow} \end{matrix}$	$y\bar{c} : \begin{matrix} \bar{u}^{\bar{c}}_{\Downarrow\downarrow} \\ u^{c}_{\Uparrow\downarrow} \end{matrix}$	$zrgb : \begin{matrix} \nu_{\Downarrow\uparrow} \\ \bar{\nu}_{\Uparrow\uparrow} \end{matrix}$	$yzrgb : \begin{matrix} \nu_{\Uparrow\uparrow} \\ \bar{\nu}_{\Downarrow\uparrow} \end{matrix}$
	$z : \begin{matrix} \bar{e}_{\Downarrow\downarrow} \\ e_{\Uparrow\downarrow} \end{matrix}$	$yz : \begin{matrix} \bar{e}_{\Uparrow\downarrow} \\ e_{\Downarrow\downarrow} \end{matrix}$	$rgb : \begin{matrix} e_{\Uparrow\uparrow} \\ \bar{e}_{\Downarrow\uparrow} \end{matrix}$	$yrgb : \begin{matrix} e_{\Downarrow\uparrow} \\ \bar{e}_{\Uparrow\uparrow} \end{matrix}$	
	$c : \begin{matrix} d^{c}_{\Uparrow\uparrow} \\ \bar{d}^{\bar{c}}_{\Downarrow\uparrow} \end{matrix}$	$yc : \begin{matrix} d^{c}_{\Downarrow\uparrow} \\ \bar{d}^{\bar{c}}_{\Uparrow\uparrow} \end{matrix}$	$z\bar{c} : \begin{matrix} \bar{d}^{\bar{c}}_{\Downarrow\downarrow} \\ d^{c}_{\Uparrow\downarrow} \end{matrix}$	$yz\bar{c} : \begin{matrix} \bar{d}^{\bar{c}}_{\Uparrow\downarrow} \\ d^{c}_{\Downarrow\downarrow} \end{matrix}$	
		$zc : \begin{matrix} u^{c}_{\Downarrow\uparrow} \\ \bar{u}^{\bar{c}}_{\Uparrow\uparrow} \end{matrix}$	$yzc : \begin{matrix} u^{c}_{\Uparrow\uparrow} \\ \bar{u}^{\bar{c}}_{\Downarrow\uparrow} \end{matrix}$		

(7)

whereas in the original Spin(10) chart (5) each entry was a 2-component Weyl spinor, in the Spin(11, 1) chart (7) the 2 components of each Weyl spinor appear in bit-flipped entries. For example, the right-handed electron e_R of the original chart is replaced by $e_{\Uparrow\uparrow}$, and its spatially rotated partner $e_{\Downarrow\downarrow}$ of the same chirality appears in the all-bit-flipped entry. Each entry still has two components, but in the Spin(11, 1) chart those two components differ by their t-bit; the upper component has t-bit up, the lower t-bit down. The net number of degrees of freedom remains the same, $2^6 = 64$.

In the unified Spin(11, 1) algebra, the Dirac boost and spin of a fermion are woven into the algebra, no longer dissociated from Spin(10). The Dirac boost $\Uparrow$ or $\Downarrow$ is the eigenvalue of the weak chiral operator $\varkappa_{tyz}$, which counts whether the number of tyz up-bits is odd or even. The Dirac spin $\uparrow$ or $\downarrow$ is the eigenvalue of the color chiral operator $\varkappa_{rgb}$, which counts whether the number of color rgb up-bits is odd or even. The weak and color chiral operators $\varkappa_{tyz}$ and $\varkappa_{rgb}$ are equal to weak and color pseudoscalars I_{tyz} and I_{rgb} modified by a phase factor to make their eigenvalues real:

$$I_{tyz} \equiv -i\gamma_t^+\gamma_t^-\gamma_y^+\gamma_y^-\gamma_z^+\gamma_z^- = -\varkappa_{tyz} \equiv -\gamma_t \wedge \gamma_{\bar{t}} \wedge \gamma_y \wedge \gamma_{\bar{y}} \wedge \gamma_z \wedge \gamma_{\bar{z}}\ , \tag{8a}$$

$$I_{rgb} \equiv \gamma_r^+\gamma_r^-\gamma_g^+\gamma_g^-\gamma_b^+\gamma_b^- = -i\varkappa_{rgb} \equiv -i\gamma_r \wedge \gamma_{\bar{r}} \wedge \gamma_g \wedge \gamma_{\bar{g}} \wedge \gamma_b \wedge \gamma_{\bar{b}}\ . \tag{8b}$$

The 12-dimensional pseudoscalar J is the product of the boost operator I_{tyz} and the spin operator I_{rgb},

$$J \equiv I_{tyz} I_{rgb} = -i \gamma_t^+ \gamma_t^- \gamma_y^+ \gamma_y^- \gamma_z^+ \gamma_z^- \gamma_r^+ \gamma_r^- \gamma_g^+ \gamma_g^- \gamma_b^+ \gamma_b^-$$
$$= i\varkappa_{12} \equiv i \gamma_t \wedge \gamma_{\bar{t}} \wedge \gamma_y \wedge \gamma_{\bar{y}} \wedge \gamma_z \wedge \gamma_{\bar{z}} \wedge \gamma_r \wedge \gamma_{\bar{r}} \wedge \gamma_g \wedge \gamma_{\bar{g}} \wedge \gamma_b \wedge \gamma_{\bar{b}} \, . \tag{9}$$

In the Dirac algebra, the charge of a chiral fermion is ambiguous: a fermion and its antifermion partner, which have opposite charges, are linear combinations of the same chiral components, Eq. (2). The t-bit removes the ambiguity, specifying whether a fermion is going forwards or backwards in time. The charge of a fermion is determined unambiguously by its 6 $tyzrgb$ bits. In Spin(10), the standard-model charges of a fermion can be read off from its 5 $yzrgb$ bits. In Spin(11, 1), the standard-model charges are equal to Spin(10) charges multiplied by the color chiral operator $\varkappa_{rgb}$, as is evident from the fact that the spinors in the Spin(10) chart (5) are fermions (unbarred) or antifermions (barred) depending on whether their color chirality is odd or even.

In Spin(10), the 5 standard-model charges are eigenvalues of the 5 diagonal bivector generators of Spin(10),

$$\tfrac{1}{2} \gamma_i^+ \wedge \gamma_i^- = \tfrac{i}{2} \gamma_i \wedge \gamma_{\bar{\imath}} \, , \quad i = y, z, r, g, b \, . \tag{10}$$

In Spin(11, 1), standard-model charges are eigenvalues of the 5 diagonal bivectors (10) multiplied by the color chiral operator $\varkappa_{rgb}$. A consistent way to implement this modification, that leaves the bivector algebra of the standard model unchanged, is to multiply all imaginary bivectors $\gamma_i^+ \gamma_j^-$ in the Spin(10) geometric algebra by $\varkappa_{rgb}$, while leaving all real bivectors $\gamma_i^+ \gamma_j^+$ and $\gamma_i^- \gamma_j^-$ unchanged,

$$\gamma_i^+ \gamma_j^- \to \gamma_i^+ \gamma_j^- \varkappa_{rgb} \quad i, j = y, z, r, g, b \, . \tag{11}$$

Equivalently, replace the imaginary i in all Spin(10) bivectors by the color pseudoscalar $-I_{rgb} = i\varkappa_{rgb}$, Eq. (8b). A key point that allows this adjustment to be made consistently is that $\varkappa_{rgb}$ commutes with all standard-model bivectors. Note that $\varkappa_{rgb}$ does not commute with SU(5) bivectors that transform between leptons and quarks; but that is fine, because SU(5) is not an unbroken symmetry of the standard model.

The definitive proof that unification in Spin(11, 1) is consistent comes from expressing the 4 orthonormal vectors γ_m, $m = 0, 1, 2, 3$, of the Dirac algebra in terms of the 12 orthonormal vectors $\gamma_i^\pm$, $i = t, y, z, r, g, b$ of the Spin(11, 1) algebra:

$$\gamma_0 = i \gamma_t^- \, , \tag{12a}$$
$$\gamma_1 = \gamma_y^- \gamma_z^- \gamma_r^+ \gamma_g^+ \gamma_b^+ \, , \tag{12b}$$
$$\gamma_2 = \gamma_y^- \gamma_z^- \gamma_r^- \gamma_g^- \gamma_b^- \, , \tag{12c}$$
$$\gamma_3 = \gamma_t^+ \gamma_y^+ \gamma_y^- \gamma_z^+ \gamma_z^- \, . \tag{12d}$$

The Dirac vectors (12) all have grade 1 mod 4 in the Spin(11, 1) algebra. The multiplication rules for the Dirac vectors γ_m given by Eq. (12) agree with the usual multiplication rules for Dirac γ-matrices: the vectors γ_m anticommute, and their scalar products form the Minkowski metric. All the spacetime vectors γ_m commute with all standard-model generators modified per (11). The Dirac pseudoscalar I coincides with the Spin(11, 1) pseudoscalar J, Eq. (9),

$$I \equiv \gamma_0\gamma_1\gamma_2\gamma_3 = J \ . \tag{13}$$

Thus the Dirac and standard-model algebras are subalgebras of the Spin(11, 1) geometric algebra, such that all Dirac generators commute with all standard-model generators modified per (11), consistent with the Coleman-Mandula theorem.

The time dimension (12a) is just a simple vector in the Spin(11, 1) algebra, but the 3 spatial dimensions (12b)–(12d) are all 5-dimensional. The spatial dimensions share a common 2-dimensional factor $\gamma_y^-\gamma_z^-$. Aside from that common factor, each of the 3 spatial dimensions is itself 3-dimensional: $\gamma_r^+\gamma_g^+\gamma_b^+$, $\gamma_r^-\gamma_g^-\gamma_b^-$, and $\gamma_t^+\gamma_y^+\gamma_z^+$.

5 Predictions of the Spin(11,1) Theory

A first response to any new theory is, Does it make any predictions? Much of [1] is devoted to answering this question. The specific question is, what predictions can be made if the Grand Unified group is Spin(11, 1) and no additional ingredients are admitted? The condition of no additional ingredients is highly restrictive.

The end result is that the theory predicts the following sequence of symmetry breakings, at energies determined by the running of coupling parameters:

$$\begin{aligned}\mathrm{Spin}(11,1) \xrightarrow[??]{} \mathrm{Spin}(10,1) \xrightarrow[10^{15}\,\mathrm{GeV}]{} \mathrm{Spin}(4) \times \mathrm{Spin}(6) \xrightarrow[10^{12}\,\mathrm{GeV}]{} \\ \mathrm{U}_Y(1) \times \mathrm{SU}_L(2) \times \mathrm{SU}(3) \xrightarrow[100\,\mathrm{GeV}]{} \mathrm{U}_{\mathrm{em}}(1) \times \mathrm{SU}(3) \ .\end{aligned} \tag{14}$$

The top line of the sequence (14) is the prediction, while the bottom line is the standard model.

The addition of the 6th bit, the time bit t, to the 5 $yzrgb$ bits of Spin(10) adjoins to the bivectors of Spin(10) additional bivectors involving either or both of the two extra dimensions $\gamma_t^\pm$. Of those bivectors, four commute with all the Dirac vectors γ_m defined by Eq. (12), and could therefore potentially play a role in the standard model. The four happen to have precisely the properties of the 4-component electroweak Higgs multiplet required by the Weinberg [3] model of electroweak symmetry breaking, motivating the identification of the electroweak Higgs field $\boldsymbol{H}$ as (with the bivectors being understood to be modified per (11) as usual)

$$\boldsymbol{H} \equiv H^{i^\pm}\gamma_t^+\gamma_i^\pm \ , \quad i = y, z \ . \tag{15}$$

Electroweak symmetry breaking occurs when the Higgs field acquires a vacuum expectation value $\langle \boldsymbol{H} \rangle$ proportional to $\gamma_t^+\gamma_y^-$,

$$\langle \boldsymbol{H} \rangle = \langle H \rangle \gamma_t^+\gamma_y^- \varkappa_{rgb} \tag{16}$$

(the factor of $\varkappa_{rgb}$ from the modification (11), omitted from (15), is included here to avoid possible confusion). The electroweak Higgs field (16) carries y-charge, breaks y-symmetry, and generates masses for fermions by flipping their y-bit. The three remaining components of the Higgs multiplet are absorbed into the longitudinal components of the electroweak $W^{\pm}$ and Z bosons, giving them mass, while leaving the photon massless.

As long as spacetime is 4-dimensional, as in today's world, any intermediate gauge group on the path to grand unification must commute with all the Dirac vectors (12). The largest subgroup of Spin(11, 1) whose bivector generators, modified per (11), all commute with the Dirac vectors (12) is a product of weak and color groups Spin(5) × Spin(6) generated by, respectively, the ten bivectors formed from γ_t^+ and $\gamma_i^{\pm}$, $i = y, z$, and the fifteen bivectors formed from $\gamma_i^{\pm}$, $i = r, g, b$. However, the subset of four Spin(5) bivectors $\gamma_t^+\gamma_i^{\pm}$ fail to commute with the field (18) that mediates grand symmetry breaking, so those bivectors are already eliminated as gauge fields (but not as scalar fields) at grand symmetry breaking. Thus the largest possible group on the path to grand unification is the product of extended weak and color groups, the Pati-Salam [11] group

$$\mathrm{Spin}(4) \times \mathrm{Spin}(6) . \tag{17}$$

The running of the three coupling parameters of the standard model indicates that unification to Spin(4) × Spin(6) should happen at 10^{12} GeV, so that unification does in fact happen. The energy 10^{12} GeV is comparable to that of the most energetic cosmic rays observed [22,23].

The general principles underlying symmetry breaking by the Higgs mechanism are: the Higgs field before symmetry breaking must be a scalar (spin 0) multiplet of the unbroken symmetry; one component of the Higgs multiplet must acquire a non-zero vacuum expectation value; components of the Higgs multiplet whose symmetry is broken are absorbed into longitudinal components of the broken gauge (spin 1) fields, giving those gauge fields mass; and unbroken components of the Higgs field persist as scalar fields, potentially available to mediate the next level of symmetry breaking.

In the sequence (14) of symmetry breakings, the primordial Higgs field is a scalar 66-component bivector multiplet of Spin(11, 1). The primordial Higgs field is the parent of all the other Higgs fields.

The field that breaks grand symmetry proves to be the Majorana-Higgs field $\langle \boldsymbol{T} \rangle$ proportional to the bivector $\gamma_t^+\gamma_t^-$,

$$\langle \boldsymbol{T} \rangle = -i\langle T \rangle \gamma_t^+\gamma_t^- \varkappa_{rgb} , \tag{18}$$

the imaginary i coming from the time vector being timelike, $\gamma_0 = i\gamma_t^-$, and the factor $\varkappa_{rgb}$ from the modification (11). The Majorana-Higgs field (18) has the property that it commutes with all Spin(4) × Spin(6) fields, and fails to commute with all Spin(10) fields not in Spin(4) × Spin(6).

The Majorana-Higgs field $\langle \boldsymbol{T} \rangle$ carries t-charge, and is able to flip the t-bit of the right-handed neutrino, flipping the neutrino between itself and its left-handed antineutrino partner of opposite boost, giving the right-handed neutrino a so-called Majorana mass. Only the right-handed neutrino can acquire a Majorana mass, because only the right-handed neutrino possesses no conserved standard-model charge. A large Majorana mass for the right-handed neutrino can generate a small mass for the left-handed neutrino by the well-known see-saw mechanism proposed by [24].

The Majorana-Higgs field $\langle \boldsymbol{T} \rangle$ is available to drive cosmological inflation at the GUT scale. The running of weak and color coupling parameters implies that grand unification occurs at an energy of $3 \times 10^{14}\,\text{GeV}$. This unification energy is well within the upper limit on the energy scale $\mu_{\text{inflation}}$ of cosmological inflation inferred from the upper limit to B-mode polarization power in the cosmic microwave background measured by the Planck satellite [25, eq. (26)],

$$\mu_{\text{inflation}} \leq 2 \times 10^{16}\,\text{GeV} \ . \tag{19}$$

The first step in the symmetry-breaking sequence (14) is $\text{Spin}(11,1) \to \text{Spin}(10,1)$. The problem is that the $\text{Spin}(11,1)$ bivectors $\gamma_t^+ \gamma_i^\pm$ cannot be generators of a gauge (spin 1) field after grand symmetry breaking, because if they were, then their Higgs scalar (spin 0) counterparts would be absorbed into the gauge field after grand symmetry breaking, whereas the scalar counterparts apparently persist in the form of the electroweak Higgs multiplet (15).

The vector γ_t^+, the spatial vector companion to the time vector $\gamma_0 = i\gamma_t^-$, stands out as the only spatial vector missing from the Spin(10) algebra. The solution to γ_t^+ not generating any gauge symmetry is to assert that it behaves as a scalar dimension prior to grand symmetry breaking, so that the grand unified group is $\text{Spin}(10,1)$, not $\text{Spin}(11,1)$. Why this should be so is unclear. Possibly a non-trivial quantum field theory in higher dimensions requires 10+1 dimensions, as in M theory. Spin algebras live naturally in even dimensions, and one way to accommodate a spin algebra in 10+1 dimensions is to embed it in one extra dimension, 11+1 dimensions, and to treat the extra dimension, here γ_t^+, as a scalar. The scalar dimension γ_t^+, which anticommutes with the other 11 dimensions, plays the role of a time-reversal operator, essential to a consistent quantum field theory. It remains to be seen whether the $\text{Spin}(11,1)$ model can in fact be accommodated in M theory.

References

1. Hamilton, A.J.S., McMaken, T.: Unification of the four forces in the Spin(11,1) geometric algebra. Phys. Scr. **98**, 085306 (2023). https://doi.org/10.1088/1402-4896/acdaff
2. Hamilton, A.J.S.: The Supergeometric algebra: the square root of the geometric algebra. In: ENGAGE 2022 workshop at CGI 2022 (2022, accepted)
3. Weinberg, S.: A model of leptons. Phys. Rev. Lett. **19**, 1264–1266 (1967). https://doi.org/10.1103/PhysRevLett.19.126
4. Englert, F., Brout, R.: Broken symmetry and the mass of gauge vector mesons. Phys. Rev. Lett. **13**, 321–323 (1964). https://doi.org/10.1103/PhysRevLett.13.321

5. Higgs, P.W.: Broken symmetries and the masses of gauge bosons. Phys. Rev. Lett. **13**, 508–509 (1964) https://doi.org/10.1016/0550-3213(74)90038-8
6. Aad, G., et al.: (ATLAS Collaboration, 2934 authors): Observation of a new particle in the search for the Standard Model Higgs boson with the ATLAS detector at the LHC. Phys. Lett. B **716**, 1–29 (2012). https://doi.org/10.1016/j.physletb.2012.08.020
7. Chatrchyan, S., et al. (CMS Collaboration, 2885 authors): Observation of a new boson at a mass of 125 GeV with the CMS experiment at the LHC. Phys. Lett. B **716**, 30–61 (2012). arXiv:1207.7235, https://doi.org/10.3847/1538-4357/ab042c, https://doi.org/10.1016/j.physletb.2012.08.021
8. Georgi, H.: In Particles and Fields - 1974. In: Carlson, C.E. (ed.) Proceedings of Meeting of the APS Division of Particles and Fields, Williamsburg, Virginia (AIP, New York) (1975)
9. Fritzsch, H., Minkowski, P.: Unified interactions of leptons and hadrons. Ann. Phys. **93**, 193–266 (1975)
10. Georgi, H., Glashow, S.: Unity of all elementary-particle forces. Phys. Rev. Lett. **32**, 438–441 (1974)
11. Pati, J.C., Salam, A.: Lepton number as the fourth color. Phys. Rev. D **10**, 275–289 (1974)
12. Wilczek, F.: SO(10) marshals the particles. Nature **394**, 15 (1998). DOIurlhttps://doi.org/10.1038/27761
13. Baez, J.C., Huerta, J.: The algebra of grand unified theories. Bull. Am. Math. Soc. **47**, 483–552 (2010). arXiv:0904.1556
14. Ivezić, Ž., et al. (LSST Collaboration, 313 authors): LSST: from science drivers to reference design and anticipated data products. Astrophys. J. **873**, 11 (2019). arXiv:0805.2366, https://doi.org/10.3847/1538-4357/ab042c, arXiv:1806.00612 , https://doi.org/10.1140/epjc/s10052-018-5844-7
15. Atiyah, M.F., Bott, R., Shapiro, A.: Clifford modules. Topology **3**, 3–38 (1964)
16. Coleman, S., Mandula, J.: All possible symmetries of the S matrix. Phys. Rev. **159**, 1251–1256 (1967)
17. Mandula, J.E.: Coleman-Mandula theorem. Scholarpedia **10**(6), 7416 (1015). https://doi.org/10.4249/scholarpedia.7476
18. Pelc, O., Horwitz, L.P.: Generalization of the Coleman-Mandula theorem to higher dimension. J. Math. Phys. **38**, 139–172 (1997). https://doi.org/10.1063/1.531846
19. Eduard Study & Élie Cartan, Nombres complexes. Encyclopédie des sciences mathématiques; tome 1, volume 1, fascicule 3, pp. 353–411 (1908)
20. Bott, R.: The periodicity theorem for the classical groups and some of its applications. Adv. Math. **4**, 353–411 (1970). https://doi.org/10.1016/0001-8708(70)90030-7
21. Coquereaux, R.: Modulo 8 periodicity of real Clifford algebras and particle physics. Phys. Lett. B **115**, 389–395 (1982). https://doi.org/10.1016/0370-2693(82)90524-X
22. Telescope Array Collaboration: 126 authors, Indications of intermediate-scale anisotropy of cosmic rays with energy greater than 57 EeV in the Northern Sky measured with the surface detector of the telescope array experiment, Astrophys. J. **790**, L21 (2014). arXiv:1404.5890, https://doi.org/10.1088/2041-8205/790/2/l21
23. Batista, R.A., et al. (16 authors): Open questions in cosmic-ray research at ultra-high energies. Front. Astron. Space Sci. **6**, 23 (2019). arXiv:1903.06714, https://doi.org/10.3389/fspas.2019.00023, https://doi.org/10.1103/physrevd.95.012004

24. Gell-Mann, M., Ramond, P., Slansky, R.: Complex spinors and unified theories. In: Freedman, D.Z., van Nieuwenhuizen, P. (eds.) Supergravity, pp. 315–321. North Holland, Amsterdam (1979). arXiv:1306.46694
25. Planck Collaboration (246 authors): Planck 2015 results. XX. Constraints on inflation, Astron. Astrophys. **594**, A20 (2016). arXiv:1502.02114, https://doi.org/10.1051/0004-6361/201525898

Geometric Algebra Speaks Quantum Esperanto, I

Sebastian Xambó-Descamps(✉)

IMTech/UPC & BSC/CNS, Jordi Girona 1-3, 08034 Barcelona, Spain
sebastia.xambo@upc.edu

Abstract. The fundamental Stern-Gerlach (SG) experiments suggest that the (pure) states of a q-bit are the points of the unit sphere S^2 (in some suitable system of units), with a distinguished vector corresponding to the direction of the magnetic field. The goal of this paper is to elucidate the Hermitian structure of the algebra of geometric quaternions $\mathbf{H} = \mathcal{G}_3^+$ (that is, the even algebra of the geometric algebra of the Euclidean 3D space) which allows to regard it as the Hilbert space of the q-bit. The main results are phrased in terms of an explicit *ket map* $\kappa : \mathbf{H} \to E_3$ such that $|\kappa(\mathfrak{q})| = |\mathfrak{q}|$ for all $\mathfrak{q} \in \mathbf{H}$, and include: that $\kappa(\mathfrak{q}') = \kappa(\mathfrak{q})$ if and only if $\mathfrak{q}' \equiv \mathfrak{q}$ (this relation denotes that the two quaternions differ by a phase factor –a unit *geometric* complex number); that κ is onto; a check that the computed probabilities obey the statistics of the SG experiments; and a recall of the relations between the multiplicative group $\mathbf{H}^\times$ and the rotation group $\mathrm{SO}(E_3)$. A sequel paper will explore other facets of the proposed analysis, including the study of the polarization states of electromagnetic waves and more complex spin systems. In conclusion:

Jes, geometria algebro povas paroli kvantan Esperanton.

Keywords: Hermitian spaces · Quantum Esperanto · Geometric algebra · Ket map · Spin statistics

1 The Grammar Rules

> The language of mathematics makes the world of Maxwell fields and the world of quantum processes equally transparent. [...] Each of the interpretations of quantum mechanics is an attempt to describe quantum mechanics in a language that lacks the appropriate concepts. The battles between the rival interpretations continue unabated and no end is in sight.
>
> FREEMAN J. DYSON, [1].

Since geometric algebras are finite-dimensional, in principle their use in quantum mechanical questions cannot go beyond systems whose Hilbert spaces are finite-dimensional. For such systems, the mathematics of their Hilbert spaces reduces to the simpler mathematics of the Hermitian spaces, a simplicity that here we metaphorically call 'quantum Esperanto' (QE)—after the wonderfully

D. W. Silva et al. (Eds.): ICACGA 2022, LNCS 13771, pp. 174–185, 2024.
https://doi.org/10.1007/978-3-031-34031-4_15

simple international language Esperanto invented by the ophthalmologist Ludwik L. Zamenhof. A convenient introduction of the QE notions that we will use is through four *axioms*. The first two, QE_1 and QE_2, will be introduced in this section and are sufficient for our present discussions, which will be focused on the bearing of geometric algebra to model the *geometry* of a q-bit. The other two, QE_3 (unitary evolution) and QE_4 (state vectors of a composite system), and their fundamental role in areas such as quantum computing, will be inspected with a similar lens in a sequel manuscript [2].

The reader can find the notions about Hermitian spaces to be used here in the Appendix, page 10. In particular, the Hermitian scalar product $\langle x|x'\rangle$, the norm $|x|$ of a vector ($|x|^2 = \langle x|x\rangle$), and the notation $\hat{x} = x/|x|$ for the *normalization* of a non-zero vector x (cf. §12). It is important to note the distinctive features of the Hermitian angle between two vectors (P.15) as compared to the Euclidean angle (P.16), which ultimately explain why the Hermitian angle between two state vectors of a q-bit is half the angle between the corresponding states (for instance, the state vectors of two antipodal points of S^2 are orthogonal, P.8).

QE_1. (*a*) ***Quantum systems, state vectors and states***
In QE, there are three ingredients defining a *quantum system*: (1) A Hermitian space $\mathcal{H}$, whose non-zero elements are called *state vectors*, sometimes also *wave functions*; (2) A set Σ, whose elements are called (pure) *states*; and (3) an onto map $\mathcal{H} - \{0\} \to \Sigma$, $x \mapsto |x\rangle$ (Dirac's *ket* notation) such that $|x\rangle = |x'\rangle$ if and only if $x' \sim x$ (a shorthand for 'x and x' are equal up to a complex factor').

The third property just says that Σ can be seen as the set of classes of non-zero state vectors by the relation $\sim$. This set is the *projective space* associated to $\mathcal{H}$ and is denoted by $[\mathcal{H}]$ or $\mathbf{P}\mathcal{H}$. In geometry, the elements of this space are called (projective) *points* and the point associated to a non-zero vector x is denoted by $[x]$. Thus, by definition, the relation $[x] = [x']$ is equivalent to $x' \sim x$ for non-zero vectors $x, x' \in \mathcal{H}$

The fact that $|x\rangle$ and $[x]$ obey the same rule is not a coincidence, for although Dirac never mentioned projective geometry explicitly in his research papers, later in his life he acknowledged having used it in his reasonings all along. This is a fascinating story, a bit mysterious, for which we can only refer to the literature, for instance the biography [3] and the references there, particularly [4]. For our purposes, the clearest connection appears in the definition of superposition of states, as we will see in next paragraph.

QE_1. (*b*) ***Quantum superposition***
Given two different states, $X = |x\rangle$ and $X' = |x'\rangle$, each state of the projective line XX' is said to be a (quantum) *superposition* of X and X'. By definition, such states have the form $|\xi x + \xi' x'\rangle$, with $\xi, \xi' \in \mathbb{C}$ and $\xi x + \xi' x' \neq 0$.

Dirac's ket notation is usually abused by writing $\xi|x\rangle + \xi'|x'\rangle$ instead of $|\xi x + \xi' x'\rangle$. Although $\xi|x\rangle + \xi'|x'\rangle$ does not make sense mathematically, as Σ is not a vector space (and even less a complex vector space), in practice it is understood that the state expressed by $|x\rangle$ "remembers" the vector x that has been used to represent it and thus calculations can usually be interpreted unambiguously. For example, if $e_1, \ldots, e_n$ is a basis of $\mathcal{H}$ and $x = \Sigma\lambda_j e_j$, then custom favors to write the expression $\Sigma\lambda_j|e_j\rangle$, which in this case is unambiguously decoded as $|\Sigma\lambda_j e_j\rangle = |x\rangle$.

*QE_2. **Quantoscopes, measurements and observables***
We define a *quantoscope*[1] (to "observe" or "measure" the system) as a set of pairs $A = \{(a_1, V_1), \dots, (a_r, V_r)\}$ such that:

1. The a_j are distinct *real* numbers. The set $\{a_1, \dots, a_r\}$ is the *dial* of the quantoscope, and we assume it is ordered; and
2. The V_j are non-zero vector subspaces of $\mathcal{H}$ such that $V_j \perp V_k$ for $j \neq k$ (orthogonality condition), and $\mathcal{H} = \oplus_j V_j$. The latter means that any $x \in \mathcal{H}$ can be written in a unique way as $x = x_1 + \cdots + x_r$ with $x_j \in V_j$, and the orthogonality condition implies the *Pythagoras theorem*:
$$|x|^2 = |x_1|^2 + \cdots + |x_r|^2.$$
Note also that $x_j = P_{V_j}(x)$ (the orthogonal projection of x on V_j). For a unit vector $u \in \mathcal{H}$, we have
$$1 = |u_1|^2 + \cdots + |u_r|^2,$$
which means that the quantities $p_j = |u_j|^2$ form a *probability distribution* on the set $\{1, \dots, r\}$.

An *observation* or *measure* with the quantoscope A, assuming that the system is in the state $|u\rangle$ (u unitary), consists in carrying out the following two operations:

(*i*) to select at random a value a_j with probability $p_j = |u_j|^2$, where $u_j = P_{V_j}(u)$ is the orthogonal projection of u to V_j (we say that a_j is the *result* or *outcome* of the observation), and
(*ii*) to update the state of the system to $|u_j\rangle$. Note that $p_j \neq 0$ if a_j is selected and hence $u_j \neq 0$.

If $u \in V_j$, then $u_j = u$ and $p_j = 1$, which means that the outcome a_j of the measurement is certain and that the system's state does not change.

Let us associate to each quantoscope A the operator $\widehat{A} = \Sigma_j a_j P_{V_j}$. This operator is selfadjoint and $A \mapsto \widehat{A}$ is a one-to-one map of the set of quantoscopes to the space of selfadjoint operators of $\mathcal{H}$. Conversely, given a selfadjoint operator A', its eigenvalues $a_1, \dots, a_r$ are real (we assume that they are distinct and ordered according to the criterion used to order the quantoscopes' dials), and the corresponding eigenspaces $V_1, \dots, V_r$ are an orthogonal decomposition of $\mathcal{H}$ (these statements are the conclusions of the diagonalization theorem for self-adjoint operators). Thus we see that $A = \{(a_1, V_1), \dots, (a_r, V_r)\}$ is a quantoscope, and we will say that it is the quantoscope *associated to* (or *defined by*) A'.

[1] We introduce this notion to mediate between the intuitive notion of quantum measurement (like a Stern-Gerlach experiment) and the less tangible realization that the observables of the system can be identified with the selfadjoint operators of $\mathcal{H}$.

Example. The quantoscope associated to the orthogonal projection P_V of $\mathcal{H}$ onto the subspace V is $\{(1, V), (0, V^\perp)\}$. Assuming that the system is in the state $|u\rangle$, u unitary, a measurement with this quantoscope selects 1 or 0 at random with probabilities $|P_V(u)|^2$ and $|P_{V^\perp}(u)|^2$, while resetting the state to $|P_V(u)\rangle$ or $|P_{V^\perp}(u)\rangle$, respectively.

Observables. To concur with the conventional terminology, henceforth we will refer to self-adjoint operators as *observables* of the system. By a *measurement* of an observable we understand a measurement with the associated quantoscope. As specified before, such a measurement supplies, if the state of the system is $|u\rangle$ (u unitary), a random eigenvalue a_j of the observable with probability $p_j = |u_j|^2$, $u_j = P_{V_j}(u)$, and resets the state of the system to $|u_j\rangle$. Note that in general the vector u_j is not unitary, which illustrates the resilience of working with state vectors that are not necessarily unitary.

2 q-Bits

The quantum behavior of *spin*, or intrinsic angular momentum, was discovered for the first time by the Stern-Gerlach (SG) experiments (see the Wikipedia article Stern-Gerlach_experiment for an apt presentation, including references to the three original papers by O. Stern and W. Gerlach published in 1922, and a discussion of the telling outcomes of composite SG procedures; for an epistemological analysis of its significance, see [5]). In general, the values j observed in an SG experiment have the form (in appropriate units) $j = -s, -s+1, \ldots, s-1, s$, with s an non-negative integer multiple of $1/2$. In all cases, the number of values j is $2s + 1$.

Excluding $s = 0$, corresponding to a spinless system, the simplest case (as in the original SG experiments) is $s = 1/2$, with two possible values: $j = \pm\frac{1}{2}$. It is such kind of systems, called *q-bits*, that we will consider in this paper. In this case, the experimental results of the SG experiments suggest that we may construe the (pure) states of a q-bit as ordinary vectors of norm 1 (in suitable units). In other words, we may take the unit sphere $S^2 \subset E_3$ as the space of (pure) states of a q-bit: $\Sigma = S^2$. Here E_3 is the ordinary Euclidean space and in what follows $u_x, u_y, u_z \in E_3$ will be an orthonormal basis, with u_z aligned with the magnetic field.

To elicit the Hermitian space, two old geometric constructions are handy. On one hand, $S^2 \simeq \widehat{\mathbb{C}} = \mathbb{C} \sqcup \{\infty\}$, via the stereographic projection of S^2 from the north pole u_z onto the equatorial plane $z = 0$ (which we identify with $\mathbb{C}$), and with $u_z \mapsto \infty$ (see Fig. 1). This is the *Riemann sphere*.

On the other hand, $\widehat{\mathbb{C}} \simeq [\mathbb{C}^2]$, via the map $\xi \mapsto [(1, \xi)]$ ($\xi \in \mathbb{C}$) and $\infty \mapsto [(0, 1)] = [e_1]$. By composing both bijections, we have a bijection $S^2 \simeq [\mathbb{C}^2]$, and so, according to $Q_1(a)$, we can take the (Pauli) *spinor space* $\mathbb{C}^2$ as the space of state vectors of the q-bit. In Table 1 we collect a detailed description of how these bijections work in both directions.

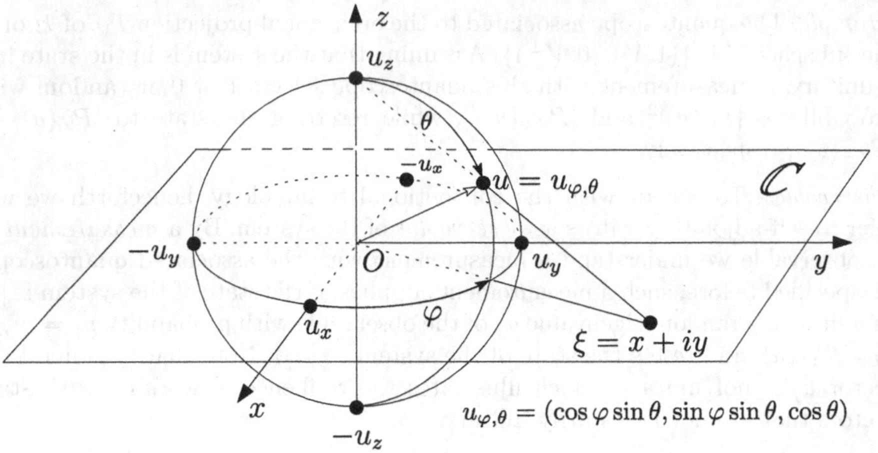

Fig. 1. Stereographic projecction of S^2 to $\widehat{\mathbb{C}} = \mathbb{C} \sqcup \{\infty\}$.

Table 1. Bijections $S^2 \simeq \widehat{\mathbb{C}} \simeq [\mathbb{C}^2]$. In the fourth row, $k = 2/(a^2+b^2+1)$ and $c = (a^2+b^2-1)/2)$

S^2		$\widehat{\mathbb{C}}$		$\mathbf{P}\mathbb{C}^2 = [\mathbb{C}^2]$
$(x,y,z) \neq (0,0,1)$	$\to$	$\frac{x}{1-z} + \frac{y}{1-z}i = \xi$	$\to$	$[(1,\xi)]$
$u_z = (0,0,1)$	$\to$	∞	$\to$	$[(0,1)] = [e_1]$
$k(a,b,c)$	$\leftarrow$	$\xi = a+bi = \xi_1/\xi_0$	$\leftarrow$	$[(\xi_0,\xi_1)]$ $(\xi_0 \neq 0)$
$u_z = (0,0,1)$	$\leftarrow$	∞	$\leftarrow$	$[(0,\xi_1)] = [(0,1)] = [e_1]$

Let us use Dirac's notation $|\xi_0,\xi_1\rangle \in S^2$ to denote the state corresponding to $(\xi_0,\xi_1) \in \mathbb{C}^2$. If $\xi_0 \neq 0$, $|\xi_0,\xi_1\rangle = |1,\xi\rangle$ $(\xi = \xi_1/\xi_0)$, which we will abridge to $|\xi\rangle$. In particular, $|0\rangle = |1,0\rangle = |e_0\rangle$ is the point $-u_z = (0,0,-1)$ (the south pole of S^2). The state $|\infty\rangle = |0,1\rangle = |e_1\rangle$ is $(0,0,1) = u_z$, the north pole of S^2.

Next statement provides explicit expressions for the coordinates x, y, z of $|\xi_0,\xi_1\rangle$. First published in [6], we derive them by means of the stereographic projection.

P.1. (Representation of S^2 by spinors) If $(x,y,z) = |\xi_0,\xi_1\rangle$, then

$$x = (\xi_1\bar{\xi}_0 + \xi_0\bar{\xi}_1)/r^2,\ y = i(\xi_0\bar{\xi}_1 - \xi_1\bar{\xi}_0)/r^2,\ z = (\xi_1\bar{\xi}_1 - \xi_0\bar{\xi}_0)/r^2,\ r^2 = \xi_0\bar{\xi}_0 + \xi_1\bar{\xi}_1.$$

Proof. The formula for the stereographic projection $\xi \in \mathbb{C}$ of $(x,y,z) \in S^2$ shows that $x + iy = (1-z)\xi$. So $x - iy = (1-z)\bar{\xi}$ and hence

$$x = \tfrac{1}{2}(\xi + \bar{\xi})(1-z),\ y = -\tfrac{i}{2}(\xi - \bar{\xi})(1-z).$$

We also have

$$1 - z^2 = x^2 + y^2 = (1-z)^2\xi\bar{\xi}, \quad \text{or} \quad 1 + z = (1-z)\xi\bar{\xi},$$

from which it follows that $1 - z = \dfrac{2}{\xi\bar{\xi} + 1}$ and so

$$x = \frac{\xi + \bar{\xi}}{\xi\bar{\xi} + 1}, \quad y = -i\frac{\xi - \bar{\xi}}{\xi\bar{\xi} + 1}, \quad z = \frac{\xi\bar{\xi} - 1}{\xi\bar{\xi} + 1}.$$

In terms of spinors (cf. Table 1), $\xi = \xi_1/\xi_0$, and we get the expressions in the statement. □

Next statement can be obtained via a little play with the expressions in Table 1 or applying the formulas in P.1.

P.2. Let $\widetilde{u}_{\varphi,\theta} = (e^{-i\varphi/2}\sin\frac{\theta}{2}, e^{i\varphi/2}\cos\frac{\theta}{2}) \in \mathbb{C}^2$. Then $|\widetilde{u}_{\varphi,\theta}\rangle = u_{\varphi,\theta}$. □

Notice that Dirac's notation allows us to write $|\xi_0, \xi_1\rangle = \xi_0|e_0\rangle + \xi_1|e_1\rangle$. In other words, any state of a q-bit is a superposition of the states $|e_0\rangle$ and $|e_1\rangle$. The states $|e_1\rangle$ and $|e_0\rangle$ can be described as "spin up" (parallel to the magnetic field) and "spin down" (antiparallel to the magnetic field), and are sometimes denoted by $|+\rangle$ and $|-\rangle$, or $|\uparrow\rangle$ and $|\downarrow\rangle$.

Given $s = (\xi_0, \xi_1) \in \mathbb{C}^2$, we set $s^\perp = (-\bar{\xi}_1, \bar{\xi}_0)$. This expression is appropriate, as $\langle s|s^\perp\rangle = 0$. The map $s \mapsto s^\perp$ is *antilinear* and satisfies $|s| = |s^\perp|$, hence $s, s^\perp$ is an orthonormal basis of $\mathbb{C}^2$ if s is a unit vector. Note also that $s^{\perp\perp} = -s$.

3 Let the Geometric Quaternions Speak up

Instead of $\mathbb{C}^2$, whose relation to E_3 is quite artificial (these structures speak different languages), we could try to go over to Hamilton's quaternions $\mathbb{H}$ by means of the $\mathbb{C}$-isomorphism $\mathsf{j} : \mathbb{C}^2 \to \mathbb{H}$ given by $\mathsf{j}((\xi_0, \xi_1)) = \xi_0 + \xi_1\mathsf{j}$, where $1, \mathsf{i}, \mathsf{j}, \mathsf{k}$ is the usual basis of $\mathbb{H}$ as a real vector space, and use $\mathbf{P}_\mathbb{C}\mathbb{H}$ as state space. To do that, the basic requirement would be to express the Hermitian scalar product inherited from $\mathbb{C}^2$, via j, purely in terms of $\mathbb{H}$. But even with this we would still have to rely on the space of pure quaternions $\langle \mathsf{i}, \mathsf{j}, \mathsf{k}\rangle$ as an artificial substitute of E_3.

These considerations clearly suggest that the optimal structure on which to base the geometric theory of the q-bit is the algebra of geometric quaternions, $\mathbf{H} = \mathcal{G}_3^+ \subset \mathcal{G}_3$, as it is constructed directly on the geometry of E_3, and besides it keeps a free copy of E_3, namely $E_3 = \mathcal{G}_3^1$ (the grade 1 elements). There is a little work we have to do, but, as we show next, it may be found to be worthwhile.

As a real vector space, $\mathbf{H} = \langle 1, \mathsf{i}, \mathsf{j}, \mathsf{k}\rangle$, where $\mathsf{i} = u_y u_z$, $\mathsf{j} = u_x u_z$ and $\mathsf{k} = u_x u_y$. *Formally* we have not moved from $\mathbb{H}$, as (the new) $1, \mathsf{i}, \mathsf{j}, \mathsf{k}$ satisfy Hamilton's relations, but now these objects are meaningful bivectors (unit areas) of E_3. We also have to ask a crucial question: what *complex* structure of $\mathbf{H}$ shall we use? A convenient choice, as warranted by what follows, is $\mathbf{C} = \mathbf{C}_\mathsf{i} = \langle 1, \mathsf{i}\rangle$, the (geometric) complex numbers of the form $a + b\,\mathsf{i}$, and denote by $\mathbf{H}_\mathsf{i}$ the algebra $\mathbf{H}$ when considered as a $\mathbf{C}_\mathsf{i}$-vector space. It is worth to keep in mind that there are very many quaternions $\mathfrak{l}$ such that $\mathfrak{l}^2 = -1$ (unit pure quaternions satisfy this condition, but they are not the only ones) and hence $\mathbf{C}_\mathfrak{l} = \langle 1, \mathfrak{l}\rangle = \{a + b\mathfrak{l} :$

$a, b \in \mathbf{R}\}$ can be used as a field of complex numbers within $\mathbf{H}$, and in that case $\mathbf{H}_\mathsf{i}$ is $\mathbf{H}$ regarded as a $\mathbf{C}_\mathsf{i}$-vector space.

Notice that $\mathsf{q} = a + b\mathsf{i} + c\mathsf{j} + d\mathsf{k} \in \mathbf{H}$ can be written in the form $\mathsf{q} = \xi_0 + \xi_1\mathsf{j}$, where $\xi_0 = a + b\mathsf{i}$ and $\xi_1 = c + d\mathsf{i}$. The question of how ξ_0 and ξ_1 can be retrieved from q by operations involving *only* $\mathbf{H}$ is answered in the following proposition.

P.3. If $\mathsf{q} = \xi_0 + \xi_1\mathsf{j}$, $\xi_0, \xi_1 \in \mathbf{C}$, then $\xi_0 = \frac{1}{2}(\mathsf{q} - \mathsf{iqi})$ and $\xi_1 = -\frac{1}{2}(\mathsf{qj} + \mathsf{iqk})$.

Proof. Since the expression $\frac{1}{2}(\mathsf{q} - \mathsf{iqi})$ is $\mathbf{R}$-linear in q, it is enough to check that it supplies q for $\mathsf{q} = 1, \mathsf{i}$ and 0 if $\mathsf{q} = \mathsf{j}, \mathsf{k}$. In fact, if q commutes with i, as is the case for the elements of $\mathbf{C}$, the formula supplies q, and if q anti-commutes with i, as is the case for all elements of $\langle \mathsf{j}, \mathsf{k} \rangle$, it supplies 0. The second part follows similarly: $-\frac{1}{2}(\mathsf{qj} + \mathsf{iqk})$ yields 0 for $\mathsf{q} = 1, \mathsf{i}$ and q for $\mathsf{q} = \mathsf{j}, \mathsf{k}$. □

Now we seek a Hermitian scalar product in $\mathbf{H}$, again defined only in terms of $\mathbf{H}$, that plays the role of the Hermitian scalar product of $\mathbb{C}^2$. The next proposition provides the answer. We will use the map $\mathsf{i} : \mathbf{H} \to \mathbf{C}$ defined by the relation $\mathsf{i}(\mathsf{q}) = \frac{1}{2}(\mathsf{q} - \mathsf{iqi})$. The reverse involution in $\mathbf{H}$ will be denoted by $\bar{\mathsf{q}}$ (it coincides with the Clifford involution). The customary symbol $\tilde{x}$ for the reverse involution will be used for other purposes below.

P.4. (The hidden Hermitian structure of $\mathbf{H}$) The scalar product $\mathbf{H} \times \mathbf{H} \to \mathbf{C}$, denoted by $\langle \mathsf{q} | \mathsf{q}' \rangle$ and defined by the formula $\langle \mathsf{q} | \mathsf{q}' \rangle = \mathsf{i}(\mathsf{q}'\bar{\mathsf{q}})$ is Hermitian and for $\xi_0, \xi_1, \xi_0', \xi_1' \in \mathbf{C}$ we have $\langle \xi_0 + \xi_1\mathsf{j} | \xi_0' + \xi_1'\mathsf{j} \rangle = \bar{\xi}_0\xi_0' + \bar{\xi}_1\xi_1'$. Moreover, $\langle \mathsf{q} | \mathsf{q} \rangle = |\mathsf{q}|^2$ for any $\mathsf{q} \in \mathbf{H}$, so that the Euclidean norm of q coincides with the Hermitian norm. In the special case in which $\mathsf{q} = v\,\boldsymbol{i}$ is the dual of a vector $v \in E_3$, $|\mathsf{q}|^2 = |v|^2$. Here $\boldsymbol{i}$ is the pseudoscalar of $\mathcal{G}_3$, namely $\boldsymbol{i} = u_x u_y u_z$.

Proof. Since $\bar{\mathsf{q}} = \bar{\xi}_0 - j\bar{\xi}_1$, the terms in the expansion of $\mathsf{q}'\bar{\mathsf{q}}$ not involving j and k are $\xi_0'\bar{\xi}_0 + (\xi_1' j)(-j\bar{\xi}_1) = \xi_0'\bar{\xi}_0 + \xi_1'\bar{\xi}_1$, as stated. If $\mathsf{q}' = \mathsf{q}$, then $\mathsf{q}\bar{\mathsf{q}} = |\mathsf{q}|^2$ is real and the claim follows. For the last point, note that for $\mathsf{q} = v\,\boldsymbol{i}$ we have $\bar{\mathsf{q}} = -\boldsymbol{i}\,v$ and $\mathsf{q}\bar{\mathsf{q}} = v^2 = |v|^2$. □

Next question is how to realize the state space of $\mathbf{H}$, which is the (abstract) sphere $\mathbf{PH}_\mathsf{i}$, as the sphere $S^2 \subset E_3$. In other words, if $S^3 = \{\mathsf{q} \in \mathbf{H} : |\mathsf{q}|^2 = 1\}$, we are seeking a map $S^3 \to S^2$, $\mathsf{q} \mapsto |\mathsf{q}\rangle$, that is onto and such that $|\mathsf{q}\rangle = |\mathsf{q}'\rangle$ if and only if $\mathsf{q}' \equiv \mathsf{q}$ (this map is usually called the *Hopf fibration*).

For that, let us define a more convenient generalized *ket map* $\kappa : \mathbf{H} \to E_3$ that induces, as we shall see in a moment, an onto map $S_r^3 \to S_r^2$ for any radius r with the property that $\kappa(\mathsf{q}) = \kappa(\mathsf{q}')$ if and only if $\mathsf{q}' \equiv \mathsf{q}$. If $\mathsf{q} = \xi_0 + \xi_1\mathsf{j}$ ($\xi_0, \xi_1 \in \mathbf{C}$), let $\kappa(\mathsf{q}) = 0$ if $\mathsf{q} = 0$, and otherwise define it as the map in P.1, but with denominator r instead of r^2. In other words, let $r = |q|$ and $\kappa(\mathsf{q}) = \alpha u_x + \beta u_y + \gamma u_z$, where

$$\alpha = (\xi_1\bar{\xi}_0 + \xi_0\bar{\xi}_1)/r, \quad \beta = i(\xi_0\bar{\xi}_1 - \xi_1\bar{\xi}_0)/r, \quad \gamma = (\xi_1\bar{\xi}_1 - \xi_0\bar{\xi}_0)/r.$$

We know that $\alpha^2 + \beta^2 + \gamma^2 = r^2$ when $|\mathsf{q}|^2 = r^2$, so indeed $\kappa : S_r^3 \to S_r^2$. For $r = 1$, we clearly have $\kappa(\mathsf{q}) = |\mathsf{q}\rangle$ for any $\mathsf{q} \in S^3$.

P.5. (Examples) (1) $\kappa(e^{i\varphi}\mathsf{q}) = \kappa(\mathsf{q})$, for any $\varphi \in \mathbb{R}$. Thus $\kappa(\mathsf{q}') = \kappa(\mathsf{q})$ if $\mathsf{q}' \equiv \mathsf{q}$. (2) $\kappa(1) = \kappa(\mathsf{i}) = -u_z$ (the south pole of S^2) and $\kappa(\mathsf{j}) = \kappa(\mathsf{k}) = u_z$ (the north pole of S^2). Note that $\mathsf{i} \equiv 1$ and $\mathsf{k} = \mathsf{ij} \equiv \mathsf{j}$, so it is enough to check that $\kappa(1) = -u_z$ and $\kappa(\mathsf{j}) = u_z$. (3) If $\kappa(\mathsf{q}) = u_z$ then $\mathsf{q} \equiv \mathsf{j}$. □

P.6. The map $\kappa : S_r^3 \to S_r^2$ is surjective and for $\mathsf{q}, \mathsf{q}' \in S_r^3$ we have $\kappa(\mathsf{q}') = \kappa(\mathsf{q})$ if and only if $\mathsf{q}' \equiv \mathsf{q}$.

Proof. From the definitions if follows easily that it is enough to prove the case $r = 1$. Let $u = \alpha u_x + \beta u_y + \gamma u_z \in S^2$ (so $\alpha^2 + \beta^2 + \gamma^2 = 1$). We are seeking $\mathsf{q} = a + b\,\mathsf{i} + c\,\mathsf{j} + d\,\mathsf{k} \in S^3$ such that $\kappa(\mathsf{q}) = u$. By P.5, we may assume that $u \neq u_z$ (that is, $\gamma \neq 1$). We may further assume that $b = 0$ (if φ is the phase of $a + bi$, it is enough to replade q by $e^{-i\varphi}\mathsf{q}$). In other words, we may assume that $\mathsf{q} = a + (c + di)\mathsf{j} = a + c\mathsf{j} + d\mathsf{k}$, which means, with the notations above, that $\xi_0 = a$ and $\xi_1 = c + d\,\mathsf{i}$. Since $\xi_0 = a$ is real, the expressions that define $\kappa(\mathsf{q})$ are $a(\xi_1 + \bar{\xi}_1) = 2ac$, $ia(\bar{\xi}_1 - \xi_1) = 2ad$, and $c^2 + d^2 - a^2 = 1 - 2a^2$, so the condition $\kappa(\mathsf{q}) = u$ is equivalent to the relations

$$\alpha = 2ac, \quad \beta = 2ad, \quad \gamma = 1 - 2a^2.$$

The third relation gives $a = \pm\sqrt{(1-\gamma)/2}$, and $a \neq 0$ because $\gamma \neq 1$. For each of the two solutions, we get $c = \alpha/2a$ and $d = \beta/2a$. This shows that κ is surjective. And the argument also shows that if $\kappa(\mathsf{q}') = \kappa(\mathsf{q})$, then $\mathsf{q}' \equiv \mathsf{q}$. □

Now in order to find the probabilities of an observation event when the state is $u = \alpha u_x + \beta u_y + \gamma u_z \in S^2$, it is convenient to define $\tilde{u} \in \mathbf{H}$ to be j if $u = u_z$ and otherwise $\tilde{u} = a + (c + d\,\mathsf{i})\mathsf{j}$, where $a = \sqrt{(1-\gamma)/2}$, $c = \alpha/2a$ and $d = \beta/2a$. Thus we have $\kappa(\tilde{u}) = u$ in all cases. *Note*: this definition of $\tilde{u}$ is not the same as the one introduced in P.2, but, as shown in next statement, they differ (as they should) by a phasor factor.

P.7. (1) For $u \neq u_z$, $\tilde{u} = \frac{1}{\sqrt{2(1-\gamma)}}(1 - \gamma + \alpha\mathsf{j} + \beta\mathsf{k}) \in S^3$.

(2) If we use spherical coordinates for the state $u = \alpha u_x + \beta u_y + \gamma u_z$, that is (see Fig. 1), $\alpha = \cos\varphi \sin\theta$, $\beta = \sin\varphi \sin\theta$, $\gamma = \cos\theta$, then

$$\begin{aligned}\tilde{u} &= \sin\tfrac{\theta}{2} + \cos\varphi\cos\tfrac{\theta}{2}\,\mathsf{j} + \sin\varphi\cos\tfrac{\theta}{2}\,\mathsf{k}\\ &= \sin\tfrac{\theta}{2} + e^{\mathsf{i}\varphi}\cos\tfrac{\theta}{2}\,\mathsf{j}\\ &\equiv e^{-\mathsf{i}\varphi/2}\sin\tfrac{\theta}{2} + e^{\mathsf{i}\varphi/2}\cos\tfrac{\theta}{2}\,\mathsf{j}.\end{aligned}$$

This in particular shows that when $\theta \to 0$ (so u approaches u_z, the north pole) we get, in the limit, $e^{\mathsf{i}\varphi}\,\mathsf{j} \equiv \mathsf{j}$, which is how we have defined $\tilde{u}_z$.

(3) We also have that

$$\widetilde{-u} = \frac{1}{\sqrt{2(1+\gamma)}}(1 + \gamma - \alpha j - \beta k).$$

In particular $\widetilde{-u_z} = 1$, in agreement with the fact that $\kappa(1) = -u_z$. □

P.8. (Hermitian angle of two anti-podal states)If $u \in S^2$, then $\tilde{u}$ and $\widetilde{-u}$ are orthogonal with respect to the Hermitian metric of $\mathbf{H}$. □

Now let us answer the question about probabilities. Let $u \in S^2$ be a state. This defines the quantoscope $A_u = \{(1, \langle\tilde{u}\rangle), (-1, \langle\widetilde{-u}\rangle)\}$. What is the probability of obtaining 1 if before measurement with A_u the state is $v \in S^2$? The answer is given by the following result, which is in agreement with the experimental observations. The result also says that the Hermitian angle between state vectors is half the Euclidean angle between the corresponding states, which is a more general statement than P.8.

P.9. The probability of observing 1 with A_u, if the state before measurement is $v \in S^2$, is given by the expression $p_u(v) = \cos^2(\alpha/2)$, where α is the Euclidean angle between u and v (that is, $\cos(\alpha) = u \cdot v$).

Proof. According to the QE prescriptions, $p_u(v) = |\langle\tilde{u}|\tilde{v}\rangle|^2$. To compute this, let us use spherical coordinates for u and v, so that $\tilde{u} = \sin\frac{\theta}{2} + e^{\mathrm{i}\varphi}\cos\frac{\theta}{2}\mathrm{j}$ and likewise $\tilde{v} = \sin\frac{\theta'}{2} + e^{\mathrm{i}\varphi'}\cos\frac{\theta'}{2}\mathrm{j}$. Now we know that $\langle\tilde{u}|\tilde{v}\rangle = \mathrm{i}(\tilde{v}\bar{\tilde{u}})$. On expanding $\tilde{v}\bar{\tilde{u}}$ and retaining only the terms that are not scalar multiples of j and k, we find:

$$\begin{aligned}\mathrm{i}(\tilde{v}\bar{\tilde{u}}) &= \mathrm{i}\left(\left(\sin\tfrac{\theta'}{2} + e^{\mathrm{i}\varphi'}\cos\tfrac{\theta'}{2}\mathrm{j}\right)\left(\sin\tfrac{\theta}{2} - \mathrm{j}\,e^{-\mathrm{i}\varphi}\cos\tfrac{\theta}{2}\right)\right)\\ &= \sin\tfrac{\theta'}{2}\sin\tfrac{\theta}{2} + e^{\mathrm{i}(\varphi'-\varphi)}\cos\tfrac{\theta'}{2}\cos\tfrac{\theta}{2}.\end{aligned}$$

Now $p_u(v)$ is the modulus squared of this expression and a little joggling with trigonometric expressions yields that it is $\cos^2(\frac{\alpha}{2})$, where, by definition of α, $\cos(\alpha) = \cos\theta\cos\theta' + \sin\theta\sin\theta'\cos(\varphi' - \varphi)$. □

P.10. (The group $\mathrm{SU}(\mathbf{H})$)

(1) Any $\mathsf{q} \in \mathbf{H}_1$ defines a map $U_\mathsf{q} : \mathbf{H} \to \mathbf{H}$ by the formula $U_\mathsf{q}(x) = x\mathsf{q}$. This map is clearly $\mathbf{H}$-linear, hence also $\mathbf{C}_\mathrm{i}$-linear, and satisfies $|U_\mathsf{q}(x)| = |x|$. So (see P20) $U_\mathsf{q} \in \mathrm{U}(\mathbf{H})$.
(2) $U_\mathsf{q} \in \mathrm{SU}(\mathcal{H})$.
(3) The map $\mathbf{H}_1 \to \mathrm{SU}(\mathbf{H})$, $\mathsf{q} \mapsto U_\mathsf{q}$, is a group anti-isomorphism. By taking matrices with respect to $\{1, \mathrm{j}\}$, we are led to $\mathbf{H}_1 \simeq \mathrm{SU}_2$.

Proof. (1) $|U_\mathsf{q}(x)|^2 = |x\mathsf{q}|^2 = (x\mathsf{q})(\bar{\mathsf{q}}\bar{x}) = x\bar{x} = |x|^2$.

(2) It is enough to see that the matrix A_q of U_q with respect to the basis $\{1, \mathrm{j}\}$, which is obtained from $U_\mathsf{q}(1) = \mathsf{q}$, $U_\mathsf{q}(\mathrm{j}) = \mathrm{j}\mathsf{q} = \mathsf{q}^\perp$, has determinant 1.
(3) From the definition of U_q it follows that $(U_{\mathsf{q}'}U_\mathsf{q})(x) = (x\mathsf{q})\mathsf{q}' = x(\mathsf{q}\mathsf{q}') = U_{\mathsf{q}\mathsf{q}'}(x)$. And this implies that $A_{\mathsf{q}\mathsf{q}'} = A_\mathsf{q}A_{\mathsf{q}'}$.

P.11. (Symmetries). (1) A quaternion $\mathsf{q} \in \mathbf{H}^\times$ defines a map $R_\mathsf{q} : E_3 \to E_3$ by the formula $R_\mathsf{q}(v) = \mathsf{q}v\mathsf{q}^{-1}$, and $R_\mathsf{q} \in \mathrm{SO}(E_3)$.
(2) The map $\mathbf{H}^\times \to \mathrm{SO}(E_3)$, $\mathsf{q} \mapsto R_\mathsf{q}$, is a group homomorphism.

(3) We have $\ker(\rho) = \mathbf{R}^{\times}$. In other words, $\rho_{\mathfrak{q}} = Id$ if and only if $\mathfrak{q}$ is a non zero real number. If $\mathfrak{q} \in \mathbf{H}_1$, then $\rho_{\mathfrak{q}} = Id$ if and only if $\mathfrak{q} = \pm 1$.
(4) The rotation $\rho_{v,\alpha}$ of amplitude α about $v \in S^2$ is given by $\rho_{v,\alpha} = R_{\mathfrak{q}}$, where $\mathfrak{q} = e^{-\boldsymbol{i}\frac{\alpha}{2}v} \in H_1$. It follows that the homomorphism $R : \mathbf{H}_1 \to \mathrm{SO}(E_3)$ is onto.

Proof. This is a well-known result, but cast here in terms of only the structures of geometric quaternions. Let us just sketch a proof of (4). Since v commutes with $\mathfrak{q}$, we have $R_{\mathfrak{q}}(v) = v$. So $R_{\mathfrak{q}}$ is a rotation about v. To find its amplitude, pick any $v' \in S^2$ orthogonal to v. Since v' anticommutes with $\mathfrak{q}$, we have $R_{\mathfrak{q}}(v') = v'\bar{\mathfrak{q}}^2 = v'e^{\boldsymbol{i}\alpha v} = v'\cos\alpha + v'\,\boldsymbol{i}\,v\sin(\alpha)$, where we have used that $(\boldsymbol{i}\,v)^2 = -1$. If we set $v'' = v \times v'$, then v', v'', v is a positively oriented orthornormal basis of E_3 and, in particular, $\boldsymbol{i} = v'v''v$, so that $R_{\mathfrak{q}}(v') = v'\cos\alpha + v''\sin\alpha$.

Appendix: Hermitian spaces

The aim of this appendix is to review the basic notions needed in the preceding sections. Proofs are omitted, but the interested reader can find them in the paper [7] or in the references therein.

12. A *Hermitian* vector space is a *complex* vector space $\mathcal{H}$ endowed with a scalar product $\langle x|x'\rangle \in \mathbb{C}$ $(x, x' \in \mathcal{H})$ satisfying the following properties:

(1) $\langle x|x'\rangle = \overline{\langle x'|x\rangle}$ (the overline means complex conjugation). In particular, $\langle x|x\rangle$ is self-conjugate for any x and therefore it is a real number.
(2) It is $\mathbb{C}$-linear in x'. This property and (1) imply that the scalar product is conjugate-linear (same as anti-linear) in x.
(3) $\langle x|x\rangle > 0$ if $x \neq 0$. We say that the scalar product is positive-definite.

For any $x \in \mathcal{H}$, we set $|x| = \sqrt{\langle x|x\rangle}$ (*norm* or *length* of x; it is also customary to denote it by $\|x\|$). If $|x| = 1$, we say that x is *unitary*, or a *unit vector*. For example, $\widehat{x} = x/|x|$ (*normalization* of x) is unitary for any $x \neq 0$.

We say that $x, x' \in \mathcal{H}$ are *orthogonal* if $\langle x|x'\rangle = 0$, and we will write $x \perp x'$ to denote this relation.

A basis $e_1, \ldots, e_n$ of $\mathcal{H}$ is said to be *orthonormal* if $\langle e_j|e_k\rangle = \delta_{jk}$ for any j, k. This means that the e_j are unit vectors such that $e_j \perp e_k$ for $j \neq k$. The components $\lambda_j \in \mathbb{C}$ of a vector $x \in \mathcal{H}$ with respect to an orthonormal basis $e_1, \ldots, e_n$ are given by $\lambda_j = \langle e_j|x\rangle$, so that $x = \langle e_1|x\rangle e_1 + \cdots + \langle e_n|x\rangle e_n$.

We will often use the relations $x \sim x'$ and $x \equiv x'$ $(x, x' \in \mathcal{H})$. The first is a shorthand for stating that $x' = \lambda x$, for some non-zero $\lambda \in \mathbb{C}$. For example, we have $x \sim \hat{x}$ for any non-zero $x \in \mathcal{H}$ (in this case λ is real). The second relation is a shorthand for stating that $x' = \lambda x$ for some $\lambda \in \mathbb{C}$ such that $|\lambda| = 1$. In other words, $x' = e^{i\varphi}x$ for some $\varphi \in \mathbb{R}$. □

13. (Example) $\mathbb{C}^n$ with the scalar product

$$\langle(\xi_1, \ldots, \xi_n)|(\xi_1', \ldots, \xi_n')\rangle = \bar{\xi}_1\xi_1' + \cdots + \bar{\xi}_n\xi_n'$$

is a Hermitian space. For the treatment of q-bits, the basic space we initially need is $\mathbb{C}^2$. For historical reasons, the elements of this space are called (Pauli) *spinors* and we will re-index them as (ξ_0, ξ_1). Thus, in this case the Hermitian scalar product reads $\langle(\xi_0,\xi_1)|(\xi_0',\xi_1')\rangle = \bar{\xi}_0\xi_0' + \bar{\xi}_1\xi_1'$. Moreover, we will use the notation $e_0 = (1,0)$ and $e_1 = (0,1)$. □

P.14. (Cauchy-Schwarz inequality for Hermitian spaces) Let $\mathcal{H}$ be a Hermitian vector space. Then, for all $x, x' \in \mathcal{H}$, $|\langle x|x'\rangle| \leq |x||x'|$. □

P.15 (Hermitian angle between two no-zero vectors) If $x, x' \in \mathcal{H}$ are non-zero vectors, P.14 tells us that $0 \leq |\langle x|x'\rangle|/|x||x'| \leq 1$ and hence there is a unique real number $\beta = \beta(x,x') \in [0,\pi/2]$ such that $\cos(\beta) = |\langle x|x'\rangle|/|x||x'|$. Moreover, $\beta = \pi/2$ precisely when $\langle x|x'\rangle = 0$ (that is, precisely when $x \perp x'$), and $\beta = 0$ if and only if $x' \sim x$. Finally, $\beta(x,x') = \beta(y,y')$ when $y \sim x$ and $y' \sim x'$. □

P.16. (Cauchy-Schwarz inequality for Euclidean spaces) For Euclidean spaces, where the (real symmetric) inner product is denoted $x \cdot x'$, the Cauchy-Schwarz inequality says that $|x \cdot x'| \leq |x||x'|$. Since $x \cdot x'$ is *real*, this is equivalent to the inequalities $-|x||x'| \leq x \cdot x' \leq |x||x'|$. This implies that $-1 \leq (x \cdot x')/|x||x'| \leq 1$ if $x, x' \neq 0$, and therefore there exists a unique real number $\alpha = \alpha(x,x') \in [0,\pi]$ such that $\cos(\alpha) = x{\cdot}x'/|x||x'|$. This α is the (Euclidean) *angle* between x and x'. With a bit more attention, it can be seen that $\alpha = 0$ if and only if $x' = tx$, with $t > 0$, and that $\alpha = \pi$ if and only if $x' = tx$, with $t < 0$. In any case, the angle does not vary if we rescale the vectors: if t, t' are positive real numbers, then $\alpha(tx, t'x') = \alpha(x,x')$. But note that $\alpha(-x,x') = \alpha(x,-x') = \pi - \alpha(x,x')$. □

17. The $\mathbb{C}$-endomorphisms of $\mathcal{H}$ are usually called *operators*. If L is an operator, its *adjoint*, denoted $L^\dagger$, is defined as the unique endomorphism of $\mathcal{H}$ such that

$$\langle L^\dagger y|x\rangle = \langle y|Lx\rangle.$$

The map $L \mapsto L^\dagger$ is conjugate-linear. If $L^\dagger = L$, we say that L is *selfadjoint* or *Hermitian*.

Example. Let F be vector subspace of $\mathcal{H}$. The *orthogonal projection* $P_F : \mathcal{H} \to \mathcal{H}$ (defined as $P_F x = x'$ if $x = x' + x''$ with $x' \in F$ and $x'' \in F^\perp$) is selfadjoint, as the expressions $\langle P_F y|x\rangle$ and $\langle y|P_F x\rangle$ are both equal to $\langle y'|x'\rangle$.

If $v \in \mathcal{H}$ is a non-zero vector, instead of $P_{\langle v\rangle}$ we will simply write P_v.

P.18. (Computation of P_F). If we know an orthonormal basis $e_1, \ldots, e_r$ of F, then $P_F(x) = \langle e_1|x\rangle e_1 + \cdots + \langle e_r|x\rangle e_r$. In particular we have, for any non-zero vector $v \in \mathcal{H}$, that $P_v(x) = \langle v|x\rangle v$ if v is unitary, or, in general, $P_v(x) = \frac{1}{|v|^2}\langle v|x\rangle v$. □

P.19. (Matrix of an operator relative to an orthonormal basis). Let A be the matrix of an operator L of the Hermitian space $\mathcal{H}$ with respect to an orthonormal basis $e_1, \ldots, e_n$, that is, $A = (a_{ij})$, where $L(e_i) = a_{i1}e_1 + \cdots + a_{in}e_n$. Then the matrix of $L^\dagger$, with respect to the same basis, is $A^\dagger$: $L^\dagger(e_i) = \bar{a}_{1i}e_1 + \cdots + \bar{a}_{ni}e_n$. This implies that L is self-adjoint if and only if the matrix A is self-adjoint (that is, $A^\dagger = A$).

P.20. (Unitary operators). An operator U is said to be *unitary* if $U^{\dagger}U = I$ (the identity operator). With the notations of the preceding exercise, we see that U is unitary if and only if its matrix A is a unitary matrix ($A^{\dagger}A = I_n$). Equivalently, U is unitary if and only if $\langle Ux|Ux'\rangle = \langle x|x'\rangle$ for all $x, x' \in \mathcal{H}$. This relation implies that $|Ux| = |x|$ for all $x \in \mathcal{H}$. This condition is also *sufficient*, because of the identity $4\langle x|x'\rangle = |x+x'|^2 - |x-x'|^2 + i|ix+x'|^2 - i|-ix+x'|^2 = \sum_{\nu=0}^{\nu=3} i^{\nu}|i^{\nu}x+x'|$. The unitary operators of $\mathcal{H}$ form a group with the composition operation (the *unitary group* of $\mathcal{H}$). It is denoted by $\mathrm{U}(\mathcal{H})$. If $\mathcal{H}$ has dimension n, $\mathrm{U}(\mathcal{H}) \simeq \mathrm{U}_n$ (the group of unitary matrices of order n). $\mathrm{SU}(\mathcal{H}) = \{U \in \mathrm{U}(\mathcal{H}) : \det(U) = 1\}$ is the *special unitary group* of $\mathcal{H}$. Clearly, $\mathrm{SU}(\mathcal{H}) \simeq \mathrm{SU}_n$ if $\mathcal{H}$ has dimension n.

References

1. Dyson, F.J.: Why is Maxwell's theory so hard to understand?. In: James Clerk Maxwell Commemorative Booklet, pp. 1–6. James Clerk Maxwell Foundation, Edinburgh (1999)
2. Xambó-Descamps, S.: Geometric Algebra Speaks Quantum Esperanto, II (2022, in preparation). Integrated in [7]
3. Farmelo, G.: The Strangest Man: The Hidden Life of Paul Dirac. Faber & Faber (2009)
4. Galison, P.: The suppressed drawing: Paul Dirac's hidden geometry. Representations **72**, 145–166 (2000). https://datascience.iq.harvard.edu/files/andrewhsmith/files/2902912.pdf
5. Weinert, F.: Wrong theory-right experiment: the significance of the Stern-Gerlach experiments. Stud. History Philos. Sci. Part B: Stud. History Philos. Mod. Phys. **26**(1), 75–86 (1995)
6. vand der Waerden, B.: Spinoranalyse. Nachrichten von der Gesellschaft der Wissenschaften zu Göttingen, Mathematisch-Physikalische Klasse, vol. 2, pp. 100–109 (1929)
7. Xambó-Descamps, S.: Geometric algebra speaks quantum Esperanto. To appear in Advances in Applied Clifford Algebras, Topical Collection ICACGA 2022 (2022)

On Some Lie Groups in Degenerate Geometric Algebras

Ekaterina Filimoshina[1(✉)] and Dmitry Shirokov[1,2]

[1] HSE University, 101000 Moscow, Russia
erfilimoshina@edu.hse.ru, dshirokov@hse.ru
[2] Institute for Information Transmission Problems of the Russian Academy of Sciences, 127051 Moscow, Russia
shirokov@iitp.ru

Abstract. In this paper, we introduce and study several Lie groups in degenerate (Clifford) geometric algebras. These Lie groups preserve the even and odd subspaces under the adjoint representation and the twisted adjoint representation. The considered Lie groups are interesting for the study of spin groups and their generalizations in degenerate case.

Keywords: Geometric algebra · Clifford algebra · Lie group · Degenerate geometric algebra

1 Introduction

In this paper, we study degenerate (Clifford) geometric algebras $\mathcal{G}_{p,q,r}$ of arbitrary dimension and signature. Degenerate geometric algebras are important for applications in geometry, computer science, engineering, signal and image processing, physics, etc. For instance, projective geometric algebra (PGA) $\mathcal{G}_{p,0,1}$ is useful for computations with flat objects and is applied in computer graphics and vision, robotics, motion capture, dynamics simulations [4,6,14,19,20]. PGA can be realized as a subalgebra of conformal geometric algebra (CGA) [13,22,25,26,28], which has applications in pose estimation, robotics, computer animation, machine learning, neural networks, etc. [15–17,23,24,33]. The algebras $\mathcal{G}_{3,0,1}$, $\mathcal{G}_{0,3,1}$, even subalgebras $\mathcal{G}^{(0)}_{3,0,1}$ (known as the motor algebra), $\mathcal{G}^{(0)}_{0,3,1}$, $\mathcal{G}^{(0)}_{6,0,2}$, $\mathcal{G}^{(0)}_{6,0,6}$ are applied in robotics and computer vision [4,5,31].

We introduce and study several Lie groups in degenerate geometric algebras. These groups are closely related to the degenerate spin groups, and that is why they are interesting for consideration. These groups preserve the even subspace $\mathcal{G}^{(0)}_{p,q,r}$ and the odd subspace $\mathcal{G}^{(1)}_{p,q,r}$ under the adjoint representation and the twisted adjoint representation. The twisted adjoint representation was introduced in the classic paper [3], and it is an important mathematical notion that is used to describe two-sheeted coverings of orthogonal groups by spin groups. The Lie groups introduced in this paper are important for studying spin groups and their generalizations in degenerate case. Degenerate spin groups, degenerate

D. W. Silva et al. (Eds.): ICACGA 2022, LNCS 13771, pp. 186–198, 2024.
https://doi.org/10.1007/978-3-031-34031-4_16

orthogonal groups, and their applications in physics are discussed in the papers [1,7–9,11,12]. This paper generalizes the results of the papers [18,32] on Lie groups in non-degenerate geometric algebras $\mathcal{G}_{p,q}$ to the case of degenerate geometric algebras $\mathcal{G}_{p,q,r}$.

In Sect. 2, we discuss degenerate geometric algebras, the Jacobson radical, and some auxiliary statements. In Sect. 3, we present some statements on the adjoint representation and the twisted adjoint representation in degenerate geometric algebras $\mathcal{G}_{p,q,r}$. In Sect. 4, we introduce the four groups $\mathrm{P}^{\pm}_{p,q,r}$, $\mathrm{P}_{p,q,r}$, $\mathrm{P}^{\pm\Lambda}_{p,q,r}$, $\mathrm{P}^{\Lambda}_{p,q,r}$ and discuss several equivalent definitions of these groups. In Sect. 5, we prove that the considered four groups preserve the even subspace and the odd subspace under the adjoint representation and the twisted adjoint representation in $\mathcal{G}_{p,q,r}$. Theorems 1, 2, and 3 are new. We consider the Lie algebras of the four Lie groups in Sect. 6. The conclusions follow in Sect. 7.

2 Degenerate Geometric Algebra and the Jacobson Radical

Let us consider the (Clifford) geometric algebra [21,29,30] $\mathcal{G}(V) = \mathcal{G}_{p,q,r}$, $p+q+r = n \geq 1$, over a vector space V with a symmetric bilinear form g. We consider the real case $V = \mathbb{R}^{p,q,r}$ and the complex case $V = \mathbb{C}^{p+q,r}$. We use $\mathbb{F}$ to denote the field of real numbers $\mathbb{R}$ in the first case and the field of complex numbers $\mathbb{C}$ in the second case respectively. In this paper, we concentrate on the degenerate geometric algebras with $r \neq 0$, but all the following statements are true for arbitrary $r \geq 0$.

We denote the identity element of the algebra $\mathcal{G}_{p,q,r}$ by e, the generators by e_a, $a = 1, \ldots, n$. In the case of the real geometric algebra $\mathcal{G}(\mathbb{R}^{p,q,r})$, the generators satisfy $e_a e_b + e_b e_a = 2\eta_{ab} e$, $a, b = 1, \ldots, n$, where $\eta = (\eta_{ab})$ is the diagonal matrix with p times 1, q times -1 and r times 0 on the diagonal. In the case of the complex geometric algebra $\mathcal{G}(\mathbb{C}^{p+q,r})$, the generators satisfy the same conditions but with the diagonal matrix η with $p+q$ times 1 and r times 0 on the diagonal. Let us denote by $\Lambda_r := \mathcal{G}_{0,0,r}$ the subalgebra of $\mathcal{G}_{p,q,r}$, which is the Grassmann (exterior) algebra [11,13,29].

Consider the subspaces $\mathcal{G}^k_{p,q,r}$ of grades $k = 0, 1, \ldots, n$, which elements are linear combinations of the basis elements $e_{a_1 \ldots a_k} := e_{a_1} \cdots e_{a_k}$, $a_1 < \cdots < a_k$, with ordered multi-indices of length k. Note that the subspace $\mathcal{G}^0_{p,q,r}$ of grade 0 does not depend on the signature of the algebra, so we denote it by $\mathcal{G}^0$ without the lower indices p, q, r.

The grade involute of the element $U \in \mathcal{G}_{p,q,r}$ is denoted by $\widehat{U}$. This operation has the following well-known property: $\widehat{UV} = \widehat{U}\widehat{V}$ for any $U, V \in \mathcal{G}_{p,q,r}$. Consider the even $\mathcal{G}^{(0)}_{p,q,r}$ and odd $\mathcal{G}^{(1)}_{p,q,r}$ subspaces: $\mathcal{G}^{(k)}_{p,q,r} = \{U \in \mathcal{G}_{p,q,r} : \quad \widehat{U} = (-1)^k U\} = \bigoplus_{j=k \bmod 2} \mathcal{G}^j_{p,q,r}$, $k = 0, 1$, with the property

$$\mathcal{G}^{(k)}_{p,q,r} \mathcal{G}^{(l)}_{p,q,r} \subset \mathcal{G}^{(k+l) \bmod 2}_{p,q,r}, \qquad k, l = 0, 1. \tag{1}$$

Let us consider the Jacobson radical rad $\mathcal{G}_{p,q,r}$ of the algebra $\mathcal{G}_{p,q,r}$. Let A, B, C be ordered multi-indices with the non-zero length and $e_A = e_{a_1} \cdots e_{a_k}$

with $\{a_1,\ldots,a_k\} \subseteq \{1,\ldots,p\}$, $e_B = e_{b_1}\cdots e_{b_l}$ with $\{b_1,\ldots,b_l\} \subseteq \{p+1,\ldots,p+q\}$, $e_C = e_{c_1}\cdots e_{c_m}$ with $\{c_1,\ldots,c_m\} \subseteq \{p+q+1,\ldots,n\}$. An arbitrary element $y \in \operatorname{rad} \mathcal{G}_{p,q,r}$ has the form $y = \sum_C v_C e_C + \sum_{A,C} v_{AC} e_A e_C + \sum_{B,C} v_{BC} e_B e_C + \sum_{A,B,C} v_{ABC} e_A e_B e_C$, where $v_C, v_{AC}, v_{BC}, v_{ABC} \in \mathbb{F}$.

Remark 1. Any element of the Jacobson radical is non-invertible (see [27]). The Jacobson radical of the Grassmann algebra $\mathcal{G}_{0,0,n} = \Lambda_n$ is the direct sum of the subspaces of grades $1,\ldots,n$:

$$\operatorname{rad} \mathcal{G}_{0,0,n} = \mathcal{G}^1_{0,0,n} \oplus \mathcal{G}^2_{0,0,n} \oplus \cdots \oplus \mathcal{G}^n_{0,0,n}, \quad \mathcal{G}_{0,0,n} = \mathcal{G}^0 \oplus \operatorname{rad} \mathcal{G}_{0,0,n}.$$

The non-degenerate algebra $\mathcal{G}_{p,q,0}$ is semi-simple and $\operatorname{rad} \mathcal{G}_{p,q,0} = \{0\}$ ([1,11, 27]).

We need the following well-known (see, for example, [2,27]) lemma.

Lemma 1. *The element $e + xy$ is invertible for any $y \in \operatorname{rad} \mathcal{G}_{p,q,r}$, $x \in \mathcal{G}_{p,q,r}$.*

The subset of invertible elements of any set is denoted with $\times$. For example, we denote the group of invertible elements of the algebra $\mathcal{G}_{p,q,r}$ by $\mathcal{G}^{\times}_{p,q,r}$.

Lemma 2. *The element $T \in \mathcal{G}^0 \oplus \operatorname{rad} \mathcal{G}_{p,q,r}$ is invertible if and only if its projection on grade 0 is non-zero:*

$$T \in \mathcal{G}^0 \oplus \operatorname{rad} \mathcal{G}_{p,q,r}, \quad \langle T \rangle_0 \neq 0 \qquad \Leftrightarrow \qquad T \in (\mathcal{G}^0 \oplus \operatorname{rad} \mathcal{G}_{p,q,r})^{\times}.$$

Proof. Suppose $\langle T \rangle_0 \neq 0$ for some $T = \alpha e + W = \alpha(e + \frac{1}{\alpha} W)$, where $\alpha \in \mathbb{F}^{\times}$, $W \in \operatorname{rad} \mathcal{G}_{p,q,r}$. We have $e + \frac{1}{\alpha} W \in \mathcal{G}^{\times}_{p,q,r}$ by Lemma 1; thus, $T \in (\mathcal{G}^0 \oplus \operatorname{rad} \mathcal{G}_{p,q,r})^{\times}$.

Suppose $T \in (\mathcal{G}^0 \oplus \operatorname{rad} \mathcal{G}_{p,q,r})^{\times}$. Assume $\langle T \rangle_0 = 0$; then $T \in \operatorname{rad} \mathcal{G}^{\times}_{p,q,r}$, and we get a contradiction by Remark 1. □

Remark 2. The inverse of any invertible $T = \alpha e + \beta e_{1\ldots n} \in (\mathcal{G}^0 \oplus \mathcal{G}^n_{p,q,r})^{\times}$, where $\alpha, \beta \in \mathbb{F}$, has the form $T^{-1} = \alpha e - \beta e_{1\ldots n} \in (\mathcal{G}^0 \oplus \mathcal{G}^n_{p,q,r})^{\times}$, since $(\alpha e + \beta e_{1\ldots n})(\alpha e - \beta e_{1\ldots n}) = \alpha^2 e - \beta^2 (e_{1\ldots n})^2 \in \mathcal{G}^0$.

3 Adjoint and Twisted Adjoint Representations in $\mathcal{G}_{p,q,r}$

Consider the adjoint representation ad and the twisted adjoint representation $\check{\mathrm{ad}}$ acting on the group of all invertible elements ad, $\check{\mathrm{ad}} : \mathcal{G}^{\times}_{p,q,r} \to \operatorname{Aut}\mathcal{G}_{p,q,r}$ as $T \mapsto \mathrm{ad}_T$ and $T \mapsto \check{\mathrm{ad}}_T$ respectively, where $\mathrm{ad}_T(U) = TUT^{-1}$, $\check{\mathrm{ad}}_T(U) = \widehat{T} U T^{-1}$, $U \in \mathcal{G}_{p,q,r}$. It is well-known (see, for example, [1,9]) that the center of $\mathcal{G}_{p,q,r}$ is

$$\mathrm{Z}_{p,q,r} = \begin{cases} \Lambda^{(0)}_r \oplus \mathcal{G}^n_{p,q,r} & \text{if } n \text{ is odd}, \\ \Lambda^{(0)}_r & \text{if } n \text{ is even}. \end{cases} \tag{2}$$

Lemma 3. *We have*

$$\ker(\mathrm{ad}) = \mathrm{Z}^{\times}_{p,q,r} = \begin{cases} (\Lambda^{(0)}_r \oplus \mathcal{G}^n_{p,q,r})^{\times} & \text{if } n \text{ is odd}, \\ \Lambda^{(0)\times}_r & \text{if } n \text{ is even}, \end{cases} \qquad \ker(\check{\mathrm{ad}}) = \Lambda^{(0)\times}_r. \tag{3}$$

Proof. We obtain the statement on ker(ad) from (2). Let us prove $\Lambda_r^{(0)\times} \subseteq \ker(\check{\mathrm{ad}})$. Suppose $T \in \Lambda_r^{(0)\times}$; then $TUT^{-1} = U$ for any $U \in \mathcal{G}_{p,q,r}$. Since T is even, we have $\widehat{T} = T$; therefore, $\widehat{T}UT^{-1} = U$ for any $U \in \mathcal{G}_{p,q,r}$.

Let us prove $\ker(\check{\mathrm{ad}}) \subseteq \Lambda_r^{(0)\times}$. Suppose $T \in \mathcal{G}^{\times}_{p,q,r}$ satisfies $\widehat{T}UT^{-1} = U$ for any $U \in \mathcal{G}$. Substituting the element $U = e$, we obtain $\widehat{T} = T$; hence, $T \in \mathcal{G}^{(0)\times}_{p,q,r}$ and $TUT^{-1} = U$ for any $U \in \mathcal{G}_{p,q,r}$. In other words, $T \in \mathcal{G}^{(0)\times}_{p,q,r} \cap \ker(\mathrm{ad})$. Using (3), we obtain $T \in \mathcal{G}^{(0)\times}_{p,q,r} \cap (\Lambda_r^{(0)} \oplus \mathcal{G}^n_{p,q,r})^{\times} = \Lambda_r^{(0)\times}$ in the case of odd n, $T \in \Lambda_r^{(0)\times}$ in the case of even n, and the proof is completed. □

Lemma 4. *We have* $\{X \in \mathcal{G}_{p,q,r}: \quad \widehat{X}V = VX \quad \forall V \in \mathcal{G}^1_{p,q,r}\} = \Lambda_r$.

Lemma 5. *We have*

$$\{X \in \mathcal{G}_{p,q,r}: \quad XV = VX \quad \forall V \in \mathcal{G}^{(0)}_{p,q,r}\} = \Lambda_r \oplus \mathcal{G}^n_{p,q,r}, \tag{4}$$

$$\{X \in \mathcal{G}_{p,q,r}: \quad \widehat{X}V = VX \quad \forall V \in \mathcal{G}^{(0)}_{p,q,r}\} = \begin{cases} \Lambda_r^{(0)} & \text{if } n \text{ is odd,} \\ \Lambda_r^{(0)} \oplus \mathcal{G}^n_{p,q,r} & \text{if } n \text{ is even.} \end{cases} \tag{5}$$

Lemma 6. *Consider an arbitrary element* $X \in \mathcal{G}_{p,q,r}$ *and an arbitrary fixed subset* H *of the set* $\mathcal{G}^{(0)}_{p,q,r} \cup \mathcal{G}^{(1)}_{p,q,r}$. *We have*

$$\widehat{X}U = UX \quad \forall U \in \mathrm{H} \quad \Rightarrow \quad \widehat{X}(U_1 \cdots U_m) = (U_1 \cdots U_m)X \quad \forall U_1, \ldots, U_m \in \mathrm{H}$$

for any odd natural number m.

Proof. Lemmas 4–6 are proved in the similar way as the formula (2) and Lemma 3. The proof of these lemmas will be provided in the extended version of this paper. □

4 The Groups $\mathrm{P}^{\pm}_{p,q,r}$, $\mathrm{P}_{p,q,r}$, $\mathrm{P}^{\pm\Lambda}_{p,q,r}$, and $\mathrm{P}^{\Lambda}_{p,q,r}$

Let us denote by $\mathrm{S}_{p,q,r}$ the following subset of the center $\mathrm{Z}_{p,q,r}$ (2):

$$\mathrm{S}_{p,q,r} := \begin{cases} \mathcal{G}^0 \oplus \mathcal{G}^n_{p,q,r} & \text{if } n \text{ is odd,} \\ \mathcal{G}^0 & \text{if } n \text{ is even.} \end{cases} \tag{6}$$

Note that $\mathrm{S}_{p,q,r} \oplus (\Lambda_r^{(0)} \setminus \mathcal{G}^0) = \mathrm{Z}_{p,q,r}$. In the case of the non-degenerate algebra $\mathcal{G}_{p,q,0}$, we have $\mathrm{S}_{p,q,0} = \mathrm{Z}_{p,q,0}$. Let us consider the groups $\mathrm{P}^{\pm}_{p,q,r}$ and $\mathrm{P}_{p,q,r}$:

$$\mathrm{P}^{\pm}_{p,q,r} := \mathcal{G}^{(0)\times}_{p,q,r} \cup \mathcal{G}^{(1)\times}_{p,q,r}, \tag{7}$$

$$\mathrm{P}_{p,q,r} := \mathrm{P}^{\pm}_{p,q,r}\mathrm{Z}^{\times}_{p,q,r} = \begin{cases} (\mathcal{G}^{(0)\times}_{p,q,r} \cup \mathcal{G}^{(1)\times}_{p,q,r})(\Lambda_r^{(0)} \oplus \mathcal{G}^n_{p,q,r})^{\times} & \text{if } n \text{ is odd,} \\ (\mathcal{G}^{(0)\times}_{p,q,r} \cup \mathcal{G}^{(1)\times}_{p,q,r})\Lambda_r^{(0)\times} & \text{if } n \text{ is even,} \end{cases} \tag{8}$$

$$= \mathrm{P}^{\pm}_{p,q,r}\mathrm{S}^{\times}_{p,q,r} = \begin{cases} (\mathcal{G}^{(0)\times}_{p,q,r} \cup \mathcal{G}^{(1)\times}_{p,q,r})(\mathcal{G}^0 \oplus \mathcal{G}^n_{p,q,r})^{\times} & \text{if } n \text{ is odd,} \\ \mathcal{G}^{(0)\times}_{p,q,r} \cup \mathcal{G}^{(1)\times}_{p,q,r} & \text{if } n \text{ is even,} \end{cases} \tag{9}$$

where we get (9) by Lemma 7. In the particular case $\mathcal{G}_{p,q,0}$, we obtain the groups from the paper [32]:

$$\mathrm{P}^{\pm}_{p,q,0} = \mathrm{P}^{\pm} = \mathcal{G}^{(0)\times}_{p,q,0} \cup \mathcal{G}^{(1)\times}_{p,q,0}, \qquad \mathrm{P}_{p,q,0} = \mathrm{P} = \mathrm{Z}^{\times}_{p,q,0}(\mathcal{G}^{(0)\times}_{p,q,0} \cup \mathcal{G}^{(1)\times}_{p,q,0}). \quad (10)$$

Lemma 7. *In the case of arbitrary n, we have*

$$(\mathcal{G}^{(0)\times}_{p,q,r} \cup \mathcal{G}^{(1)\times}_{p,q,r})\Lambda^{(0)\times}_r = \mathcal{G}^{(0)\times}_{p,q,r} \cup \mathcal{G}^{(1)\times}_{p,q,r}, \quad (11)$$
$$(\mathcal{G}^{(0)\times}_{p,q,r} \cup \mathcal{G}^{(1)\times}_{p,q,r})(\Lambda^{(0)}_r \oplus \mathcal{G}^{n}_{p,q,r})^{\times} = (\mathcal{G}^{(0)\times}_{p,q,r} \cup \mathcal{G}^{(1)\times}_{p,q,r})(\mathcal{G}^0 \oplus \mathcal{G}^{n}_{p,q,r})^{\times}. \quad (12)$$

Proof. The statement (11) is true by (1). The proof of the statement (12) in the case $r = 0$ is trivial, since $\Lambda^{(0)}_0 = \mathcal{G}^0$. Consider the case $r \neq 0$. The right set in (12) is a subset of the left one. Let us prove that the left set in (12) is a subset of the right one. Suppose $T = AW$, where $A \in \mathcal{G}^{(0)\times}_{p,q,r} \cup \mathcal{G}^{(1)\times}_{p,q,r}$ and $W = \alpha e + X + \beta e_{1\ldots n} \in (\Lambda^{(0)}_r \oplus \mathcal{G}^n_{p,q,r})^{\times}$, $\alpha, \beta \in \mathbb{F}$ and $X \in \Lambda^{(0)}_r \setminus \mathcal{G}^0$. Since W is invertible, $\alpha \neq 0$ by Lemma 2. Then we get $W = (e + \frac{1}{\alpha}X)(\alpha e + \beta e_{1\ldots n}) \in \Lambda^{(0)\times}_r(\mathcal{G}^0 \oplus \mathcal{G}^n_{p,q,r})^{\times}$, where the first factor is invertible by Lemma 2. Hence, $T = AW = A(e + \frac{1}{\alpha}X)(\alpha e + \beta e_{1\ldots n}) \in (\mathcal{G}^{(0)\times}_{p,q,r} \cup \mathcal{G}^{(1)\times}_{p,q,r})\Lambda^{(0)\times}_r(\mathcal{G}^0 \oplus \mathcal{G}^n_{p,q,r})^{\times} = (\mathcal{G}^{(0)\times}_{p,q,r} \cup \mathcal{G}^{(1)\times}_{p,q,r})(\mathcal{G}^0 \oplus \mathcal{G}^n_{p,q,r})^{\times}$, and the proof is completed. □

Also let us consider the groups $\mathrm{P}^{\pm\Lambda}_{p,q,r}$ and $\mathrm{P}^{\Lambda}_{p,q,r}$:

$$\mathrm{P}^{\pm\Lambda}_{p,q,r} := \Lambda^{\times}_r \mathrm{P}^{\pm}_{p,q,r} = \mathrm{P}^{\pm}_{p,q,r}\Lambda^{\times}_r = (\mathcal{G}^{(0)\times}_{p,q,r} \cup \mathcal{G}^{(1)\times}_{p,q,r})\Lambda^{\times}_r, \quad (13)$$
$$\mathrm{P}^{\Lambda}_{p,q,r} := \Lambda^{\times}_r \mathrm{P}_{p,q,r} = \mathrm{P}_{p,q,r}\Lambda^{\times}_r \quad (14)$$
$$= \mathrm{P}^{\pm\Lambda}_{p,q,r}\mathrm{Z}^{\times}_{p,q,r} = \begin{cases} (\mathcal{G}^{(0)\times}_{p,q,r} \cup \mathcal{G}^{(1)\times}_{p,q,r})(\Lambda^{(0)}_r \oplus \mathcal{G}^n_{p,q,r})^{\times}\Lambda^{\times}_r & \text{if } n \text{ is odd,} \\ \mathrm{P}^{\pm\Lambda}_{p,q,r} = (\mathcal{G}^{(0)\times}_{p,q,r} \cup \mathcal{G}^{(1)\times}_{p,q,r})\Lambda^{\times}_r & \text{if } n \text{ is even,} \end{cases} \quad (15)$$
$$= \mathrm{P}^{\pm\Lambda}_{p,q,r}\mathrm{S}^{\times}_{p,q,r} = \begin{cases} (\mathcal{G}^{(0)\times}_{p,q,r} \cup \mathcal{G}^{(1)\times}_{p,q,r})(\mathcal{G}^0 \oplus \mathcal{G}^n_{p,q,r})^{\times}\Lambda^{\times}_r & \text{if } n \text{ is odd,} \\ (\mathcal{G}^{(0)\times}_{p,q,r} \cup \mathcal{G}^{(1)\times}_{p,q,r})\Lambda^{\times}_r & \text{if } n \text{ is even,} \end{cases} \quad (16)$$
$$= \begin{cases} (\mathcal{G}^{(0)\times}_{p,q,r} \cup \mathcal{G}^{(1)\times}_{p,q,r})(\Lambda_r \oplus \mathcal{G}^n_{p,q,r})^{\times} & \text{if } n \text{ is odd,} \\ (\mathcal{G}^{(0)\times}_{p,q,r} \cup \mathcal{G}^{(1)\times}_{p,q,r})\Lambda^{\times}_r & \text{if } n \text{ is even,} \end{cases} \quad (17)$$

where we get (16) and (17) by Lemma 8.

Lemma 8. *In the case of arbitrary n, we have*

$$(\Lambda^{(0)}_r \oplus \mathcal{G}^n_{p,q,r})^{\times}\Lambda^{\times}_r = (\mathcal{G}^0 \oplus \mathcal{G}^n_{p,q,r})^{\times}\Lambda^{\times}_r = (\Lambda_r \oplus \mathcal{G}^n_{p,q,r})^{\times}. \quad (18)$$

Proof. In the case $r = 0$, the proof of the equalities is trivial, since $\Lambda_0 = \Lambda^{(0)}_0 = \mathcal{G}^0$. Consider the case $r \neq 0$. By multiplying the factors in the first and the second sets in (18), we get that each of these sets is a subset of the third one in (18).

Let us show that the third set in (18) is a subset of the first two ones. Suppose $T = \alpha e + X + \beta e_{1\ldots n} \in (\Lambda_r \oplus \mathcal{G}^n_{p,q,r})^{\times}$, where $\alpha, \beta \in \mathbb{F}$ and $X \in \Lambda_r \setminus \mathcal{G}^0$.

Since T is invertible, $\alpha \neq 0$ by Lemma 2. Then $T = (\alpha e + \beta e_{1\ldots n})(e + \frac{1}{\alpha}X) \in (\mathcal{G}^0 \oplus \mathcal{G}^n_{p,q,r})^{\times}\Lambda^{\times}_r \subseteq (\Lambda^{(0)}_r \oplus \mathcal{G}^n_{p,q,r})^{\times}\Lambda^{\times}_r$, where the second factor is invertible by Lemma 2, and the proof is completed. □

Remark 3. The groups $\mathrm{P}^{\pm}_{p,q,r}$, $\mathrm{P}_{p,q,r}$, $\mathrm{P}^{\pm\Lambda}_{p,q,r}$, and $\mathrm{P}^{\Lambda}_{p,q,r}$ are related as follows:

$$\mathrm{P}_{p,q,r} = \mathrm{P}^{\pm}_{p,q,r}\mathrm{Z}^{\times}_{p,q,r} = \mathrm{P}^{\pm}_{p,q,r}\mathrm{S}^{\times}_{p,q,r}, \qquad \mathrm{P}^{\pm\Lambda}_{p,q,r} = \mathrm{P}^{\pm}_{p,q,r}\Lambda^{\times}_r, \tag{19}$$

$$\mathrm{P}^{\Lambda}_{p,q,r} = \mathrm{P}^{\pm}_{p,q,r}\mathrm{Z}^{\times}_{p,q,r}\Lambda^{\times}_r = \mathrm{P}^{\pm}_{p,q,r}\mathrm{S}^{\times}_{p,q,r}\Lambda^{\times}_r, \tag{20}$$

$\mathrm{P}^{\pm}_{p,q,r}$ is a subgroup of the groups $\mathrm{P}_{p,q,r}$, $\mathrm{P}^{\Lambda}_{p,q,r}$, $\mathrm{P}^{\pm\Lambda}_{p,q,r}$, and the groups $\mathrm{P}^{\pm}_{p,q,r}$, $\mathrm{P}_{p,q,r}$, $\mathrm{P}^{\pm\Lambda}_{p,q,r}$ are subgroups of $\mathrm{P}^{\Lambda}_{p,q,r}$.

Remark 4. In the particular case of the algebra $\mathcal{G}_{p,q,0}$, the groups $\mathrm{P}^{\pm\Lambda}_{p,q,r}$ and $\mathrm{P}^{\Lambda}_{p,q,r}$ coincide with the groups $\mathrm{P}^{\pm}$ and P respectively:

$$\mathrm{P}^{\pm\Lambda}_{p,q,0} = \mathrm{P}^{\pm}_{p,q,0} = \mathrm{P}^{\pm} \subseteq \mathrm{P}^{\Lambda}_{p,q,0} = \mathrm{P}_{p,q,0} = \mathrm{P}, \tag{21}$$

moreover, if $n = p + q$ is even, all the considered groups coincide.

Let us give some examples on the groups $\mathrm{P}^{\pm}_{p,q,r}$, $\mathrm{P}_{p,q,r}$, $\mathrm{P}^{\pm\Lambda}_{p,q,r}$, and $\mathrm{P}^{\Lambda}_{p,q,r}$ in the cases of the low-dimensional degenerate geometric algebras. We use that the degenerate geometric algebra can be embedded into the non-degenerate geometric algebra of larger dimension (see Clifford–Jordan–Wigner representation [10]), which is isomorphic to the matrix algebra (see, for example, [29, 30]).

Example 1. Consider the algebra $\Lambda_1 = \mathcal{G}_{0,0,1}$, which can be embedded into $\mathcal{G}_{1,1,0} \cong \mathrm{Mat}(2, \mathbb{F})$. We obtain $\mathrm{P}^{\pm}_{0,0,1} = \Lambda^{(0)\times}_1 = \mathcal{G}^{0\times}$ and

$$\mathrm{P}_{0,0,1} = \mathrm{P}^{\pm\Lambda}_{0,0,1} = \mathrm{P}^{\Lambda}_{0,0,1} = \Lambda^{\times}_1 \cong \{\begin{pmatrix} x_0 + x_1 & x_1 \\ -x_1 & x_0 - x_1 \end{pmatrix} : \quad x_0, x_1 \in \mathbb{F}, \; x_0 \neq 0\}.$$

Example 2. Since $\Lambda_2 = \mathcal{G}_{0,0,2}$ can be embedded into $\mathcal{G}_{2,2,0} \cong \mathrm{Mat}(4, \mathbb{F})$, we get

$$\mathrm{P}^{\pm\Lambda}_{0,0,2} = \mathrm{P}^{\Lambda}_{0,0,2} = \Lambda^{\times}_2 \cong \{\begin{pmatrix} x_0 & 0 & 0 & 0 \\ x_1 & x_0 & 0 & 0 \\ x_2 & 0 & x_0 & 0 \\ x_3 & -x_2 & x_1 & x_0 \end{pmatrix} : \quad x_0, x_1, x_2, x_3 \in \mathbb{F}, \quad x_0 \neq 0\},$$

$$\mathrm{P}^{\pm}_{0,0,2} = \mathrm{P}_{0,0,2} = \Lambda^{(0)\times}_2 \cong \{\begin{pmatrix} x_0 & 0 & 0 & 0 \\ 0 & x_0 & 0 & 0 \\ 0 & 0 & x_0 & 0 \\ x_3 & 0 & 0 & x_0 \end{pmatrix} : \quad x_0, x_3 \in \mathbb{F}, \quad x_0 \neq 0\}.$$

Remark 5. In the case of the Grassmann algebra $\mathcal{G}_{0,0,n} = \Lambda_n$, we have

$$\mathrm{P}^{\pm}_{0,0,n} = \mathrm{P}_{0,0,n} = \Lambda^{(0)\times}_n \subset \mathrm{P}^{\pm\Lambda}_{0,0,n} = \mathrm{P}^{\Lambda}_{0,0,n} = \Lambda^{\times}_n, \qquad n \text{ is even;}$$
$$\mathrm{P}^{\pm}_{0,0,n} = \Lambda^{(0)\times}_n \subset \mathrm{P}_{0,0,n} = (\Lambda^{(0)}_n \oplus \Lambda^n_n)^{\times} \subset \mathrm{P}^{\pm\Lambda}_{0,0,n} = \mathrm{P}^{\Lambda}_{0,0,n} = \Lambda^{\times}_n, \quad n \geq 3 \text{ is odd.}$$

The statements $\mathrm{P}^{\pm}_{0,0,n} = \Lambda^{(0)\times}_n$ and $\mathrm{P}_{0,0,n} = \Lambda^{(0)\times}_n(\mathcal{G}^0 \oplus \Lambda^n_n)^{\times}$ follow from Lemma 2, since any invertible element of $\Lambda^{\times}_n$ has the non-zero projection on grade 0 and, consequently, is not odd.

In Theorems 1 and 2, we give the equivalent definitions of the groups $\mathrm{P}_{p,q,r}$, $\mathrm{P}^{\pm}_{p,q,r}$, $\mathrm{P}^{\Lambda}_{p,q,r}$, and $\mathrm{P}^{\pm\Lambda}_{p,q,r}$. We use these definitions to prove Theorem 3. Note that $\mathcal{G}^0 \subseteq \Lambda_r^{(0)} \subseteq \mathcal{G}^0 \oplus \mathrm{rad}\, \mathcal{G}^{(0)}_{p,q,r}$ in (23)–(25) and (27)–(29).

Theorem 1. *We have the following equivalent definitions of the group* $\mathrm{P}_{p,q,r}$*:*

$$\mathrm{P}_{p,q,r} = \begin{cases} (\mathcal{G}^{(0)\times}_{p,q,r} \cup \mathcal{G}^{(1)\times}_{p,q,r})(\mathcal{G}^0 \oplus \mathcal{G}^n_{p,q,r})^\times, & n \text{ is odd,} \\ \mathcal{G}^{(0)\times}_{p,q,r} \cup \mathcal{G}^{(1)\times}_{p,q,r}, & n \text{ is even,} \end{cases} \tag{22}$$

$$= \{T \in \mathcal{G}^\times_{p,q,r} : \quad \widehat{T^{-1}}T \in \mathrm{S}^\times_{p,q,r} = \begin{cases} (\mathcal{G}^0 \oplus \mathcal{G}^n_{p,q,r})^\times, & n \text{ is odd,} \\ \mathcal{G}^{0\times}, & n \text{ is even,} \end{cases} \} \tag{23}$$

$$= \{T \in \mathcal{G}^\times_{p,q,r} : \widehat{T^{-1}}T \in \ker(\mathrm{ad}) = \begin{cases} (\Lambda_r^{(0)} \oplus \mathcal{G}^n_{p,q,r})^\times, & n \text{ is odd,} \\ \Lambda_r^{(0)\times}, & n \text{ is even,} \end{cases} \} \tag{24}$$

$$= \begin{cases} \{T \in \mathcal{G}^\times_{p,q,r} : \widehat{T^{-1}}T \in (\mathcal{G}^0 \oplus \mathrm{rad}\, \mathcal{G}^{(0)}_{p,q,r} \oplus \mathcal{G}^n_{p,q,r})^\times\}, & n \text{ is odd,} \\ \{T \in \mathcal{G}^\times_{p,q,r} : \widehat{T^{-1}}T \in (\mathcal{G}^0 \oplus \mathrm{rad}\, \mathcal{G}^{(0)}_{p,q,r})^\times\}, & n \text{ is even,} \end{cases} \tag{25}$$

and the group $\mathrm{P}^{\pm}_{p,q,r}$*:*

$$\mathrm{P}^{\pm}_{p,q,r} = \mathcal{G}^{(0)\times}_{p,q,r} \cup \mathcal{G}^{(1)\times}_{p,q,r} \tag{26}$$

$$= \{T \in \mathcal{G}^\times_{p,q,r} : \widehat{T^{-1}}T \in \mathcal{G}^{0\times}\} \tag{27}$$

$$= \{T \in \mathcal{G}^\times_{p,q,r} : \widehat{T^{-1}}T \in \ker(\check{\mathrm{ad}})\} = \{T \in \mathcal{G}^\times_{p,q,r} : \widehat{T^{-1}}T \in \Lambda_r^{(0)\times}\} \tag{28}$$

$$= \{T \in \mathcal{G}^\times_{p,q,r} : \widehat{T^{-1}}T \in (\mathcal{G}^0 \oplus \mathrm{rad}\, \mathcal{G}^{(0)}_{p,q,r})^\times\}. \tag{29}$$

Proof. First let us prove (22)–(25). Let us prove that the set (22) is a subset of the set (23). Suppose $T = AB \in \mathrm{P}_{p,q,r}$, where $A \in \mathcal{G}^{(0)\times}_{p,q,r} \cup \mathcal{G}^{(1)\times}_{p,q,r}$ and $B \in (\mathcal{G}^0 \oplus \mathcal{G}^n_{p,q,r})^\times$ in the case of odd n, $B = e$ in the case of even n. Then $\widehat{T^{-1}}T = (\widehat{AB})^{-1}(AB) = \widehat{B^{-1}}\widehat{A^{-1}}AB = \pm\widehat{B^{-1}}A^{-1}AB = \pm\widehat{B^{-1}}B$. We have $\widehat{B^{-1}}B \in (\mathcal{G}^0 \oplus \mathcal{G}^n_{p,q,r})^\times$ in the case of odd n (see Remark 2) and $\widehat{B^{-1}}B = e \in \mathcal{G}^{0\times}$ in the case of even n, and the proof is completed. The set (23) is a subset of the set (24), which is a subset of the set (25), since $\mathcal{G}^0 \subseteq \Lambda_r^{(0)} \subseteq \mathcal{G}^0 \oplus \mathrm{rad}\, \mathcal{G}^{(0)}_{p,q,r}$.

Let us prove that the set (25) is a subset of the set (22) in the case of even n. Suppose $T \in \mathcal{G}^\times$ satisfies $\widehat{T^{-1}}T = W_0 \in (\mathcal{G}^0 \oplus \mathrm{rad}\, \mathcal{G}^{(0)}_{p,q,r})^\times$; then $T = \widehat{T}W_0$. Suppose $T = T_0 + T_1$, where $T_0 \in \mathcal{G}^{(0)}_{p,q,r}$, $T_1 \in \mathcal{G}^{(1)}_{p,q,r}$. Then $T_0 + T_1 = (T_0 - T_1)W_0$, i.e. $T_0(e - W_0) = 0$, $T_1(e + W_0) = 0$. If at least one of the elements $e - W_0$ and $e + W_0$ is invertible, then we get either $T_0 = 0$ or $T_1 = 0$. Thus, $T \in \mathcal{G}^{(0)\times}_{p,q,r} \cup \mathcal{G}^{(1)\times}_{p,q,r} = \mathrm{P}^{\pm}_{p,q,r}$, and the proof is completed. Let us show that at least one of the elements $e - W_0$ and $e + W_0$ is invertible. Note that at least one of these elements has the non-zero projection on grade 0, since otherwise we can sum the equations $\langle e - W_0 \rangle_0 = 0$, $\langle e + W_0 \rangle_0 = 0$ and get $\langle 2e \rangle_0 = 0$, i.e. a contradiction, where we use the linearity of the projection operator. Then we obtain either $e - W_0 \in (\mathcal{G}^0 \oplus \mathrm{rad}\, \mathcal{G}^{(0)}_{p,q,r})^\times$ or $e + W_0 \in (\mathcal{G}^0 \oplus \mathrm{rad}\, \mathcal{G}^{(0)}_{p,q,r})^\times$ by Lemma 2.

Let us prove that the set (25) is a subset of the set (22) in the case of odd n. This statement is proved in the particular case $\mathcal{G}_{p,q,0}$ in the paper [32] (see Theorem 3.2). Consider the case $r \neq 0$. Suppose $\widehat{T^{-1}}T = W_0 + \beta e_{1\ldots n} \in (\mathcal{G}^0 \oplus \mathrm{rad}\, \mathcal{G}^{(0)}_{p,q,r} \oplus \mathcal{G}^n_{p,q,r})^\times$, where $W_0 \in \mathcal{G}^0 \oplus \mathrm{rad}\, \mathcal{G}^{(0)}_{p,q,r}$ and $\beta \in \mathbb{F}$. Then $T = \widehat{T}(W_0 + \beta e_{1\ldots n})$. Suppose $T = T_0 + T_1$, where $T_0 \in \mathcal{G}^{(0)}_{p,q,r}$, $T_1 \in \mathcal{G}^{(1)}_{p,q,r}$. Then we get $T_0 = T_0 W_0 - \beta T_1 e_{1\ldots n}$, $T_1 = -T_1 W_0 + \beta T_0 e_{1\ldots n}$; therefore, $T_0(e - W_0) = -\beta T_1 e_{1\ldots n}$, $T_1(e + W_0) = \beta T_0 e_{1\ldots n}$. As shown above, at least one of the elements $e - W_0$ and $e + W_0$ is invertible; hence, we obtain one of the two following equations:

$$T_0 = -\beta T_1 e_{1\ldots n}(e - W_0)^{-1}, \qquad T_1 = \beta T_0 e_{1\ldots n}(e + W_0)^{-1}. \tag{30}$$

Therefore, either $T_0 = \lambda T_1 e_{1\ldots n}$ or $T_1 = \mu T_0 e_{1\ldots n}$, where $\lambda, \mu \in \mathbb{F}$, and we use that $(e - W_0)^{-1}, (e + W_0)^{-1} \in \mathcal{G}^0 \oplus \mathrm{rad}\, \mathcal{G}^{(0)}_{p,q,r}$ and $e_{1\ldots n}\mathrm{rad}\, \mathcal{G}^{(0)}_{p,q,r} = 0$. Then we have either $T = T_0 + T_1 = T_1(e + \lambda e_{1\ldots n}) \in \mathcal{G}^{(1)\times}_{p,q,r}(\mathcal{G}^0 \oplus \mathcal{G}^n_{p,q,r})^\times$ or $T = T_0(e + \mu e_{1\ldots n}) \in \mathcal{G}^{(0)\times}_{p,q,r}(\mathcal{G}^0 \oplus \mathcal{G}^n_{p,q,r})^\times$, where in both cases, the second factor is invertible by Lemma 2. Thus, $T \in (\mathcal{G}^{(0)\times}_{p,q,r} \cup \mathcal{G}^{(1)\times}_{p,q,r})(\mathcal{G}^0 \oplus \mathcal{G}^n_{p,q,r})^\times$.

Now let us prove (26)–(29). The set (26) is a subset of the set (27), since we obtain $\widehat{T^{-1}}T = \pm T^{-1}T = \pm e \in \mathcal{G}^{0\times}$ for any $T \in \mathcal{G}^{(0)\times}_{p,q,r} \cup \mathcal{G}^{(1)\times}_{p,q,r}$. The set (27) is a subset of the set (28), which is a subset of the set (29), since $\mathcal{G}^0 \subseteq \Lambda^{(0)}_r \subseteq \mathcal{G}^0 \oplus \mathrm{rad}\, \mathcal{G}^{(0)}_{p,q,r}$.

Let us prove that the set (29) is a subset of the set (26). In the case of even n, we have proved $\{T \in \mathcal{G}^\times : \quad \widehat{T^{-1}}T \in (\mathcal{G}^0 \oplus \mathrm{rad}\, \mathcal{G}^{(0)}_{p,q,r})^\times\} = \mathcal{G}^{(0)\times}_{p,q,r} \cup \mathcal{G}^{(1)\times}_{p,q,r}$ (see (25) and (22)). Consider the case of odd n. Suppose $\widehat{T^{-1}}T = W_0 + \beta e_{1\ldots n} \in \mathcal{G}^0 \oplus \mathrm{rad}\, \mathcal{G}^{(0)}_{p,q,r}$, where $\beta = 0$, $W_0 \in \mathcal{G}^0 \oplus \mathrm{rad}\, \mathcal{G}^{(0)}_{p,q,r}$. As shown above, we obtain one of the Eqs. (30). Since $\beta = 0$, we get either $T_0 = 0$ or $T_1 = 0$; thus, $T \in \mathcal{G}^{(0)\times}_{p,q,r} \cup \mathcal{G}^{(1)\times}_{p,q,r}$ and the proof is completed. □

Theorem 2. *We have the following equivalent definitions of the group* $\mathrm{P}^{\Lambda}_{p,q,r}$:

$$\mathrm{P}^{\Lambda}_{p,q,r} = \begin{cases} (\mathcal{G}^{(0)\times}_{p,q,r} \cup \mathcal{G}^{(1)\times}_{p,q,r})(\mathcal{G}^0 \oplus \mathcal{G}^n_{p,q,r})^\times \Lambda^\times_r, & n \text{ is odd,} \\ (\mathcal{G}^{(0)\times}_{p,q,r} \cup \mathcal{G}^{(1)\times}_{p,q,r})\Lambda^\times_r, & n \text{ is even,} \end{cases} \tag{31}$$

$$= \begin{cases} \{T \in \mathcal{G}^\times_{p,q,r} : \quad \widehat{T^{-1}}T \in (\Lambda_r \oplus \mathcal{G}^n_{p,q,r})^\times\}, & n \text{ is odd,} \\ \{T \in \mathcal{G}^\times_{p,q,r} : \quad \widehat{T^{-1}}T \in \Lambda^\times_r\}, & n \text{ is even,} \end{cases} \tag{32}$$

$$= \{T \in \mathcal{G}^\times_{p,q,r} : \quad \widehat{T^{-1}}T \in (\Lambda_r \oplus \mathcal{G}^n_{p,q,r})^\times\} \tag{33}$$

and the group $\mathrm{P}^{\pm\Lambda}_{p,q,r}$:

$$\mathrm{P}^{\pm\Lambda}_{p,q,r} = (\mathcal{G}^{(0)\times}_{p,q,r} \cup \mathcal{G}^{(1)\times}_{p,q,r})\Lambda^\times_r = \{T \in \mathcal{G}^\times_{p,q,r} : \quad \widehat{T^{-1}}T \in \Lambda^\times_r\} \tag{34}$$

$$= \begin{cases} \{T \in \mathcal{G}^\times_{p,q,r} : \quad \widehat{T^{-1}}T \in \Lambda^\times_r\}, & n \text{ is odd,} \\ \{T \in \mathcal{G}^\times_{p,q,r} : \quad \widehat{T^{-1}}T \in (\Lambda_r \oplus \mathcal{G}^n_{p,q,r})^\times\}, & n \text{ is even.} \end{cases} \tag{35}$$

Proof. This theorem is proved in the similar way as Theorem 1. The proof of this theorem will be provided in the extended version of this paper. □

5 The Groups Preserving the Subspaces of Fixed Parity Under the Adjoint and Twisted Adjoint Representations

We use the following notation for the groups preserving the subspaces of fixed parity under the adjoint representation: $\Gamma^{(k)} = \{T \in \mathcal{G}^{\times}_{p,q,r} : T\mathcal{G}^{(k)}_{p,q,r}T^{-1} \subseteq \mathcal{G}^{(k)}_{p,q,r}\}$, $k = 0, 1$, and the twisted adjoint representation: $\check{\Gamma}^{(k)} = \{T \in \mathcal{G}^{\times}_{p,q,r} : \widehat{T}\mathcal{G}^{(k)}_{p,q,r}T^{-1} \subseteq \mathcal{G}^{(k)}_{p,q,r}\}$, $k = 0, 1$.

Theorem 3. $\mathrm{P}_{p,q,r} = \Gamma^{(1)} \subseteq \mathrm{P}^{\Lambda}_{p,q,r} = \Gamma^{(0)}, \quad \mathrm{P}^{\pm}_{p,q,r} = \check{\Gamma}^{(0)} \subseteq \mathrm{P}^{\pm\Lambda}_{p,q,r} = \check{\Gamma}^{(1)}$.

Proof. The statements $\mathrm{P}_{p,q,r} \subseteq \mathrm{P}^{\Lambda}_{p,q,r}$ and $\mathrm{P}^{\pm}_{p,q,r} \subseteq \mathrm{P}^{\pm\Lambda}_{p,q,r}$ follow from the definitions of the groups (22), (9), (13), and (16).

Let us prove $\mathrm{P}^{\pm}_{p,q,r} \subseteq \check{\Gamma}^{(0)}$. Suppose $T \in \mathrm{P}^{\pm}_{p,q,r}$ (22). If $T \in \mathcal{G}^{(0)\times}_{p,q,r}$, then $\widehat{T} = T$ and $T^{-1} \in \mathcal{G}^{(0)\times}_{p,q,r}$. If $T \in \mathcal{G}^{(1)\times}_{p,q,r}$, then $\widehat{T} = -T$ and $T^{-1} \in \mathcal{G}^{(1)\times}_{p,q,r}$. In both cases, we obtain $\widehat{T}\mathcal{G}^{(0)}_{p,q,r}T^{-1} \subseteq \mathcal{G}^{(0)}_{p,q,r}$ by (1). Thus, $T \in \check{\Gamma}^{(0)}$. Let us prove $\mathrm{P}_{p,q,r} \subseteq \Gamma^{(1)}$. Suppose $T = XW \in \mathrm{P}_{p,q,r} = \mathrm{P}^{\pm}_{p,q,r}\mathrm{Z}^{\times}_{p,q,r}$ (15), where $X \in \mathcal{G}^{(0)\times}_{p,q,r} \cup \mathcal{G}^{(1)\times}_{p,q,r}$ and $W \in \mathrm{Z}^{\times}_{p,q,r}$. Then we get $T\mathcal{G}^{(1)}_{p,q,r}T^{-1} = XW\mathcal{G}^{(1)}_{p,q,r}W^{-1}X^{-1} = X\mathcal{G}^{(1)}_{p,q,r}WW^{-1}X^{-1} = X\mathcal{G}^{(1)}_{p,q,r}X^{-1} \subseteq \mathcal{G}^{(1)}_{p,q,r}$, where we use (1). Thus, $T \in \Gamma^{(1)}$.

Let us prove $\mathrm{P}^{\Lambda}_{p,q,r} \subseteq \Gamma^{(0)}$. Suppose $T = XW \in \mathrm{P}^{\Lambda}_{p,q,r}$ (31), where $X \in \mathcal{G}^{(0)\times}_{p,q,r} \cup \mathcal{G}^{(1)\times}_{p,q,r}$, $W \in (\Lambda_r \oplus \mathcal{G}^{n}_{p,q,r})^{\times}$ in the case of odd n and $W \in \Lambda^{\times}_r$ in the case of even n. We obtain $T\mathcal{G}^{(0)}_{p,q,r}T^{-1} = XW\mathcal{G}^{(0)}_{p,q,r}W^{-1}X^{-1} = X\mathcal{G}^{(0)}_{p,q,r}WW^{-1}X^{-1} = X\mathcal{G}^{(0)}_{p,q,r}X^{-1} \subseteq \mathcal{G}^{(0)}_{p,q,r}$, where we use the property (1) and that $W\mathcal{G}^{(0)}_{p,q,r} = \mathcal{G}^{(0)}_{p,q,r}W$ by Lemma 5. Thus, $T \in \Gamma^{(0)}$. Let us prove $\mathrm{P}^{\pm\Lambda}_{p,q,r} \subseteq \check{\Gamma}^{(1)}$. Suppose $T = XW \in \mathrm{P}^{\pm\Lambda}_{p,q,r}$ (13), where $X \in \mathcal{G}^{(0)\times}_{p,q,r} \cup \mathcal{G}^{(1)\times}_{p,q,r}$ and $W \in \Lambda^{\times}_r$. Since $\widehat{W}e_a = e_aW$ for any generator e_a, $a = 1, \ldots, n$, by Lemma 4 and since any odd basis element can be represented as the product of an odd number of generators, we get $\widehat{W}\mathcal{G}^{(1)}_{p,q,r} = \mathcal{G}^{(1)}_{p,q,r}W$ by Lemma 6. Then we obtain $\widehat{T}\mathcal{G}^{(1)}_{p,q,r}T^{-1} = \widehat{X}\widehat{W}\mathcal{G}^{(1)}_{p,q,r}W^{-1}X^{-1} = \pm X\mathcal{G}^{(1)}_{p,q,r}WW^{-1}X^{-1} = \pm X\mathcal{G}^{(1)}_{p,q,r}X^{-1} \subseteq \mathcal{G}^{(1)}_{p,q,r}$ by (1). Thus, $T \in \check{\Gamma}^{(1)}$.

Let us prove $\Gamma^{(1)} \subseteq \mathrm{P}_{p,q,r}$. Suppose $T \in \mathcal{G}^{\times}_{p,q,r}$ satisfies $T\mathcal{G}^{(1)}_{p,q,r}T^{-1} \subseteq \mathcal{G}^{(1)}_{p,q,r}$; then we obtain $TUT^{-1} = -(TUT^{-1})\widehat{\ } = \widehat{T}\widehat{U}\widehat{T^{-1}}$ for any $U \in \mathcal{G}^{(1)}_{p,q,r}$. Multiplying both sides of this equation on the left by $\widehat{T^{-1}}$, on the right by T, we get

$$(\widehat{T^{-1}}T)U = U(\widehat{T^{-1}}T), \qquad \forall U \in \mathcal{G}^{(1)}_{p,q,r}. \tag{36}$$

In particular, (36) is true for any generator $U = e_a \in \mathcal{G}^{(1)}_{p,q,r}$, $a = 1, \ldots, n$. Since the identity element $U = e \in \mathcal{G}^0$ satisfies (36) as well, we get $\mathrm{ad}_{\widehat{T^{-1}}T}(U) = U$ for any $U \in \mathcal{G}_{p,q,r}$. Therefore, $\widehat{T^{-1}}T \in \ker(\mathrm{ad})$. Thus, $T \in \mathrm{P}_{p,q,r}$ by Theorem 1.

Let us prove $\check{\Gamma}^{(1)} \subseteq \mathrm{P}^{\pm\Lambda}_{p,q,r}$. Suppose $T \in \mathcal{G}^{\times}_{p,q,r}$ satisfies $\widehat{T}\mathcal{G}^{(1)}_{p,q,r}T^{-1} \subseteq \mathcal{G}^{(1)}_{p,q,r}$. Then we get $\widehat{T}UT^{-1} = -(\widehat{T}UT^{-1})\widehat{\ } = T\widehat{U}\widehat{T^{-1}}$ for any $U \in \mathcal{G}^{(1)}_{p,q,r}$. Multiplying both sides of the equation on the left by T^{-1}, on the right by T, we obtain $T^{-1}\widehat{T}U = U\widehat{T^{-1}T}$, i.e. $(\widehat{\widehat{T^{-1}}T})U = U(\widehat{T^{-1}}T)$ for any $U \in \mathcal{G}^{(1)}_{p,q,r}$. In particular, this equation is true for any generator $U = e_a \in \mathcal{G}^1_{p,q,r}$, $a = 1,\ldots,n$. Using Lemma 4, we get $\widehat{T^{-1}}T \in \Lambda^\times_r$; hence, $T \in \mathrm{P}^{\pm\Lambda}_{p,q,r}$ by Theorem 2.

Let us prove $\Gamma^{(0)} \subseteq \mathrm{P}^{\Lambda}_{p,q,r}$. Suppose $T \in \mathcal{G}^\times_{p,q,r}$ satisfies $T\mathcal{G}^{(0)}_{p,q,r}T^{-1} \subseteq \mathcal{G}^{(0)}_{p,q,r}$. Then we get $TUT^{-1} = (TUT^{-1})\widehat{\ } = \widehat{T}U\widehat{T^{-1}}$ for any $U \in \mathcal{G}^{(0)}_{p,q,r}$. Multiplying both sides of this equation on the left by $\widehat{T^{-1}}$, on the right by T, we obtain $(\widehat{T^{-1}}T)U = U(\widehat{T^{-1}}T)$ for any $U \in \mathcal{G}^{(0)}_{p,q,r}$. Using Lemma 5, we have $\widehat{T^{-1}}T \in \Lambda_r \oplus \mathcal{G}^n_{p,q,r}$. Thus, $T \in \mathrm{P}^{\Lambda}_{p,q,r}$ by Theorem 2.

Let us prove $\check{\Gamma}^{(0)} \subseteq \mathrm{P}^{\pm}_{p,q,r}$. This statement is proved in the case $r = 0$ in the paper [18]. Consider the case $r \neq 0$. Suppose $T \in \mathcal{G}^\times_{p,q,r}$ satisfies $\widehat{T}\mathcal{G}^{(0)}_{p,q,r}T^{-1} \subseteq \mathcal{G}^{(0)}_{p,q,r}$. Then $\widehat{T}UT^{-1} = (\widehat{T}UT^{-1})\widehat{\ } = TU\widehat{T^{-1}}$ for any $U \in \mathcal{G}^{(0)}_{p,q,r}$. Multiplying both sides of this equation on the left by T^{-1}, on the right by T, we obtain $T^{-1}\widehat{T}U = U\widehat{T^{-1}}T$, i.e. $(\widehat{\widehat{T^{-1}}T})U = U(\widehat{T^{-1}}T)$ for any $U \in \mathcal{G}^{(0)}_{p,q,r}$. Using (5), we get $\widehat{T^{-1}}T \in (\Lambda^{(0)}_r \oplus \mathcal{G}^n_{p,q,r})^\times$ in the case of even n and $\widehat{T^{-1}}T \in \Lambda^{(0)\times}_r$ in the case of odd n. Therefore, $T \in \mathrm{P}^{\pm}_{p,q,r}$ by (28) in the case of odd n and by (29) in the case of even n, since $\Lambda^{(0)}_r \oplus \mathcal{G}^n_{p,q,r} \subseteq \mathcal{G}^0 \oplus \mathrm{rad}\,\mathcal{G}^{(0)}_{p,q,r}$. □

Remark 6. In the particular case $r = 0$, we have by (10) and (21):

$$\mathrm{P}^{\pm}_{p,q,0} = \mathrm{P}^{\pm\Lambda}_{p,q,0} = \check{\Gamma}^{(0)} = \check{\Gamma}^{(1)} \subset \mathrm{P}_{p,q,0} = \mathrm{P}^{\Lambda}_{p,q,0} = \Gamma^{(1)} = \Gamma^{(0)}, \qquad n \text{ is odd},$$
$$\mathrm{P}^{\pm}_{p,q,0} = \mathrm{P}^{\pm\Lambda}_{p,q,0} = \check{\Gamma}^{(0)} = \check{\Gamma}^{(1)} = \mathrm{P}_{p,q,0} = \mathrm{P}^{\Lambda}_{p,q,0} = \Gamma^{(1)} = \Gamma^{(0)}, \qquad n \text{ is even}.$$

Remark 7. In the particular case of the Grassmann algebra $\mathcal{G}_{0,0,n} = \Lambda_n$, we have three different groups:

$$\mathrm{P}^{\pm}_{0,0,n} = \check{\Gamma}^{(0)} = \ker(\check{\mathrm{ad}}) = \Lambda^{(0)\times}_n, \qquad \mathrm{P}^{\Lambda}_{0,0,n} = \mathrm{P}^{\pm\Lambda}_{0,0,n} = \Gamma^{(0)} = \check{\Gamma}^{(1)} = \Lambda^\times_n,$$
$$\mathrm{P}_{0,0,n} = \Gamma^{(1)} = \ker(\mathrm{ad}) = \begin{cases} (\Lambda^{(0)}_n \oplus \Lambda^n_n)^\times & \text{if } n \text{ is odd}, \\ \Lambda^{(0)\times}_n & \text{if } n \text{ is even}. \end{cases}$$

6 The Corresponding Lie Algebras

Let us denote the Lie algebras of the Lie groups $\mathrm{P}^{\pm}_{p,q,r}$, $\mathrm{P}_{p,q,r}$, $\mathrm{P}^{\pm\Lambda}_{p,q,r}$, and $\mathrm{P}^{\Lambda}_{p,q,r}$ by $\mathfrak{p}^{\pm}_{p,q,r}$, $\mathfrak{p}_{p,q,r}$, $\mathfrak{p}^{\pm\Lambda}_{p,q,r}$, and $\mathfrak{p}^{\Lambda}_{p,q,r}$ respectively.

Theorem 4. *We have the Lie algebras*

$$\mathfrak{p}^{\pm}_{p,q,r} = \mathcal{G}^{(0)}_{p,q,r}, \qquad \mathfrak{p}_{p,q,r} = \begin{cases} \mathcal{G}^{(0)}_{p,q,r} \oplus \mathcal{G}^n_{p,q,r}, & n \text{ is odd;} \\ \mathcal{G}^{(0)}_{p,q,r}, & n \text{ is even;} \end{cases}$$
$$\mathfrak{p}^{\pm\Lambda}_{p,q,r} = \mathcal{G}^{(0)}_{p,q,r} \oplus \Lambda^{(1)}_r, \qquad \mathfrak{p}^{\Lambda}_{p,q,r} = \begin{cases} \mathcal{G}^{(0)}_{p,q,r} \oplus \Lambda^{(1)}_r \oplus \mathcal{G}^n_{p,q,r}, & n \text{ is odd}, \quad r \neq n; \\ \mathcal{G}^{(0)}_{p,q,r} \oplus \Lambda^{(1)}_r, & \text{in the other cases;} \end{cases}$$

of the following dimensions

$$\dim \mathfrak{p}^{\pm}_{p,q,r} = 2^{n-1}, \quad \dim \mathfrak{p}_{p,q,r} = \begin{cases} 2^{n-1}+1, & n \text{ is odd;} \\ 2^{n-1}, & n \text{ is even;} \end{cases}$$

$$\dim \mathfrak{p}^{\pm\Lambda}_{p,q,r} = \begin{cases} 2^{n-1}+2^{r-1}, & r \geq 1; \\ 2^{n-1}, & r = 0; \end{cases} \quad \dim \mathfrak{p}^{\Lambda}_{p,q,r} = \begin{cases} 2^{n-1}+2^{r-1}+1, & n \text{ is odd, } r \neq n,\ r \geq 1; \\ 2^{n-1}+1, & n \text{ is odd, } r = 0; \\ 2^{n-1}, & n \text{ is even, } r = 0; \\ 2^{n-1}+2^{r-1}, & \text{in the other cases.} \end{cases}$$

Proof. We use the well-known facts about the relation between an arbitrary Lie group and the corresponding Lie algebra in order to prove the statements. We calculate the dimensions of the considered Lie algebras using $\dim \mathcal{G}^{(0)}_{p,q,r} = 2^{n-1}$, $\dim \Lambda^{(1)}_r = 2^{r-1}$, $\dim \mathcal{G}^n_{p,q,r} = 1$. □

Remark 8. In the particular case of the non-degenerate algebra $\mathcal{G}_{p,q,0}$, we obtain

$$\mathfrak{p}^{\pm}_{p,q,0} = \mathfrak{p}^{\pm\Lambda}_{p,q,0} = \mathcal{G}^{(0)}_{p,q,0}, \quad \mathfrak{p}_{p,q,0} = \mathfrak{p}^{\Lambda}_{p,q,0} = \begin{cases} \mathcal{G}^{(0)}_{p,q,0} \oplus \mathcal{G}^n_{p,q,0}, & n \text{ is odd;} \\ \mathcal{G}^{(0)}_{p,q,0}, & n \text{ is even.} \end{cases} \tag{37}$$

7 Conclusions

In this paper, we introduce the four Lie groups $\mathrm{P}^{\pm}_{p,q,r}$, $\mathrm{P}_{p,q,r}$, $\mathrm{P}^{\pm\Lambda}_{p,q,r}$, and $\mathrm{P}^{\Lambda}_{p,q,r}$ in the real and complex degenerate geometric algebras $\mathcal{G}_{p,q,r}$ of arbitrary dimension and signature. We give the equivalent definitions of these groups in Theorems 1 and 2. We prove that these groups preserve the even and odd subspaces under the adjoint representation and the twisted adjoint representation in Theorem 3. We study the corresponding Lie algebras in Theorem 4.

The groups introduced in this paper are closely related to the degenerate spin groups, and that is why they are interesting for consideration. We thank the anonymous reviewers for the other important directions of the further research[1]. In the extended version of this paper, we are going to consider the groups preserving the subspaces of fixed grades and the subspaces determined by the grade involution and the reversion under the adjoint and twisted adjoint representations in degenerate geometric algebras $\mathcal{G}_{p,q,r}$. These groups are generalizations of the groups studied in the special case of the non-degenerate algebra $\mathcal{G}_{p,q,0}$ in the papers [32] and [18]. Moreover, we will study the normalized subgroups of these groups, which can be interpreted as generalizations of the spin groups in degenerate case and can be used in applications.

Acknowledgements. This work is supported by the Russian Science Foundation (project 21-71-00043), https://rscf.ru/en/project/21-71-00043/.

The authors are grateful to the three anonymous reviewers for their careful reading of the paper and helpful comments on how to improve the presentation.

[1] It would be of interest to study the relation between the results of this paper and such concepts as the Classification Scheme of Lie groups, the root systems of the Lie groups in degenerate and non-degenerate cases, the Universal enveloping algebras of the Lie algebras.

References

1. Ablamowicz, R.: Structure of spin groups associated with degenerate Clifford algebras. J. Math. Phys. **27**(1), 1–6 (1986)
2. Ablamowicz, R., Lounesto, P.: Primitive idempotents and indecomposable left ideals in degenerate Clifford algebras. In: Chisholm, J.S.R., Common, A.K. (eds.) Clifford Algebras and Their Applications in Mathematical Physics. ASIC, vol. 183, pp. 61–65. Springer, Dordrecht (1986). https://doi.org/10.1007/978-94-009-4728-3_5
3. Atiyah, M., Bott, R., Shapiro, A.: Clifford modules. Topology **3**, 3–38 (1964)
4. Bayro-Corrochano, E.: Geometric Algebra Applications Vol. I: Computer Vision, Graphics and Neurocomputing. Springer, Cham (2019). https://doi.org/10.1007/978-3-319-74830-6
5. Bayro-Corrochano, E., Daniilidis, K., Sommer, G.: Motor algebra for 3D kinematics: the case of the hand-eye calibration. J. Math. Imaging Vis. **13**, 79–100 (2000)
6. Bayro-Corrochano, E., Sobczyk, G.: Geometric Algebra with Applications in Science and Engineering. Birkhäuser, Boston (2001)
7. Brooke, J.: A Galileian formulation of spin: I. Clifford algebras and spin groups. J. Math. Phys. **19**, 952–959 (1978)
8. Brooke, J.A.: Spin groups associated with degenerate orthogonal spaces. In: Chisholm, J.S.R., Common, A.K. (eds.) Clifford Algebras and Their Applications in Mathematical Physics. ASIC, vol. 183, pp. 93–102. Springer, Dordrecht (1986). https://doi.org/10.1007/978-94-009-4728-3_8
9. Brooke, J.: Clifford Algebras, Spin Groups and Galilei Invariance - New Perspectives. Thesis, U. of Alberta (1980)
10. Catto, S., Choun, Y., Gurcan, Y., Khalfan, A., Kurt, L.: Grassmann numbers and Clifford-Jordan-Wigner representation of supersymmetry. In: Journal of Physics: Conference Series, vol. 411 (2013)
11. Crumeyrolle, A.: Orthogonal and Symplectic Clifford Algebras, 1st edn. Springer, Dordrecht (1990). https://doi.org/10.1007/978-94-015-7877-6
12. Crumeyrolle, A.: Algebres de Clifford degenerees et revetements des groupes conformes affines orthogonaux et symplectiques. Ann. Inst. H. Poincare **33**(3), 235–249 (1980)
13. Doran, C., Lasenby, A.: Geometric Algebra for Physicists. Cambridge University Press, Cambridge (2003)
14. Dorst, L., De Keninck, S.: A Guided Tour to the Plane-Based Geometric Algebra PGA. Version 2.0 (2022). http://bivector.net/PGA4CS.html
15. Dorst, L., Doran, C., Lasenby, J.: Applications of Geometric Algebra in Computer Science and Engineering. Birkhäuser, Boston (2002)
16. Dorst, L., Fontijne, D., Mann, S.: Geometric Algebra for Computer Science: An Object-Oriented Approach to Geometry. Morgan Kaufmann Publishers Inc., Burlington (2007)
17. Dorst, L., Lasenby, J.: Guide to Geometric Algebra in Practice. Springer, London (2011)
18. Filimoshina, E., Shirokov, D.: On generalization of Lipschitz groups and spin groups. Math. Methods Appl. Sci. 1–26 (2022)
19. Gunn, C.: Geometric algebras for Euclidean geometry. Adv. Appl. Clifford Algebras **27** (2017)
20. Gunn, C.: Doing Euclidean plane geometry using projective geometric algebra. Adv. Appl. Clifford Algebras **27** (2017)

21. Hestenes, D., Sobczyk, G.: Clifford Algebra to Geometric Calculus - A Unified Language for Mathematical Physics. Reidel Publishing Company, Dordrecht (1984)
22. Hestenes, D.: Old wine in new bottles: a new algebraic framework for computational geometry. In: Corrochano, E.B., Sobczyk, G. (eds.) Geometric Algebra with Applications in Science and Engineering, pp. 3–17. Birkhäuser, Boston (2001). https://doi.org/10.1007/978-1-4612-0159-5_1
23. Hitzer, E.: Geometric operations implemented by conformal geometric algebra neural nodes. In: Proceedings of the SICE Symposium on Systems and Information, Himeji, Japan, pp. 357–362 (2008). arXiv:1306.1358v1
24. Hildenbrand, D.: Foundations of Geometric Algebra Computing, 1st edn. Springer, Heidelberg (2013)
25. Hildenbrand, D.: The Power of Geometric Algebra Computing, 1st edn. Chapman and Hall/CRC, New York (2021)
26. Hrdina, J., Navrat, A., Vasik, P., Dorst, L.: Projective geometric algebra as a subalgebra of conformal geometric algebra. Adv. Appl. Clifford Algebras **31**(18) (2021)
27. Lam, T.: A First Course in Noncommutative Rings. Graduate Texts in Mathematics, vol. 131, 2nd edn. Springer, Heidelberg (2001). https://doi.org/10.1007/978-1-4419-8616-0
28. Li, H., Hestenes, D., Rockwood, A.: Generalized homogeneous coordinates for computational geometry. In: Sommer, G. (ed.) Geometric Computing with Clifford Algebras, pp. 27–59. Springer, Heidelberg (2001). https://doi.org/10.1007/978-3-662-04621-0_2
29. Lounesto, P.: Clifford Algebras and Spinors. Cambridge University Press, Cambridge (1997)
30. Porteous, I.: Clifford Algebras and the Classical Groups. Cambridge University Press, Cambridge (1995)
31. Selig, J., Bayro-Corrochano, E.: Rigid body dynamics using Clifford algebra. Adv. Appl. Clifford Algebras **20**, 141–154 (2010)
32. Shirokov, D.: On inner automorphisms preserving fixed subspaces of Clifford algebras. Adv. Appl. Clifford Algebras **31**(30) (2021)
33. Wareham, R., Cameron, J., Lasenby, J.: Applications of conformal geometric algebra in computer vision and graphics. In: Li, H., Olver, P.J., Sommer, G. (eds.) GIAE/IWMM 2004. LNCS, vol. 3519, pp. 329–349. Springer, Heidelberg (2005). https://doi.org/10.1007/11499251_24

Computing with the Universal Properties of the Clifford Algebra and the Even Subalgebra

Eric Wieser(✉) and Joan Lasenby

Department of Engineering, University of Cambridge, Cambridge, UK
efw27@cam.ac.uk

Abstract. Typically, Geometric Algebra (GA) is introduced via choosing an orthogonal basis, and defining how multiplication acts on this basis according to some simple rules. This works well computationally, but can obscure insight mathematically. In particular, operations defined in terms of coordinates on a multivector basis can be difficult to rigorously show to be "coordinate-free", especially in large algebras. This paper explores the use of the "universal property" to ensure that operations are "coordinate-free" by construction. To build some insight for applying the universal property, we draw parallels to the process of writing recursive programs. We then demonstrate a novel result using this approach by deriving a universal property of the even subalgebra. Armed with this second universal property, we provide an explicit construction for a well-known equivalence between any Clifford algebra and its "one-up" even subalgebra. We conclude with some remarks about formalization of these ideas in a theorem prover.

Keywords: Geometric Algebra · Clifford Algebra · Universal Property · Formalized Mathematics

1 Introduction

Conventionally when working with geometric algebra, we start by choosing an orthogonal basis for our real vector space $\{e_1^+, \cdots, e_p^+, e_1^-, \cdots, e_q^-, e_1^0, \cdots, e_r^0\}$, where the superscripts describe the real-valued square of each vector; $e_i^+ \cdot e_i^+ = 1$, $e_i^- \cdot e_i^- = -1$, $e_i^0 \cdot e_i^0 = 0$. The generators formed by this basis are then used to construct the algebraic object known as the "Clifford algebra"[1], via imposing the constraint that vectors square to scalars.

In this paper, we will work in the more general setting of an R-module over an arbitrary ring R rather than over a $\mathbb{R}$-vector space. We will use $\mathcal{G}(V, Q)$ as notation for the Clifford algebra over the R-module V with quadratic form Q. What is often written as $\mathcal{C}\ell(p, q, r)$ or $\mathcal{G}(\mathbb{R}^{p,q,r})$ elsewhere would under this

[1] In this paper, we will use "Clifford algebra" to refer to this object, and "geometric algebra" to refer to the field of study.

D. W. Silva et al. (Eds.): ICACGA 2022, LNCS 13771, pp. 199–211, 2024.
https://doi.org/10.1007/978-3-031-34031-4_17

convention be written as $\mathcal{G}(\mathbb{R}^{p+q+r}, Q)$ where $Q(e_i^+) = 1, Q(e_i^-) = -1, Q(e_i^0) = 0$ and is extended to all $\mathbb{R}^{p+q+r}$ in the way that satisfies the requirements of a quadratic form.

Initially, this might seem like it just increases the verbosity; but the goal is to avoid ever mentioning the basis in the first place. One of the core claims of geometric algebra is that it lets you perform geometric manipulation in a "coordinate-free" way. Exactly what this means typically depends on the author, but a common interpretation is "the choice of basis vectors does not affect the result of the manipulation" [5]. For clarity, we will refer to algorithms with this property as "basis-agnostic".

Let us quickly summarize some examples of operations which are and are not "basis-agnostic". Basic algebraic operations like multiplication and the wedge product have a precise geometric meaning with no mention of coordinates, so are "basis-agnostic". However, the pseudoscalar $I = \prod_i e_i$ is not basis-agnostic; choosing the basis vectors in a different order results in a change of sign. In 3D this basis-dependence is transferred to the cross product, which in GA can be expressed as $a \times b = -I(a \wedge b)$; the handedness of the cross-product depends on the handedness of the vector space, which is determined by the basis.

There are some operations which despite being "basis-agnostic", are still typically defined by making a choice of basis, then proven to be invariant with respect to that choice.

After introducing some intuition for recursion in Sect. 2, this paper shows how the "universal property" can be used as a computational tool in the place of choosing a basis, and in Sect. 3.2 demonstrates how to view this tool as a variant of recursion. In Sect. 3.3 we derive a new universal property for the even subalgebra from our first universal property, and use this in Sect. 3.4 to construct a well-known isomorphism in an unusual way. In Sect. 3.5, we show how this recursive technique can be applied to construct the left-contraction from one of its properties alone. Section 4 makes brief remarks about the formalized versions of our results that are provided throughout this paper via "" links.

1.1 Notation

In this paper, we will use the colon notation $x : R$ to say "x is in the ring R", and $v : V$ to say "v is in the vector space V". Similarly, we will use $F : V \to W$ to say "F is a function from the space V to the space W", or $Q : V \to R$ to describe our quadratic form. When referring to the value of $F : V \to W$, we will use $F = (v \mapsto w)$. Whether $\to$ refers to a function, a linear map, or some other type of morphism is left to the prose. For functions of two variables, we have two choices of notation; $F : U \times V \to W$ or $F : U \to (V \to W)$, where we will omit the parentheses. This second "curried"[2] interpretation may seem unusual, but it is convenient for us for reasons that become apparent in (2). Similarly, we shall use $F = (u \mapsto v \mapsto w)$ for writing the values of such functions, and use $f(u, v)$ and $f(u)(v)$ interchangeably.

2 So-named in reference to the mathematician Haskell Curry.

Note in particular that for commutative R, an R-bilinear map $F : U \times V \to W$ can be considered as an R-linear map from U to the space of R-linear maps from V to W, which is the $F : U \to V \to W$ spelling.

2 Recursors

In functional programming, a list is usually defined inductively; either it is empty, "[ ]", or it is an element a followed by another list l, "$a :: l$". This inductive definition provides a recursion principle, or recursor: "To define a function from a list of elements, it suffices to define its value on [ ], and define its value on $a :: l$ given its value on l". Consider computing in this way a sum of a list of elements of a ring R. In a functional programming language, we would usually do so as:

$$\text{sum} : \text{list}\, R \to R \tag{1}$$
$$\text{sum}([\,]) := 0 \tag{1a}$$
$$\text{sum}(a :: l) := a + \text{sum}(l) \tag{1b}$$

One way to describe the list recursor is as a "fold"; if we have a function $f : \alpha \to \beta \to \beta$, then $\text{fold}[f] : \text{list}\,\alpha \to \beta \to \beta$. This satisfies $\text{fold}[f]([\,], b_0) = b_0$ and $\text{fold}[f](a :: l, b_0) = f(a, \text{fold}[f](l, b_0))$. The "pattern matching" in (1) can be trivially transformed by the compiler into an application of $\text{fold}[f]$, as $\text{sum}(l) = \text{fold}[a \mapsto v \mapsto a + v](l, 0)$, where $\text{sum}(l)$ in (1b) has been replaced with v.

Sometimes implementing a recursion scheme requires keeping track of intermediate state. As an example, consider producing an accumulated sum of the elements of a list, starting from zero, such that $\text{accum}([a, b]) = [0, a, a + b]$. To implement this, we use the recursor to define an auxiliary helper function:

$$\text{accum\_from} : \text{list}\, R \to R \to \text{list}\, R \tag{2}$$
$$\text{accum\_from}([\,]) := a \mapsto [a] \tag{2a}$$
$$\text{accum\_from}(b :: l) := a \mapsto a :: \text{accum\_from}(l)(a + b) \tag{2b}$$

Note the unusual type signature in (2); for each list, it produces not a value but another function. It is this function that consumes our intermediate state $a : R$, which is the value to resume the accumulation from; allowing us to thread this value through the recursion while still sticking to the rules of our list recursor. We can recover our desired function by simply initializing this state:

$$\text{accum} : \text{list}\, R \to \text{list}\, R \tag{3}$$
$$\text{accum} := l \mapsto \text{accum\_from}(l)(0) \tag{3a}$$

Finally, let us consider the example where the running sum should be in reverse. As we recurse, we will keep track of what the first element of our list is. Instead of introducing this intermediate state into the input as we did in (2) by producing a function, we introduce it into the output by producing a pair ($\times$):

$$\begin{aligned}
\text{rev_accum}_{\text{aux}} : \text{list } R \to \quad & (R \times \text{list } R) && (4)\\
\text{rev_accum}_{\text{aux}}([\,]) := \quad & (0, [\,]) && (4a)\\
\text{rev_accum}_{\text{aux}}(a :: l) := \; & (a + b, b :: l') \text{ where } (b, l') := \text{rev_accum}_{\text{aux}}(l) && (4b)
\end{aligned}$$

Instead of initializing the state as in (3a), we post-process it:

$$\begin{aligned}
\text{rev_accum} : \quad & \text{list } R \to \text{list } R && (5)\\
\text{rev_accum} := \quad & l \mapsto a :: l' \text{ where } (a, l') := \text{rev_accum}_{\text{aux}}(l) && (5a)
\end{aligned}$$

In this section, we have seen two important tricks for taking a simple recursor and implementing more complex recursion schemes. In the rest of this paper, we will show how these principles translate to the language of universal properties.

3 Universal Properties

To state the universal property of the Clifford algebra [7, §14.4]; [1, II.1.1], we will need some of the terminology from abstract algebra. We met R-modules in the introduction as a generalization of vector spaces. An R-algebra is a ring that is also an R-module, while an "R-algebra morphism" is an R-linear map that additionally preserves multiplication and the multiplicative identity. Armed with these definitions (and deferring to [1] for the construction), we can state[3] the universal property,

Definition 1. *For every R-algebra A and R-module V, we have a one to one correspondence between linear maps $f : V \to A$ satisfying $f(v)^2 = Q(v)$, and algebra morphisms $F : \mathcal{G}(V, Q) \to A$. This correspondence is compositional; given another R-algebra A_2 and an algebra morphism $H : A \to A_2$, if f corresponds with F then $v \mapsto H(f(v))$ corresponds with $x \mapsto H(F(x))$. We write this correspondence as* $\text{lift}^+[f] = F$.

It can be helpful to present this graphically, as is done in Fig. 1.

[3] Note we call this a definition as we are defining lift^+, which has computational content just like the recursor for lists did.

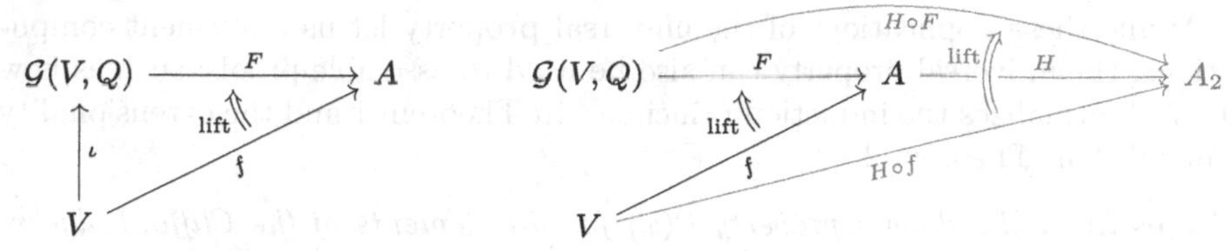

(a) via the map ι (b) via the property $\text{lift}[H \circ f] = H \circ \text{lift}[f]$

Fig. 1. Graphical representation of two equivalent ways to define the universal property, where (b) corresponds to Definition 1. Setting $F = (x \mapsto x)$ in (b) recovers (a). In this paper, we will not make explicit reference to ι, leaving it implicit wherever we turn an element of V into an element of $\mathcal{G}(V, Q)$.

This "universal property" is very similar to the perhaps more familiar concept of taking the "outermorphism" [2, §4.2], but instead of extending the linear map f to wedge products as $F(u \wedge v) = f(v) \wedge f(w)$, we extend it to geometric products using $F(uv) = f(v)f(w)$. In fact, the "outermorphism" is simply the forward direction of the universal property for the *exterior* algebra.

For one of the simplest examples of applying the universal property, consider using $f : V \to \mathcal{G}(V, Q) = (v \mapsto -v)$, which trivially satisfies $f(v)^2 = (-v)^2 = v^2 = Q(v)$; the resulting $\text{lift}[f]$ is the familiar "grade involution" operator $x \mapsto \hat{x}$. A key insight is that we can write this in the style of (1) as:

$$\text{grade\_invol} : \mathcal{G}(V, Q) \to \mathcal{G}(V, Q) \tag{6}$$
$$\text{grade\_invol}(v : V) := -v \tag{6a}$$

Note that unlike in (1), we are additionally obliged to show that (6a) is linear, and that $(-v)^2 = v^2 = Q(v)$.

A more complex example involves constructing the grade reversal operation $x \mapsto \tilde{x}$, which for example sends $e_1e_2e_3$ to $e_3e_2e_1$. For this case, we need a different choice of A than $\mathcal{G}(V, Q)$. What we choose is $\mathcal{G}(V, Q)^{\text{op}}$, where A^{op} is the algebra A but with multiplication reversed. This comes with two obvious R-linear maps, $\text{op} : A \to A^{\text{op}}$ and $\text{op}^{-1} : A^{\text{op}} \to A$, which convert between the two spaces. Note that $\text{op}(ab) = (\text{op}\, b)(\text{op}\, a)$, so these are not algebra morphisms; but we do still have $\text{op}\, 1 = 1$. Using the notation of (6), we implement this in the style of (4) as:

$$\text{grade\_rev}_{\text{aux}} : \mathcal{G}(V, Q) \to \mathcal{G}(V, Q)^{\text{op}} \tag{7}$$
$$\text{grade\_rev}_{\text{aux}}(v : V) := \text{op}\, v \tag{7a}$$

Again, we must show that (7a) is linear, and that $(\text{op}\, v)^2 = \text{op}(v^2) = \text{op}(Q(v)) = Q(v)$. To recover the reversion operator (a linear map that reverses multiplication), we simply compose this with op^{-1} to eliminate the ^{op}.

$$\text{grade\_rev} : \quad \mathcal{G}(V, Q) \to \mathcal{G}(V, Q) \tag{8}$$
$$\text{grade\_rev} := \quad x \mapsto \text{op}^{-1}(\text{grade\_rev}_{\text{aux}}(x)) \tag{8a}$$

While these applications of the universal property let us implement computations, the universal property can also be used to assemble proofs; such as how [9, §7.5] formalizes the induction principle in Theorem 1 and the extensionality principle in Theorem 2.

Theorem 1. *To show a property $P(x)$ for all elements of the Clifford algebra $\mathcal{G}(V, Q)$, it suffices to show:*

- *That $P(r)$ holds on all scalars $r : R$*
- *That $P(u)$ holds on all vectors $u : V$*
- *That $P(x + y)$ holds on all multivectors $x, y : \mathcal{G}(V, Q)$ if $P(x)$ and $P(y)$ hold*
- *That $P(xy)$ holds on all multivectors $x, y : \mathcal{G}(V, Q)$ if $P(x)$ and $P(y)$ hold*

Theorem 2. *To show that two algebra morphisms $f, g : \mathcal{G}(V, Q) \to A$ are equal, it suffices to show they agree on the generators $v : V$.*

3.1 Universal Properties as a Universal Interface

If two different representations of a Clifford algebra are available, $\mathcal{G}_1$ and $\mathcal{G}_2$, then the universal property of $\mathcal{G}_1$ provides a map between the two:

$$\text{convert} : \mathcal{G}_1(V, Q) \to \mathcal{G}_2(V, Q) \tag{9}$$

$$\text{convert}(v : V) := v \quad (= \iota_2(v)) \tag{9a}$$

In the language of software; if two libraries implement the universal property "API", then they can interoperate without direct knowledge of each other.

3.2 Universal Properties as Recursors

In (2), we saw a trick to thread extra state through our recursor by choosing our output to itself be a function. We can play a similar trick with the universal property, although we are forced to work within the functions that form an algebra. These include the endomorphism algebra *End*(R, W) (the R-linear maps of the form $W \to W$); where $1 : End(R, W)$ is the identity map and $\times : End(R, W) \to End(R, W) \to End(R, W)$ is composition. The scalars of this algebra happen to also be the "scaler"s; the endomorphisms corresponding to a uniform scaling by an element $r : R$.

This specialization to $A = End(R, W)$ allows us to apply the universal property to produce a "fold" operation (so named due to its analogy to the list version described just below (1)) by an R-bilinear map $f : V \to W \to W$ to obtain a algebra morphism into the endomorphism algebra:

$$\text{fold[f]} : \mathcal{G}(V, Q) \to \overbrace{W \to W}^{End(R,W)} \tag{10}$$

$$\text{fold[f]}(v : V) := \ w \mapsto f(v, w) \tag{10a}$$

where the $f(v)^2 = (w \mapsto f(v, f(v, w))) = Q(v)$ condition can be rewritten as

$$f(v, f(v, w)) = Q(v)w \tag{11}$$

Similarly, the fact this is an algebra morphism tells us that $\mathrm{fold}[f](r, v) = rv$ for $r : R$ and $\mathrm{fold}[f](xy, v) = \mathrm{fold}[f](x, \mathrm{fold}[f](y, v))$. As an example of what "fold"ing means in the context of a Clifford algebra, if $c + u + vw : \mathcal{G}(V, Q)$ and $x : W$ then

$$\mathrm{fold}[f](c + u + vw, x) = cx + f(u, x) + f(v, f(w, x)).$$

3.3 The Even Subalgebra

The even subalgebra $\mathcal{G}^+(V, Q)$ of a Clifford algebra $\mathcal{G}(V, Q)$ is the subalgebra consisting of the closure under addition and multiplication of all elements of the form vw where $v, w : V$; its members are known [4, (1.29)] as the "even" multivectors[4]. We will now show that this subalgebra has its own universal property, Definition 2:

Definition 2. *For every R-algebra A and R-module V, we have a one-to-one correspondence between R-bilinear maps $f : V \to V \to A$ satisfying:*

$$f(v, v) = Q(v) \tag{12}$$
$$f(u, v)f(v, w) = Q(v)f(u, w) \tag{13}$$

and algebra morphisms out of the even subalgebra $F : \mathcal{G}^+(V) \to A$. This correspondence is compositional; if f corresponds with F then $v \mapsto w \mapsto H(f(v, w))$ corresponds with $x \mapsto H(F(x))$. We write this correspondence as $\mathrm{lift}^+[f] = F$.

Again, it can be helpful to present this graphically, as is done in Fig. 2.

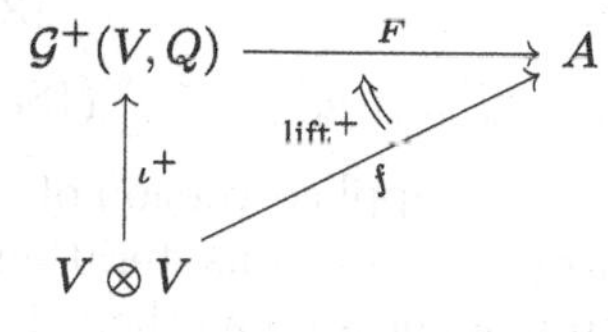

(a) via the inclusion ι^+

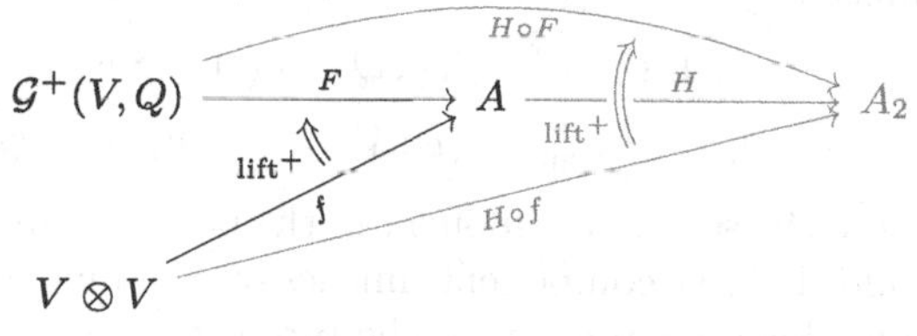

(b) via the property $\mathrm{lift}^+[H \circ f] = H \circ \mathrm{lift}[f]$

Fig. 2. The universal property of the even subalgebra. So as to resemble Fig. 1, we show the bilinear map $f : V \to V \to A$ as the equivalent linear map from the tensor product, $f : V \otimes V \to A$. Here, $\iota^+(v \otimes w) = \iota(v)\iota(w) = vw$.

[4] Although the definition in [4] needs the construction in Sect. 3.5 and therefore doesn't work in characteristic 2.

As with Definition 1 this is a definition not just a theorem; we are not just going to prove that there *is* a correspondence, but we will provide an explicit basis-agnostic computation of that correspondence. We will start by showing the reverse direction, which given the algebra morphism $F : \mathcal{G}^+(V, Q) \to A$ we choose as

$$\mathrm{lift}^{+\,-1}[F] = f = (v \mapsto w \mapsto F(vw)) \tag{14}$$

which trivially satisfies (12) and (13):

$$f(v, v) = F(vv) = F(Q(v)) = Q(v) \tag{15}$$

$$f(u, v)f(v, w) = F(uv)F(vw) = F(uvvw) = F(uQ(v)w) = Q(v)F(uw) \tag{16}$$

$$= Q(v)f(u, w) \tag{17}$$

To construct the forwards direction, we are going to use the same trick as we did in (10), setting $W = A \oplus S$ to produce an auxiliary function $\mathrm{lift}^+_{\mathrm{aux}}[f] : \mathcal{G}(V, Q) \to (A \oplus S) \to (A \oplus S)$. Here, A is our target algebra, while S is some additional state which mirrors the extra recursor state we saw in (4). $A \oplus S$ is their direct sum, which is to say it consists of pairs (a, s) with $(a_1, s_1)+(a_2, s_2) = (a_1 + a_2, s_1 + s_2)$ and $r(a, s) = (ra, rs)$. Note that for this to be an R-module as required by Definition 2, we need S to also be an R-module. We will deduce precisely what to choose for S shortly.

We want our fold to apply f on pairs of vectors $v, w : V$ at a time; that is,

$$\mathrm{lift}^+_{\mathrm{aux}}[f](vwx, (a, s)) = (f(v, w)\, \mathrm{lift}^+_{\mathrm{aux}}[f](x, (a, s)), s').$$

Using the fact that $\mathrm{lift}^+_{\mathrm{aux}}[f]$ will be an algebra morphism, this simplifies to

$$\mathrm{lift}^+_{\mathrm{aux}}[f](v, \mathrm{lift}^+_{\mathrm{aux}}[f](w, (a, s))) = (f(v, w)a, s');$$

that is, each application of $\mathrm{lift}^+_{\mathrm{aux}}[f]$ needs to apply "half" of f. An obvious choice would be to pick $S = V \to A$, the space of R-linear maps which includes the "half"-applied maps like $v \mapsto f(v, w)a$. Note that this is essentially using the trick in (2) for a second time, but instead of producing an unconstrained function we are required to produce a linear map. We can then define

$$\mathrm{lift}^+_{\mathrm{aux}}[f] : \mathcal{G}(V, Q) \to (A \oplus S) \to (A \oplus S) \tag{18}$$

$$\mathrm{lift}^+_{\mathrm{aux}}[f](v : V) := \quad (a, s) \mapsto (s(v), w \mapsto f(w, v)a), \tag{18a}$$

where the second component of the pair contains a partially-applied version of f, while the first component finishes off the invocation from the previous iteration. Note that we cannot take the product of $s(\cdot)$ and a in a single step as then this operation would cease to be linear; which is why we instead weave these terms back and forth between the left and right halves of the pair.

We now verify that our $\mathrm{lift}^+_{\mathrm{aux}}[f]$ satisfies the required property in (11) as

$$\mathrm{lift}^+_{\mathrm{aux}}[f](v, \mathrm{lift}^+_{\mathrm{aux}}[f](v, (a, s))) = \mathrm{lift}^+_{\mathrm{aux}}[f](v, (s(v), w \mapsto f(w, v)a)) \tag{19}$$

$$= (f(v, v)a, w \mapsto f(w, v)s(v))) \tag{20}$$

$$\stackrel{?}{=} (Q(v)a, w \mapsto Q(v)s(w)) \tag{21}$$

$$= Q(v)(a, s), \tag{22}$$

where $\overset{?}{=}$ is the equality to be checked. $f(v,v) = Q(v)$ was (12), so we can easily match up the first half of the pair. However, matching up the second half of the pair requires $f(w,v)s(v) = Q(v)s(w)$, which is a stronger requirement than (13) and not true for all linear maps $s : S$.

To solve this problem, we need to pick a smaller space $S^{\square}$, the module spanned by linear maps $s : V \rightarrow A$ of the form $s = (v \mapsto f(v,w)a)$ for all $w : V$ and $a : A$. Our definition of $\text{lift}^{+}_{\text{aux}}$ in (18a) trivially adapts to this definition, as $w \mapsto f(w,v)a$ lies in the new S by definition. We are now in a position to solve $f(w,v)s(v) = Q(v)s(w)$, as we can write $s = (w \mapsto \sum_i f(w,u_i)a_i)$ (for some arbitrary finite set of $u_i : V$ and $a_i : A$) to get:

$$f(w,v)\left(\textstyle\sum_i f(v,u_i)a_i\right) = \textstyle\sum_i f(w,v)f(v,u_i)a_i \tag{23}$$
$$= \textstyle\sum_i Q(v)f(w,u_i)a_i \tag{24}$$
$$= Q(v)\textstyle\sum_i f(w,u_i)a_i \tag{25}$$
$$= Q(v)s(w) \tag{26}$$

where we go from (23) to (24) using (13).

We can now extract $\text{lift}^{+}[f]$ as

$$\text{lift}^{+}[f]: \qquad \mathcal{G}^{+}(V,Q) \rightarrow A \tag{27}$$
$$\text{lift}^{+}[f] := \qquad x^{+} \mapsto a \text{ where } (a,s') = \text{lift}^{+}_{\text{aux}}[f](x, (1, v \mapsto 0)) \tag{27a}$$

This is obviously linear, as it is the composition of operations each of which is linear. We can show that $\text{lift}^{+}[f](r) = r$ by using properties of (10). To complete the proof that this is an algebra morphism, we must show that within the even subalgebra it preserves multiplication. We will do this by induction, for which we need Theorem 3$^{\square}$.

Theorem 3. *To show a property $P(x)$ for all elements of the even subalgebra $\mathcal{G}^{+}(V,Q)$, it suffices to show:*

- *That $P(r)$ holds on all scalars $r : R$*
- *That $P(x+y)$ holds on all elements $x,y : \mathcal{G}^{+}(V,Q)$ if $P(x)$ and $P(y)$ hold*
- *That $P(uvx)$ holds on all elements $u,v : V$ and $x : \mathcal{G}^{+}(V,Q)$ if $P(x)$ holds*

Proof. Follows by noting that $\mathcal{G}^{+}(V,Q)$ is the vector space spanned by all products of even numbers of vectors in V, that such products can be decomposed into pairs, and through an appropriate induction principle for spans of vectors.

We apply this principle with $P(x)$ as $\forall y, \text{lift}^{+}[f](xy) = \text{lift}^{+}[f](x)\,\text{lift}^{+}[f](y)$. The first two conditions follow trivially by $\text{lift}^{+}[f](r) = r$ and linearity. The third condition can be shown as

$$\text{lift}^{+}[f](vwy) = a \qquad \text{where } (a,s') = \text{lift}^{+}_{\text{aux}}[f](vwy, (1, v \mapsto 0)) \tag{28}$$
$$= f(v,w)a \text{ where } (a,s') = \text{lift}^{+}_{\text{aux}}[f](y, (1, v \mapsto 0)) \tag{29}$$
$$= f(v,w)\,\text{lift}^{+}[f](vwy) \tag{30}$$
$$= \text{lift}^{+}[f](vw)\,\text{lift}^{+}[f](y) \tag{31}$$

We now have $\text{lift}^+[f] : \mathcal{G}^+(V, Q) \to A$, the forward direction of the universal property. Combined with our result in (14), all that remains is to show that these two directions are inverses. Showing that this operation is a left-inverse ($\text{lift}^{+\,-1}[\text{lift}^+[f]] = f$) is straightforward. Showing that it is a right-inverse requires the extensionality principle in Theorem 4, which we use in (33).

$$\text{lift}^+[\text{lift}^{+\,-1}[F]](vw) = \text{lift}^{+\,-1}[F](v, w) = F(vw) \tag{32}$$

$$\implies \text{lift}^+[\text{lift}^{+\,-1}[F]] = F \tag{33}$$

Theorem 4. *To show that two algebra morphisms from the even subalgebra $f, g : \mathcal{G}^+(V, Q) \to A$ are equal, it suffices to show they agree on the products of two generators $v, w : V$.*

Proof. Rephrase as $\forall x, f(x) = g(x)$, apply Theorem 3, and use the properties of algebra homomorphisms.

3.4 The Isomorphism with the Even Subalgebra

Theorem 5. *$\mathcal{G}(V, Q)$ is isomorphic as an R-algebra to $\mathcal{G}^+(V \oplus R, Q')$, where $Q'((v, r)) = Q(v) - r^2$.*

Here $V \oplus R$ combines V (as elements of the form $(v, 0)$) with an extra basis vector $e = (0, 1)$ that squares to -1, and so we will write (v, r) as $v + re$. In [6, Chapter 1, Theorem 3.7], this isomorphism is evaluated by choosing a basis for V, and then copying coefficients by inspection: in the forward direction, each basis vector e_i is replaced with ee_i; and in the backwards direction[5], where all basis vectors appear in pairs, e_ie_j is left alone and e_ie is mapped back to e_i. We will proceed without choosing a basis for V, and use our pair of universal properties instead.

To construct the forward map, we can directly write down the coefficient copying approach by applying Definition 1 with $f = (v \mapsto ev)$, which satisfies $f(v)^2 = (ev)(ev) = (-ve)(ev) = -v(-1)v = Q(v)$. This gives us $F : \mathcal{G}(V, Q) \to \mathcal{G}^+(V, Q') = \text{lift}[f]$, and is exactly the approach used in [6, Chapter 1, Theorem 3.7]. This reference does not give an explicit construction for the reverse mapping, noting that to verify one exists we must "check $[F]$ on a linear basis".

The reverse map f^{-1} needs to satisfy the pair of rules above:

$$f^{-1}(0 + e,\ 0 + e) = -1 \qquad \text{(as } e^2 = -1 \text{ in } \mathcal{G}^+(V, Q')) \tag{34}$$

$$f^{-1}(0 + e, v + 0e) = v \qquad \text{(remove } e \text{ from pairs of the form } ev) \tag{35}$$

$$f^{-1}(u + 0e,\ 0 + e) = -u \qquad \text{(rewrite } ue = -eu \text{ and do the above)} \tag{36}$$

$$f^{-1}(u + 0e, v + 0e) = uv \qquad \text{(leave blades without } e \text{ untouched)} \tag{37}$$

$$\implies f^{-1}(u + re, v + se) = (u + r)(v - s) \tag{38}$$

[5] for which [6] proves only existence.

where (38) follows by linearity. (12) holds as $f^{-1}(v+se, v+se) = v^2 - s^2 = Q'(v+se)$ and (13) follows similarly. We can thus apply Definition 2 with this f to obtain $F = \mathrm{lift}^+[f^{-1}]$; or in our functional notation:

$$F^{-1} : \mathcal{G}^+(V, Q') \to \mathcal{G}(V, Q) \tag{39}$$
$$F^{-1}((u+re)(v+se)) := f^{-1}(u+re, v+se) = (u+r)(v-s) \tag{39a}$$

All that remains to conclude our construction of this isomorphism is to show that these operations are inverses, that is for all $x : \mathcal{G}(V,Q)$ and $x^+ : \mathcal{G}^+(V,Q)$,

$$F^{-1}(F(x)) = x, \qquad F(F^{-1}(x^+)) = x^+. \tag{40}$$

Rewriting (40) as equalities of functions gives

$$(x \mapsto F^{-1}(F(x))) = (x \mapsto x), \qquad (x^+ \mapsto F(F^{-1}(x^+)) = (x^+ \mapsto x^+), \tag{41}$$

which allows us to apply Theorems 2 and 4 to solve these equations:

$$\begin{aligned} &F^{-1}(F(v)) \\ &= F^{-1}(f(v)) \\ &= F^{-1}(ev) \\ &= f^{-1}(e, v) \\ &= (0+1)(v+0) \\ &= v \end{aligned}
\qquad\Bigg|\qquad
\begin{aligned} &F(F^{-1}((u+re)(v+se))) \\ &= F(f^{-1}(u+re, v+se)) \\ &= F((u+r)(v-s)) \\ &= (f(u)+r)(f(v)-s) \\ &= euev + rev - seu - rs \\ &= uv + rev - seu + rse^2 \\ &= (u+re)(v+se) \end{aligned} \tag{42}$$

3.5 The Isomorphism to the Exterior Algebra

A key result in geometric algebra is that the Clifford algebra is isomorphic as an R-module to the exterior algebra over the same vector space, as this provides the non-metric wedge product. In [3, Theorem 34], this is shown by defining[6] $v\rfloor^f$ and α^f, characterized by [3, Theorems 6 and 21] as

$$v \rfloor^f (u \otimes U) = f(v,u)U - u \otimes (v \rfloor^f U), \tag{43}$$
$$\alpha^f(u \otimes U) = u \otimes \alpha^f(U) - u\rfloor^f(\alpha^f(U)). \tag{44}$$

Here, f is a bilinear form associated with Q, that is $f(x,y) = \frac{1}{2}(Q(x+y) - Q(x) + Q(y))$; thus imposing the restriction that the ring R is not of characteristic 2. Note that these are defined on the tensor algebra; only later is it proved that these mappings can be transferred to a Clifford algebra where the $\otimes$ is simply multiplication[7]:

$$v \rfloor^f (uU) = f(v,u)U - u(v \rfloor^f U) \tag{45}$$
$$\alpha^f(uU) = u\alpha^f(U) - u\rfloor^f(\alpha^f(U)) \tag{46}$$

[6] Confusingly, [3] uses $u\llcorner U$ as notation for $u\rfloor U$, with the symbol flipped.

[7] (45) also appears as a special case of [4, (1.41a)] with $r = 1$.

Without repeating the entire proof here, we will show how to define the $v\rfloor^f$ operator satisfying (45) using the universal property, which we will write as contract$[v]$. The construction of (46) is a simple application of (10), so we shall omit it. Initially, it would seem that we cannot use the trick from "fold" in (10) here, as we not only need the current vector "u" and the result so far "$v\rfloor^f U$", but we also need the accumulation of the input so far, "U". The solution is to first apply the trick in (4), where we compute the value of U as we go along:

$$\text{contract}_{\text{aux}}[v] : \mathcal{G}(V,Q) \to \mathcal{G}(V,Q) \oplus \mathcal{G}(V,Q) \to \mathcal{G}(V,Q) \oplus \mathcal{G}(V,Q) \tag{47}$$

$$\text{contract}_{\text{aux}}[v](u:V) := \quad (U,x) \quad \mapsto \quad (uU, f(v,u)U - ux) \tag{47a}$$

This is a fold over the pairs $\mathcal{G}(V,Q) \oplus \mathcal{G}(V,Q)$, with the first entry U holding the input so far, and the second entry holding our result x. (47a) is obviously linear, and we are obliged to show

$$\begin{aligned}\text{contract}_{\text{aux}}[v](u)^2 &= (U,x) \mapsto (uuU, f(v,u)(uU) - u(f(v,u)U - ux)) &(48)\\ &= (U,x) \mapsto (Q(u)U, f(v,u)uU - f(v,u)uU + Q(u)x) &(49)\\ &= (U,x) \mapsto (Q(u)U, Q(u)x) &(50)\\ &= Q(u). &(51)\end{aligned}$$

All that remains is to initialize $(U,x) = (1,0)$ in contract$_{\text{aux}}$, then discard x' (which holds a copy of x anyway):

$$\text{contract}[v] : \mathcal{G}(V,Q) \to \mathcal{G}(V,Q) \tag{52}$$

$$\text{contract}[v] := \quad x \mapsto c \text{ where } (x',c) = \text{contract}_{\text{aux}}[v]((1,0)) \tag{52a}$$

With the aid of [3, Theorem 32] we can the show that α^{-f} is a two-sided inverse to α^f, recovering the promised isomorphism.

4 Formalization

The approaches in this paper are particularly amenable to formalization, as avoiding a basis allows them to hold in greater generality. Notably, avoiding a basis ensures our constructions continue to be valid in cases where V is not a free module, and does not have a basis at all. The results in Sects. 3.3 and 3.4 link via "" to a formalization in the Lean proving language [8], building on top of the work in [9], which can be found online at https://github.com/pygae/lean-ga. An additional construction of the (known) isomorphism between $\mathcal{G}^+(V,Q)$ and $\mathcal{G}^+(V,-Q)$ is included there, for which there was no room to describe here.

5 Conclusions

Using the universal property for anything beyond trivial constructions like (6) can appear anywhere between demanding and impossible. This paper demonstrates some essential building blocks to bring more demanding constructions within reach. While we only concerned ourselves with the universal properties related to the Clifford algebra, the strategies used apply to many other algebraic constructions with analogous properties.

Acknowledgements. The authors would like to thank David Cohoe for illuminating the ideas behind (10) that led to the rest of this paper. The first author is funded by a scholarship from the Cambridge Trust.

References

1. Chevalley, C.C.: The Algebraic Theory of Spinors. Columbia University Press (1954). https://doi.org/10.7312/chev93056
2. Dorst, L., Fontijne, D., Mann, S.: Geometric Algebra for Computer Science: an Object-Oriented Approach to Geometry. Morgan Kaufmann Series in Computer Graphics, Elsevier; Morgan Kaufmann, Amsterdam, San Francisco (2007). ISBN: 978-0-12-374942-0
3. Grinberg, D.: The Clifford algebra and the Chevalley map - a computational approach (summary version 1) (2016). http://mit.edu/~darij/www/algebra/chevalleys.pdf
4. Hestenes, D., Sobczyk, G.: Clifford Algebra to Geometric Calculus: A Unified Language for Mathematics and Physics. Fundamental Theories of Physics. Springer, Cham (1984). https://doi.org/10.1007/978-94-009-6292-7
5. Lasenby, A., Lasenby, J., Wareham, R.: A covariant approach to geometry using geometric algebra. Technical report F-INFENG/TR-483, Department of Engineering, University of Cambridge (2004). https://api.semanticscholar.org/CorpusID:124816889
6. Lawson, H., Michelsohn, M.: Spin Geometry (PMS-38). Princeton Mathematical Series, vol. 38. Princeton University Press (1989). https://books.google.co.uk/books?id=3d9JkN8w3X8C
7. Lounesto, P.: Clifford Algebras and Spinors. London Mathematical Society Lecture Note Series, 2nd edn. Cambridge University Press, Cambridge (2001). https://doi.org/10.1017/CBO9780511526022
8. de Moura, L., Kong, S., Avigad, J., van Doorn, F., von Raumer, J.: The lean theorem prover (system description). In: Felty, A.P., Middeldorp, A. (eds.) CADE 2015. LNCS (LNAI), vol. 9195, pp. 378–388. Springer, Cham (2015). https://doi.org/10.1007/978-3-319-21401-6_26
9. Wieser, E., Song, U.: Formalizing geometric algebra in lean. Adv. Appl. Clifford Algebras **32**(3), 28 (2022). https://doi.org/10.1007/s00006-021-01164-1

Binary Linear Codes via Zeon and Sym-Clifford Algebras

Melissa McLeod Price and G. Stacey Staples(✉)

Department of Mathematics and Statistics, Southern Illinois University Edwardsville, Edwardsville, IL 62026-1653, USA
{mmcleod,sstaple}@siue.edu

Abstract. Zeon algebras have proven to be useful for enumerating structures in graphs, such as paths, trails, cycles, matchings, cliques, and independent sets. Sym-Clifford algebras have been used to enumerate walks on hypercubes without the need for adjacency matrices. In the current work, zeon ("nil-Clifford") and "sym-Clifford" methods are used to reformulate essential concepts of binary linear coding theory. In particular, zeon and sym-Clifford methods are used to generate linear codes and to illustrate Clifford-algebraic formulations of encoding, decoding and error-correction.

Keywords: Binary codes · Clifford algebras · zeons

1 Introduction

Zeon algebras can be thought of as commutative analogues of fermion algebras, which are isomorphic to Clifford algebras of appropriate signature. They were first defined as subalgebras of Clifford algebras for counting self-avoiding walks (paths, cycles, trails, & circuits) in finite graphs [10]. This idea has led to numerous applications to graph problems, including routing problems in communication networks [2,4].

The "sym-Clifford" algebra $\mathcal{C}\ell_n{}^{\text{sym}}$ was first defined in [9] as a commutative subalgebra of the Clifford algebra $\mathcal{C}\ell_{n,n}$, where it was used to model random walks on hypercubes. The n-dimensional hypercube is a simple graph on 2^n vertices labeled by binary strings (i.e., "words") of length n, in which pairs of vertices are adjacent if and only if they differ in exactly one position. By associating the words with basis blades of $\mathcal{C}\ell_n{}^{\text{sym}}$, combinatorial properties of the geometric product can be used to represent random walks on hypercubes as sequences within the algebra.

In the current work, zeon and sym-Clifford methods are applied to coding theory. Here, we provide a formalism for generating, encoding, and decoding error-correcting binary linear codes using basis blades of appropriately chosen Clifford subalgebras. The symbolic approach presented here further demonstrates the applicability of Clifford algebras and their generalizations as a "unifying

D. W. Silva et al. (Eds.): ICACGA 2022, LNCS 13771, pp. 212–223, 2024.
https://doi.org/10.1007/978-3-031-34031-4_18

language" not only for physics and engineering, but for problems in discrete mathematics as well.

By reformulating binary linear codes in terms of zeon and sym-Clifford algebras, additional tools can be brought to bear in establishing theoretical results such as limit theorems and existence theorems.

1.1 Binary Linear Codes

The reader is directed to the books by Welsh [12] or Adams [1] for essential background on coding theory beyond the scope of this paper. Here, we focus only on binary linear codes.

Definition 1. *A binary linear code C of length n is a set of binary n-tuples such that the componentwise modulo 2 sum of any two codewords is contained in C.*

In this case, we say that the code's *alphabet* is the set $\{0,1\}$, with all arithmetic done modulo 2; i.e., the code's alphabet is the finite field with two elements, $GF(2)$. Since a binary linear code must contain the sum of each of its codewords, and this corresponds to what is needed to form a subspace, each binary linear code over $GF(2)$ is a subspace of the vector space $GF(2)^n$, where n is the length of each codeword.

In an $[n,k]$ linear code, messages $\mathbf{m} \in GF(2)^k$ of length k are encoded as codewords $\mathbf{c} \in GF(2)^n$ of length n using a *generator matrix* G. Encoding is done by computing $\mathbf{c} = \mathbf{m}G$. The rows of G form a basis for the linear code, so they are required to be linearly independent.

By generating all 2^k codewords, a codeword $\mathbf{c}$ can be decoded back to its original message using a lookup table, assuming $\mathbf{c}$ was received without errors in transmission. However, since errors may occur during transmission, it is vital to be able to verify that a codeword is valid. This is typically accomplished by using a parity check matrix.

Definition 2. *Given an $[n,k]$ linear code C, a* parity check matrix *for C is an $(n-k) \times n$ matrix H such that $\mathbf{c} \in C$ if and only if $\mathbf{c}H^{\intercal} = 0$.*

For the sake of brevity, details of computing the parity check matrix are omitted, but the following example illustrates the idea.

Example 1. Consider the 3×5 matrix G seen in (1).

$$G = \begin{pmatrix} 1\,0\,0\,0\,1 \\ 0\,1\,0\,1\,1 \\ 0\,0\,1\,1\,1 \end{pmatrix} \qquad H = \begin{pmatrix} 0\,1\,1\,1\,1 \\ 1\,1\,1\,0\,1 \end{pmatrix}. \tag{1}$$

By taking all products $\mathbf{m}G$, where $\mathbf{m}$ runs through all eight binary 3-tuples, the following code is generated:

$$C = \{00000, 10001, 01011, 00111, 01010, 01100, 11010, 11100\}. \tag{2}$$

The parity check matrix for G is the matrix H. Suppose $\mathbf{r_1} = 01111$ is received. Then $\mathbf{r_1}H^\intercal \neq 0$, which verifies that $\mathbf{r_1}$ is not a valid codeword. Now suppose $\mathbf{r_2} = 01010$ is received. Computing $\mathbf{r_1}H^\intercal = 0$ verifies that $\mathbf{r_2}$ is indeed a codeword, which can then be decoded to the original message 110.

Error Correction. The following questions are crucial in coding:

1. *If a received codeword contains errors, can those errors be detected?*
2. *If errors are detected, is it possible to correct them?*

In order to discuss these questions, first note that the *(Hamming) weight* of a codeword is the number of nonzero components in the codeword. For convenience, we define the following notation for the n-set: $[n] = \{1, \ldots, n\}$. Cardinality of a finite set $X = \{x_1, \ldots, x_k\}$ is denoted by $|X| = k = \sharp\{x_1, \ldots, x_k\}$.

Definition 3 (Hamming weight). *Let* $\mathbf{b} = (b_1 b_2 \cdots b_n) \in GF(2)^n$. *The* (Hamming) weight *of* b *is defined by*

$$w(\mathbf{b}) = \sharp\{i \in [n] : b_i \neq 0\}. \tag{3}$$

Now if a codeword $\mathbf{c} = (c_1 \ldots c_n)$ is sent through a "noisy" channel and the vector is $\mathbf{r} = (r_1 \ldots r_n)$ is received, then the *error vector* is defined as $\mathbf{e} = \mathbf{r} - \mathbf{c} = (e_1 \ldots e_n)$. The decoder must decide which codeword was most likely transmitted–equivalent to determining which error vector most likely occurred. Error vectors of lower weight are assumed to occur with higher probability than error vectors of higher weight, so we may use a *nearest neighbor decoding scheme*, which chooses the codeword that minimizes the distance between the received vector and possible transmitted vectors.

Definition 4. *The* Hamming distance *on* $GF(2)^n$ *is a mapping* $d : GF(2)^n \times GF(2)^n \to \mathbb{N}_0$ *defined*[1] *as follows: given* $\mathbf{x}, \mathbf{y} \in GF(2)^n$ *and writing* $\mathbf{x} = (x_1 x_2 \cdots x_n)$ *and* $\mathbf{y} = (y_1 y_2 \cdots y_n)$, *we define*

$$d(\mathbf{x}, \mathbf{y}) = \sharp\{i \in [n] : x_i \neq y_i\}. \tag{4}$$

Equivalently, $d(\mathbf{x}, \mathbf{y}) = w(\mathbf{x} + \mathbf{y})$.

It is not difficult to verify that Hamming distance establishes a well-defined metric on $GF(2)^n$. In other words, $d(\mathbf{x}, \mathbf{y})$ satisfies the following properties:

1. $d(\mathbf{x}, \mathbf{y}) = 0$ if and only if $\mathbf{x} = \mathbf{y}$;
2. $d(\mathbf{x}, \mathbf{y}) = d(\mathbf{y}, \mathbf{x})$ for all $\mathbf{x}, \mathbf{y} \in GF(2)^n$;
3. $d(\mathbf{x}, \mathbf{z}) \leq d(\mathbf{x}, \mathbf{y}) + d(\mathbf{y}, \mathbf{z})$ for all $\mathbf{x}, \mathbf{y}, \mathbf{z} \in GF(2)^n$.

Given a binary linear code C, the *minimum distance* of C is the minimum distance among distinct pairs of codewords in C. Equivalently, since the distance between codewords $\mathbf{c_1}$ and $\mathbf{c_2}$ is the weight of their binary sum, $w(\mathbf{c_1} + \mathbf{c_2})$, the

[1] The notation $\mathbb{N}_0$ denotes the nonnegative integers $\{0, 1, 2, \ldots\}$.

minimum distance $d(C)$ of a binary linear code C is equal to the weight of the lowest-weight nonzero codeword in C.

The minimum distance of a code is essential for determining the error-correcting capabilities of the code, providing methods for determining both the number of errors that can be detected and the number of errors that can be corrected.

Theorem 1. *Let C be a binary linear code having minimum distance $\mu = d(C)$. Utilizing the code C, it is possible to*

1. *detect up to s errors in any received codeword if $\mu \geq s+1$, and*
2. *correct up to t errors in any received codeword if $\mu \geq 2t+1$.*

Proof. 1. Suppose that a codeword $\mathbf{c}$ is transmitted such that s or fewer errors occur during transmission. If $d(C) \geq s+1$, the received word cannot be an element of C, since all codewords differ from $\mathbf{c}$ in at least $s+1$ places. Hence, the errors are detected.

2. Suppose $d(C) \geq 2t+1$. Suppose a codeword $\mathbf{x}$ is transmitted and that the received word, $\mathbf{r}$, contains t or fewer errors. Then $d(\mathbf{x},\mathbf{r}) \leq t$. Let $\mathbf{x}'$ be any codeword other than $\mathbf{x}$. Then $d(\mathbf{x}',\mathbf{r}) \geq t+1$, since otherwise $d(\mathbf{x}',\mathbf{r}) \leq t$ which implies that $d(\mathbf{x},\mathbf{x}') \leq d(\mathbf{x},\mathbf{r}) + d(\mathbf{x}',\mathbf{r}) \leq 2t$ (by the triangle inequality), which is impossible since $d(C) \geq 2t+1$. So $\mathbf{x}$ is the nearest codeword to $\mathbf{r}$, and $\mathbf{r}$ is decoded correctly.

1.2 Zeon and "sym-Clifford" Algebras

We begin with the essential definitions and terminology of Clifford algebras and then construct the zeon and sym-Clifford algebras as subalgebras of Clifford algebras of appropriate signature.

Definition 5. *For fixed $n \geq 0$, let V be an n-dimensional vector space having orthonormal basis $\{\mathbf{e}_{\{1\}}, \ldots, \mathbf{e}_{\{n\}}\}$. The 2^n-dimensional* Clifford algebra of signature (p,q), *where $p+q=n$, is defined as the associative algebra generated by the collection $\{\mathbf{e}_{\{i\}}\}$ along with the scalar $\mathbf{e}_{\varnothing} = 1 \in \mathbb{R}$, subject to the following multiplication rules:*

$$\mathbf{e}_{\{i\}}\,\mathbf{e}_{\{j\}} + \mathbf{e}_{\{j\}}\,\mathbf{e}_{\{i\}} = 0 \text{ for } i \neq j, \text{ and} \tag{5}$$

$$\mathbf{e}_{\{i\}}{}^2 = \begin{cases} 1, \text{ if } 1 \leq i \leq p \\ -1, \text{ if } p+1 \leq i \leq p+q = n. \end{cases} \tag{6}$$

We denote the Clifford algebra of signature (p,q) by $C\ell_{p,q}$.

Let $[n] = \{1, 2, \ldots, n\}$ and denote arbitrary, canonically ordered subsets of $[n]$ by capital Roman characters. The basis elements of $C\ell_{p,q}$ can then be indexed by these finite subsets if we write

$$\mathbf{e}_I = \prod_{j \in I} \mathbf{e}_{\{j\}}. \tag{7}$$

Arbitrary elements of $\mathcal{C}\ell_{p,q}$ then have canonical expansions of the form

$$u = \sum_{I \in 2^{[n]}} u_I \, \mathbf{e}_I, \tag{8}$$

where $u_I \in \mathbb{R}$ for each $I \in 2^{[n]}$.

Definition 6. *By the* grade *of a basis blade in $\mathcal{C}\ell_{p,q}$ we shall mean the cardinality of its multi index. That is,* $\mathrm{gr}(\mathbf{e}_I) = |I|$*. For $0 \leq k \leq n$, we define the* k-grade part *of $u \in \mathcal{C}\ell_{p,q}$ as the sum of k-grade monomials in the expansion of u. In other words,*

$$\langle u \rangle_k = \sum_{\substack{I \in 2^{[n]} \\ |I| = k}} u_I \, \mathbf{e}_I. \tag{9}$$

Introduction to the "Sym-Clifford" Algebra $\mathfrak{S}_n$. The sym-Clifford algebra first appeared in [9], where it was used to enumerate walks on hypercubes.

Definition 7. *For fixed $n > 0$, the* sym-Clifford algebra[2] $\mathfrak{S}_n$ *is defined as the 2^n-dimensional associative algebra generated by the elements $\varsigma_{\{i\}} = \mathbf{e}_{\{i\}} \, \mathbf{e}_{\{n+i\}} \in \mathcal{C}\ell_{n,n}$ for $1 \leq i \leq n$ along with the scalar $\varsigma_{\varnothing} = 1 \in \mathbb{R}$.*

It is easy to see that $\mathfrak{S}_n$ is a *commutative* graded algebra whose generators satisfy ${\varsigma_{\{i\}}}^2 = 1$ for $1 \leq i \leq n$; i.e., the generators are unipotent.

Basis elements of $\mathfrak{S}_n$ can again be indexed by canonically-ordered subsets of $[n]$ so that arbitrary elements have the form

$$u = \sum_{I \in 2^{[n]}} u_I \, \varsigma_I. \tag{10}$$

By the properties of Clifford multiplication, we see that for arbitrary $I, J \in 2^{[n]}$ we have

$$\varsigma_I \, \varsigma_J = \varsigma_{I \triangle J}, \tag{11}$$

where $I \triangle J = (I \cup J) \setminus (I \cap J)$ denotes the *set-symmetric difference* of I and J.

Remark 1. The generators ς_I of $\mathfrak{S}_n$ (disjoint bivectors in $\mathcal{C}\ell_{n,n}$) generate an Abelian multiplicative group Σ_n which is isomorphic to the group generated by reflections across orthogonal hyperplanes in the real vector space $\mathbb{R}^n$, for these also satisfy $R_i \, R_j = R_j \, R_i$ and ${R_i}^2 = \mathbb{I}$. It is equally evident that $\Sigma_n \cong (2^{[n]}, \triangle)$, the group consisting of the power set of $[n] = \{1, 2, \ldots, n\}$ with the set symmetric difference operator. These groups are also isomorphic to the additive abelian group $\underbrace{\mathbb{Z}_2 \oplus \cdots \oplus \mathbb{Z}_2}_{n\text{-times}}$. Finally, it is evident that the Cayley graph of Σ_n is the n-dimensional hypercube, a structure commonly seen in connection with coding theory.

[2] The 2^n-dimensional sym-Clifford algebra has been denoted by ${\mathcal{C}\ell_n}^{\mathrm{sym}}$ in other works, but that notation is cumbersome for this paper.

Zeon ("nil-Clifford") Algebra. For $n \in \mathbb{N}$, let $\mathfrak{Z}_n$ denote the real abelian algebra generated by the collection $\{\zeta_{\{i\}} : 1 \le i \le n\}$ along with the scalar $1 = \zeta_\varnothing$ subject to the following multiplication rules:

$$\zeta_{\{i\}}\,\zeta_{\{j\}} = \zeta_{\{i,j\}} = \zeta_{\{j\}}\,\zeta_{\{i\}} \text{ for } i \neq j, \text{ and} \tag{12}$$

$$\zeta_{\{i\}}{}^2 = 0 \text{ for } 1 \le i \le n. \tag{13}$$

It is evident that a general element $u \in \mathfrak{Z}_n$ can be expanded as $u = \sum_{I \in 2^{[n]}} u_I\,\zeta_I$, or more simply as $\sum_I u_I \zeta_I$, where $I \in 2^{[n]}$ is a subset of the *n-set*, $[n] := \{1, 2, \ldots, n\}$, used as a multi-index, $u_I \in \mathbb{R}$, and $\zeta_I = \prod_{\iota \in I} \zeta_\iota$. The algebra $\mathfrak{Z}_n$ is called the (n-particle) *zeon algebra*[3].

As a vector space, this 2^n-dimensional algebra has a canonical basis of *basis blades* of the form $\{\zeta_I : I \subseteq [n]\}$. The null-square property of the generators $\{\zeta_{\{i\}} : 1 \le i \le n\}$ guarantees that the product of two basis blades satisfies the following:

$$\zeta_I \zeta_J = \begin{cases} \zeta_{I \cup J} & I \cap J = \varnothing, \\ 0 & \text{otherwise.} \end{cases} \tag{14}$$

It should be clear that $\mathfrak{Z}_n$ is graded. For non-negative integer k, the *k-grade part* of element $u = \sum_I u_I \zeta_I$ is defined as

$$\langle u \rangle_k = \sum_{\{I:|I|=k\}} u_I \zeta_I. \tag{15}$$

It is often convenient to separate the scalar (0-grade) part of a zeon from the rest of it. To this end, for $z \in \mathfrak{Z}_n$ we write $\Re z = \langle z \rangle_0$, the *real part* of z, and $\mathfrak{D} z = z - \Re z$, the *dual part* of z (these are referred to as the body and soul of z in Neto's works [5–7]).

Remark 2. Like $\mathfrak{S}_n$, the algebra $\mathfrak{Z}_n$ can be constructed within a Clifford algebra of appropriate signature. For example, $\mathfrak{Z}_n$ is isomorphic to a subalgebra of $\mathcal{C}\ell_{2n,2n}$. To see this, begin by letting $\{\mathbf{e}_i : 1 \le i \le 4n\}$ be orthonormal generators of $\mathcal{C}\ell_{2n,2n}$. For each $j = 1, \ldots, 2n$, let $\mathbf{f}_j = (\mathbf{e}_j - e_{n+j})$ to obtain a collection of orthogonal, pairwise-anticommuting, null-square elements $\{\mathbf{f}_j : 1 \le j \le 2n\}$. Finally, define $\zeta_{\{\ell\}} = \mathbf{f}_{\{2\ell-1,2\ell\}}$ for $\ell = 1, \ldots, n$. The resulting collection of null-square bivectors $\{\mathbf{f}_{\{1,2\}}, \mathbf{f}_{\{3,4\}}, \ldots, \mathbf{f}_{\{2n-1,2n\}}\}$ is then pairwise commutative and generates the algebra $\mathfrak{Z}_n$.

Multiplicative Properties of Zeons

Since $\mathfrak{Z}_n$ is an algebra, its elements form a multiplicative semigroup. It is not difficult to establish convenient formulas for expanding products of zeons. As shown in [3], $u \in \mathfrak{Z}_n$ is invertible if and only if $\Re u \neq 0$.

[3] The n-particle zeon algebra is often denoted by $\mathcal{C}\ell_n{}^{\text{nil}}$ in other works, but that notation is cumbersome for this paper.

More important for matters at hand, the null square property of zeon generators make zeons useful for performing computations on partitions of sets.

Lemma 1 (Powers of nilpotent zeons). *Let $u = \zeta_{I_1} + \zeta_{I_2} + \cdots + \zeta_{I_m} \in \mathfrak{Z}_n$, and let k be a positive integer. Then,*

$$u^k = k! \sum_J \zeta_J, \tag{16}$$

where the sum is over multi indices J obtained from disjoint unions of k of the multi indices $I_1, \ldots, I_m$.

Proof. Proof follows from a simple application of the multinomial theorem:

$$\begin{aligned} u^k &= (\zeta_{I_1} + \zeta_{I_2} + \cdots + \zeta_{I_m})^k \\ &= \sum_{\substack{\ell_1,\ldots,\ell_m \geq 0 \\ \ell_1+\cdots+\ell_m=k}} \binom{k}{\ell_1,\ldots,\ell_m} {\zeta_{I_1}}^{\ell_1} {\zeta_{I_2}}^{\ell_2} \cdots {\zeta_{I_m}}^{\ell_m} \\ &= \sum_{\substack{\ell_1,\ldots,\ell_m \in \{0,1\} \\ \ell_1+\cdots+\ell_m=k}} \binom{k}{\ell_1,\ldots,\ell_m} {\zeta_{I_1}}^{\ell_1} {\zeta_{I_2}}^{\ell_2} \cdots {\zeta_{I_m}}^{\ell_m}. \end{aligned} \tag{17}$$

Note that the only nonzero terms of the sum correspond to pairwise-disjoint unions of k subsets.

In the next section, the exponential function will be used to generate a binary linear code. The exponential function $\exp : \mathfrak{Z}_n \to \mathfrak{Z}_n$ is defined on zeon algebras in the standard way. However, the null-square properties of zeon generators reduce the infinite power series to a finite sum, as developed in [11]:

$$\begin{aligned} \exp(u) &= \exp(\mathfrak{R}u + \mathfrak{D}u) \\ &= e^{\mathfrak{R}u} \sum_{k=0}^{n} \frac{1}{k!} (\mathfrak{D}u)^k. \end{aligned} \tag{18}$$

2 Binary $[n, k]$ Codes in $\mathfrak{Z}_k \otimes \mathfrak{S}_n$

Any codeword, consisting of a binary n-tuple, can be instead represented as a blade from $\mathfrak{S}_n$ via a mapping that takes binary strings of length n to elements of the power set $2^{[n]}$. In particular, letting $\mathbf{b} = (b_1 b_2 \ldots b_n) \in \{0,1\}^n$, the *subset representation* of $\mathbf{b}$ is given by

$$\mathbf{b} \mapsto B = \{j : b_j = 1\} \subseteq [n]. \tag{19}$$

One can then represent $\mathbf{b}$ within $\mathfrak{S}_n$ by

$$\begin{aligned} (b_1 b_2 \cdots b_n) &\mapsto \prod_{j=1}^{n} {\varsigma_{\{j\}}}^{b_j} \\ &= \varsigma_{\{j: b_j = 1\}} = \varsigma_B. \end{aligned} \tag{20}$$

Example 2. The binary string $\mathbf{b} = 100110 \in GF(2)^6$ has subset representation $B = \{1, 4, 5\}$ and sym-Clifford representation $\varsigma_{\{1,4,5\}}$.

Linear independence of code generators in generator matrices is replaced by "multiplicative independence" in $\mathfrak{S}_n$.

Definition 8. *Let $\mathcal{B} = \{\varsigma_I : I \in 2^{[n]}\}$ be the canonical basis for $\mathfrak{S}_n$. A collection $X \subset \mathcal{B}$ is said to be* multiplicatively independent *if no element of X can be written as a product of other elements of X. If an element of X can be expressed as a product of other elements, then X is* multiplicatively dependent.

It is evident from the definition that no multiplicatively independent collection can contain the scalar unit, $\varsigma_\varnothing = 1$.

Example 3. In $\mathfrak{S}_4$, the collection $\{\varsigma_{\{1\}}, \varsigma_{\{1,4\}}, \varsigma_{\{2,4\}}, \varsigma_{\{3,4\}}\}$ is multiplicatively independent, whereas the collection $\{\varsigma_{\{1,3\}}, \varsigma_{\{2,3\}}, \varsigma_{\{1,2\}}\}$ is multiplicatively dependent.

Using this notation, a binary code can be generated using an expression from $\mathfrak{S}_n$ that represents the same information contained in a code's generator matrix but in a more succinct format.

Given a multiplicatively independent collection $\mathcal{G} = \{\varsigma_{I_\ell} : 1 \leq \ell \leq k\}$, an encoding map $\varphi : 2^{[k]} \to 2^{[n]}$ is defined implicitly by

$$\varsigma_{\varphi(M)} = \prod_{\ell \in M} \varsigma_{I_\ell}. \tag{21}$$

The collection $\mathfrak{C} = \{\varsigma_{\varphi(M)} : M \in 2^{[k]}\}$ is referred to as the $[n, k]$ *sym-Clifford code generated by the collection* $\mathcal{G} = \{\varsigma_{I_\ell} : 1 \leq \ell \leq k\}$.

By combining properties of zeons and sym-Clifford algebras, the code can be generated by the exponential of a generator g in the tensor algebra $\mathfrak{Z}_k \otimes \mathfrak{S}_n$. The algebra $\mathfrak{Z}_k \otimes \mathfrak{S}_n$ is the 2^{k+n}-dimensional commutative algebra spanned by canonical basis elements $\{\zeta_I \otimes \varsigma_J : I \in 2^{[k]}, J \in 2^{[n]}\}$. For ease of notation, a typical element $u \in \mathfrak{Z}_k \otimes \mathfrak{S}_n$ will be expanded in the form

$$u = \sum_{I \in 2^{[k]}, J \in 2^{[n]}} u_{IJ} \zeta_I \varsigma_J, \tag{22}$$

where $u_{IJ} \in \mathbb{R}$ for all I, J.

Lemma 2 (Generating A Code). *Given $g = \sum_{\ell=1}^{k} \zeta_{\{\ell\}} \varsigma_{I_\ell} \in \mathfrak{Z}_k \otimes \mathfrak{S}_n$, where $\{\varsigma_{I_\ell} : 1 \leq \ell \leq k\}$ is a multiplicatively independent collection, the exponential $\exp(g) = \sum_{M \in 2^{[k]}} \zeta_M \varsigma_{\varphi(M)}$ reveals the $[n, k]$ sym-Clifford code $\mathfrak{C}$ generated by g. In particular, $\exp(g)$ is a sum over messages M, such that each summand represents a message/codeword pair.*

Proof. By nilpotent properties of zeons, the exponential of g is a finite sum representing the code $\mathfrak{C}$. Writing $g = \sum_{\ell=1}^{k} \zeta_{\{\ell\}}\varsigma_{I_\ell}$, one sees that

$$\begin{aligned}\exp(g) &= \sum_{\ell=0}^{k} \frac{g^\ell}{\ell!} \\ &= 1 + \sum_{\substack{M\in 2^k \\ M\neq\varnothing}} \zeta_M \prod_{\ell\in M} \varsigma_{I_\ell} \\ &= 1 + \sum_{\substack{M\in 2^k \\ M\neq\varnothing}} \zeta_M \varsigma_{\varphi(M)}. \end{aligned} \tag{23}$$

The canonical basis for the smallest subspace of $\mathfrak{Z}_k \otimes \mathfrak{S}_n$ containing $\exp(g)$ is

$$\mathcal{B} = \{\zeta_I \varsigma_{\varphi(I)} : I \in 2^{[k]}\}. \tag{24}$$

This basis provides all information necessary for encoding and decoding messages.

Definition 9. *An element* $g = \sum_{\ell=1}^{k} \zeta_{\{\ell\}}\varsigma_{I_\ell} \in \mathfrak{Z}_k \otimes \mathfrak{S}_n$ *is an* $[n, k]$ sym-Clifford code generator *when the collection* $\mathcal{G} = \{\varsigma_{I_\ell} : \ell = 1, \ldots, k\}$ *is multiplicatively independent.*

Inner Products. The algebra $\mathfrak{Z}_k \otimes \mathfrak{S}_n$ admits two particularly useful inner products.

Definition 10. *Define the* zeon inner product $\langle\cdot,\cdot\rangle_{\mathfrak{Z}} : \mathfrak{Z}_k \otimes \mathfrak{S}_n \times \mathfrak{Z}_k \otimes \mathfrak{S}_n \to \mathfrak{S}_n$ *by bilinear extension of*

$$\langle \zeta_I\varsigma_J, \zeta_L\varsigma_K\rangle_{\mathfrak{Z}} = \begin{cases} \varsigma_{J\Delta K} & \text{if } I = L \\ 0 & \text{otherwise.}\end{cases} \tag{25}$$

The sym-Clifford inner product $\langle\cdot,\cdot\rangle_s : \mathfrak{Z}_k \otimes \mathfrak{S}_n \times \mathfrak{Z}_k \otimes \mathfrak{S}_n \to \mathfrak{Z}_k$ *is defined by bilinear extension of*

$$\langle \zeta_I\varsigma_J, \zeta_L\varsigma_K\rangle_s = \begin{cases} \zeta_{I\cup L} & \text{if } J = K \text{ and } I\cap L = \varnothing \\ 0 & \text{otherwise.}\end{cases} \tag{26}$$

It is not difficult to see that the zeon and sym-Clifford inner products define canonical orthogonal projections $\pi_{\mathfrak{Z}}$ onto $\mathfrak{Z}_k$ and π_s onto $\mathfrak{S}_n$, respectively. More specifically, for $u \in \mathfrak{Z}_k \otimes \mathfrak{S}_n$, these projections satisfy

$$\pi_{\mathfrak{Z}}(u) = \sum_{I\in 2^{[k]}} \langle u, \zeta_I\rangle_{\mathfrak{Z}}\, \zeta_I, \quad \text{and} \tag{27}$$

$$\pi_s(u) = \sum_{I\in 2^{[n]}} \langle u, \varsigma_I\rangle_s\, \varsigma_I. \tag{28}$$

Lemma 3 (Encoding). *Let $g \in \mathfrak{Z}_k \otimes \mathfrak{S}_n$ be the generator of an $[n, k]$ code. For message $M \in 2^{[k]}$, the corresponding codeword is given by*

$$\varsigma_{\varphi(M)} = \langle \exp(g), \zeta_M \rangle_{\mathfrak{Z}}. \tag{29}$$

Definition 11. *For a sym-Clifford element $u = \sum_{I \in 2^{[n]}} u_I \varsigma_I$, it is useful to define the* minimal grade *of u by*

$$\natural u = \begin{cases} \min\{|I| : I \neq \varnothing, u_I \neq 0\} & \text{if } u \notin \mathbb{R} \\ 0 & u = u_\varnothing \in \mathbb{R}. \end{cases} \tag{30}$$

Note that $\natural u = 0$ if and only if u is a scalar. The minimal grade part *of u is defined by*

$$\langle u \rangle_\natural = \sum_{\{I : |I| = \natural u\}} u_I \varsigma_I. \tag{31}$$

Proposition 1. *Let g be the generator of an $[n, k]$ sym-Clifford code. Utilizing the code generated by g, it is possible to*

1. *detect up to $\natural(\pi_s(\exp(g))) - 1$ errors and*
2. *correct up to $(\natural(\pi_s(\exp(g))) - 1)/2$ errors in any received codeword.*

Proof. Note that letting $\mu = \natural(\pi_s(\exp(g)))$, which is the smallest nonzero grade among codewords generated by g, this proposition is equivalent to Theorem 1.

Lemma 4 (Verification). *Let $g \in \mathfrak{Z}_k \otimes \mathfrak{S}_n$ be the generator of an $[n, k]$ code, and let $\varsigma_I \in \mathfrak{S}_n$. Then,*

$$\langle \exp(g), \varsigma_I \rangle_s = \begin{cases} \zeta_{\varphi^{-1}(I)} & \text{if } I \in C, \\ 0 & \text{otherwise.} \end{cases} \tag{32}$$

Here $M = \varphi^{-1}(I)$ is the decoded message.

Proof. Let g be the generator of an $[n, k]$ code $\mathcal{G}$, and suppose ς_L is a basis blade of $\mathfrak{S}_n$. Expanding the exponential of g,

$$\exp(g) = \sum_{I \in \mathcal{G}} \zeta_{\varphi^{-1}(I)} \varsigma_I. \tag{33}$$

If ς_L is a codeword in $\mathcal{G}$, it follows that

$$\begin{aligned} \langle \exp(g), \varsigma_L \rangle_s &= \left\langle \sum_{I \in \mathcal{G}} \zeta_{\varphi^{-1}(I)} \varsigma_I, \varsigma_L \right\rangle_s \\ &= \sum_{I \in \mathcal{G}} \zeta_{\varphi^{-1}(I)} \langle \varsigma_I, \varsigma_L \rangle_s \\ &= \zeta_{\varphi^{-1}(L)} = \zeta_M, \end{aligned} \tag{34}$$

where ζ_M is the message corresponding to codeword ς_L.

On the other hand, if ς_L is not among the codewords of $\mathcal{G}$, then all inner products appearing in the sum are zero.

Lemma 4 provides a method for decoding any valid received codeword. Error correction is possible when a unique nearest-neighbor to a received codeword exists.

Proposition 2 (Decoding). *Let $g \in \mathfrak{Z}_k \otimes \mathfrak{S}_n$ be the generator of an $[n, k]$ code, and let $\varsigma_I \in \mathfrak{S}_n$. If $\langle \exp(g), \varsigma_I \rangle_s = 0$ and if $\natural(\varsigma_I \pi_s(\exp(g))) \leq (\natural(\pi_s(\exp(g))) - 1)/2$, then the message M is recovered from*

$$\zeta_M = \langle \exp(g), \varsigma_J \rangle_s, \tag{35}$$

where

$$\varsigma_J = \varsigma_I \langle \varsigma_I \pi_s(\exp(g)) \rangle_\natural. \tag{36}$$

Proof. Note that if $\langle \exp(g), \varsigma_I \rangle_s = 0$, then ς_I is not an element of the code $\mathcal{G}$ generated by g. However, if $\natural(\varsigma_I \pi_s(\exp(g))) \leq (\natural(\pi_s(\exp(g))) - 1)/2$, the minimal grade part of $\varsigma_I \pi_s(\exp(g)))$ is a unique nonzero term. In fact, this term is $\varsigma_{I \triangle J}$, where $\varsigma_J \in \mathcal{G}$ and

$$\begin{aligned} \varsigma_J &= \varsigma_I \langle \varsigma_I \pi_s(\exp(g))) \rangle_\natural \\ &= \varphi(\zeta_M). \end{aligned} \tag{37}$$

3 Example: A [7, 4]-sym-Clifford Code

In $\mathfrak{Z}_4 \otimes \mathfrak{S}_7$, let $g = \zeta_{\{1\}}\varsigma_{\{1,2,3\}} + \zeta_{\{2\}}\varsigma_{\{1,4,5\}} + \zeta_{\{3\}}\varsigma_{\{1,6,7\}} + \zeta_{\{4\}}\varsigma_{\{2,4,6\}}$. The exponential of g is then given by

$$\begin{aligned} \exp(g) = 1 &+ \zeta_{\{1\}}\varsigma_{\{1,2,3\}} + \zeta_{\{2\}}\varsigma_{\{1,4,5\}} + \zeta_{\{3\}}\varsigma_{\{1,6,7\}} + \zeta_{\{4\}}\varsigma_{\{2,4,6\}} + \zeta_{\{2,3,4\}}\varsigma_{\{2,5,7\}} \\ &+ \zeta_{\{1,3,4\}}\varsigma_{\{3,4,7\}} + \zeta_{\{1,2,4\}}\varsigma_{\{3,5,6\}} + \zeta_{\{3,4\}}\varsigma_{\{1,2,4,7\}} + \zeta_{\{2,4\}}\varsigma_{\{1,2,5,6\}} \\ &+ \zeta_{\{1,4\}}\varsigma_{\{1,3,4,6\}} + \zeta_{\{1,2,3,4\}}\varsigma_{\{1,3,5,7\}} + \zeta_{\{1,2\}}\varsigma_{\{2,3,4,5\}} + \zeta_{\{1,3\}}\varsigma_{\{2,3,6,7\}} \\ &+ \zeta_{\{2,3\}}\varsigma_{\{4,5,6,7\}} + \zeta_{\{1,2,3\}}\varsigma_{\{1,2,3,4,5,6,7\}}. \end{aligned} \tag{38}$$

The terms of the exponential reveal messages and their corresponding codewords, as illustrated in Table 1. Note that the minimum weight of the code is 3. Using this code, two errors may be detected and one error may be corrected in any received codeword.

4 Conclusion

We have formalized error-correcting binary linear codes using zeon and sym-Clifford algebraic methods. With this foundation established, generalizations (ternary codes, Reed-Solomon Codes, etc.) and extensions (cryptography) can be treated within a straightforward unified algebraic framework. Recent developments in encryption schemes using geometric algebra make these techniques look promising for future work [8].

By reformulating binary linear codes in terms of zeon and sym-Clifford algebras, previously overlooked algebraic methods can be applied to establishing theoretical results such as limit theorems and existence theorems. Moreover, these algebraic methods may provide new insight on other problems.

Table 1. A [7, 4]-sym-Clifford Code

Message	**m**	**c**	Codeword	Message	**m**	**c**	Codeword
1	0000	0000000	1	$\zeta_{\{1\}}$	1000	1110000	$\varsigma_{\{1,2,3\}}$
$\zeta_{\{1,2\}}$	1100	0111100	$\varsigma_{\{2,3,4,5\}}$	$\zeta_{\{2\}}$	0100	1001100	$\varsigma_{\{1,4,5\}}$
$\zeta_{\{1,3\}}$	1010	0110011	$\varsigma_{\{2,3,6,7\}}$	$\zeta_{\{3\}}$	0010	1000011	$\varsigma_{\{1,6,7\}}$
$\zeta_{\{1,4\}}$	1001	1011010	$\varsigma_{\{1,3,4,6\}}$	$\zeta_{\{4\}}$	0001	0101010	$\varsigma_{\{2,4,6\}}$
$\zeta_{\{2,3\}}$	0110	0001111	$\varsigma_{\{4,5,6,7\}}$	$\zeta_{\{1,2,3\}}$	1110	1111111	$\varsigma_{\{1,2,3,4,5,6,7\}}$
$\zeta_{\{2,4\}}$	0101	1100110	$\varsigma_{\{1,2,5,6\}}$	$\zeta_{\{1,2,4\}}$	1101	0010110	$\varsigma_{\{3,5,6\}}$
$\zeta_{\{3,4\}}$	0011	1101001	$\varsigma_{\{1,2,4,7\}}$	$\zeta_{\{1,3,4\}}$	1011	0011001	$\varsigma_{\{3,4,7\}}$
$\zeta_{\{1,2,3,4\}}$	1111	1010101	$\varsigma_{\{1,3,5,7\}}$	$\zeta_{\{2,3,4\}}$	0111	0100101	$\varsigma_{\{2,5,7\}}$

References

1. Adams, S.S.: Introduction to algebraic coding theory - with Gap (2008)
2. Cruz-Sanchez, H., Staples, G.S., Schott, R., Song, Y.Q.: Operator calculus approach to minimal paths: precomputed routing in a store and forward satellite constellation. In: 2012 IEEE Global Communications Conference, GLOBECOM 2012, Anaheim, CA, USA, 3–7 December 2012, pp. 3431–3436. IEEE (2012). https://doi.org/10.1109/GLOCOM.2012.6503645
3. Dollar, L.M., Staples, G.S.: Zeon roots. Adv. Appl. Clifford Algebras **27**(2), 1133–1145 (2016). https://doi.org/10.1007/s00006-016-0732-4
4. Nefzi, B., Schott, R., Song, Y.Q., Staples, G.S., Tsiontsiou, E.: An operator calculus approach for multi-constrained routing in wireless sensor networks. In: Proceedings of the 16th ACM International Symposium on Mobile Ad Hoc Networking and Computing, MobiHoc 2015, pp. 367–376. Association for Computing Machinery, New York (2015). https://doi.org/10.1145/2746285.2746301
5. Neto, A.F.: Higher order derivatives of trigonometric functions, Stirling numbers of the second kind, and zeon algebra. J. Integer Sequences **17**, Article 14.9.3 (2014)
6. Neto, A.F.: Carlitz's identity for the Bernoulli numbers and zeon algebra. J. Integer Sequences **18**, Article 15.5.6 (2015)
7. Neto, A.F., dos Anjos, P.H.R.: Zeon algebra and combinatorial identities. SIAM Rev. **56**(2), 353–370 (2014). https://doi.org/10.1137/130906684
8. da Silva, D.W.H.A., Xavier, M.A., Chow, C.E., de Araujo, C.P.: Experiments with Clifford geometric algebra applied to cryptography. In: 2020 Joint 11th International Conference on Soft Computing and Intelligent Systems and 21st International Symposium on Advanced Intelligent Systems (SCIS-ISIS), pp. 1–8 (2020). https://doi.org/10.1109/SCISISIS50064.2020.9322757
9. Staples, G.S.: Clifford-algebraic random walks on the hypercube. Adv. Appl. Clifford Algebras **15**(2), 213–232 (2005). https://doi.org/10.1007/s00006-005-0014-z
10. Staples, G.S.: A new adjacency matrix for finite graphs. Adv. Appl. Clifford Algebras **18**, 979–991 (2008)
11. Staples, G.S., Weygandt, A.: Elementary functions and factorizations of zeons. Adv. Appl. Clifford Algebras **28**, 12 (2018). https://doi.org/10.1007/s00006-018-0836-0
12. Welsh, D.: Codes and Cryptography. Oxford University Press, Oxford (1988)

Posters

A Generalized Metric for Hypothesis Testing: A Geometric Algebra Point of View

Matthew Anderson

Data and their collection methods are changing in characteristic and dimension. Using Clifford algebras, we present a hypothesis testing framework for multidimensional vectors sampled from distributions in the real or complex field. Our procedure aims to obtain non-commutative information through the incorporation of the geometric outer product not found in traditional i.i.d estimation methods. We construct a metric that measures deviations in the geometry of the parallelepiped spanned from sampled data represented as multivectors in G(n). The outer product provides additional geometric structure of the data not captured solely by the inner product. Quadratic forms of covariance matrices and trace operators are represented using geometric algebras. Null and alternative hypothesis are formed from partitions of the geometric space generated under G(n) as well as the direction of the multivectors sampled. A test statistic is constructed to measure deviations in the geometry of the data from a model, and a decision rule is applied to outcomes of the data fit to a loss function. Our method is applied to discriminate between states of quantum information systems in the binary and multiparameter setting. Representing data collected from the quantum system as a set of multivectors with a geometric algebra, our method allows for the detection of geometric deviations from the system when compared to a hypothesized null geometry space.

D. W. Silva et al. (Eds.): ICACGA 2022, LNCS 13771, p. 227, 2024.
https://doi.org/10.1007/978-3-031-34031-4

Quantum Space Structure by Geometric Algebra Including the Hurwitz Unit Quaternion Group

Jens Erfurt Andresen

The analysis aims to find a model for the internal structure of one local indivisible spin-half fermions known from the Standard Model. The single universe idea dictates the local space structure being scale invariant in physics. The quantum approach of steady state *Angular Momentum* (AM) hides the chronometric development of internal oscillations, in which case the three-dimensional founded Geometric Algebra $G_3(R)$ is suitable. Each component *quantity* of AM is associated with the concept of constant *bivectors*, representing preserved plane areas denoted by Kepler's 2nd Law. The local perpendicular plane unit bivector *directions* make the *quaternion* basis of $G_3^+(R)$. Traditional Quantum Mechanics use the orthogonal *interconnectivity* structure of AM commutators resulting in a spin-half projection in one *direction*. As a supplement, we are using Adolf Hurwitz's (1859–1919) Number Theory of Quaternions [A. Hurwitz, Vorlesungen Über die Zahlentheorie der Quaternionen, Berlin: Verlag von Julius Springer, 1919. A. Hurwitz, "Uber die Zahlentheorie der Quaternionen," Nachrichten von der Gesellschaft der Wissenschaften zu Göttingen, Mathematisch-Physikalische Klasse, 1896, pp. 313–340.] to find a normal invariant subgroup of sixteen unit-½-quaternions by superposition of the orthonormal bivector basis. This performs a regular tetrahedron space structure of four *interconnected* non-orthogonal AM bivector *directions* in physical space. Combining these as projections in one *direction* make sixteen possibilities of one local contributing *quantity* charge for the spin-half fermions, the values are − 1, − 2/3, − 1/3, − 0, + 0, + 1/3, + 2/3, + 1. Further superposition of excitation of this tetrahedron bivector AM structure of unit-½-quaternions may construct leptons, baryons, and mesons, in just one locality. Besides the internal spin-half AM qualities, their composition also opens a port in one *direction* for integer spin-boson AM interaction with the external of each fermion locality. –This investigation shows, there is knowledge to get from using the quaternions and their bivector *directions* of $G_3^+(R)$ together with the full $G_3(R)$ Geometric Algebra when studying the particle structure of physics known from the Standard Model.

D. W. Silva et al. (Eds.): ICACGA 2022, LNCS 13771, p. 228, 2024.
https://doi.org/10.1007/978-3-031-34031-4

A New Home for Bivectors

Norm Cimon

The impetus for the work (and the poster) is this quote: "…as shown by Gel'fand's approach, we can only abstract a unique manifold if our algebra is commutative." [Hiley, B. J. and Callaghan, R. E. (2010) 'The Clifford Algebra approach to Quantum Mechanics A: The Schroedinger and Pauli Particles', *arXiv:1011.4031 [math-ph, physics:quant-ph]* [Preprint]. Available at: http://arxiv.org/abs/1011.4031 (Accessed: 18 October 2020).] Geometric algebra is non-commutative. Components of different grades can be staged on different manifolds. As operations on those elements proceed, they will effect the promotion and/or demotion of components to higher and/or lower grades, and thus to different manifolds. I've written a paper with imagery that visually displays bivector addition and rotation on a sphere. Those images were then transferred to a poster I've developed along with explanatory text. David Hestenes interpreted the vector product or rotor in two-dimensions: "as a directed arc of fixed length that can be rotated at will on the unit circle, just as we interpret a vectora as a directed line segment that can be translated at will without changing its length or direction…" [Hestenes, D. (2003) 'Oersted Medal Lecture 2002: Reforming the mathematical language of physics', *American Journal of Physics*, 71(2), pp. 104–121. Available at: https://doi.org/10.1119/1.1522700]. Rotors, it turns out, can be used to develop addition and multiplication of bivectors on a sphere. For those rotational dynamics, rotors of length $\pi/2$ are the basis elements. The geometric algebra of bivectors – Hamilton's "pure quaternions" – is shown to transparently reside on a spherical manifold.

D. W. Silva et al. (Eds.): ICACGA 2022, LNCS 13771, p. 229, 2024.
https://doi.org/10.1007/978-3-031-34031-4

Typing Gesture for One-Time Authentication Using Smart Wearable

Kristina Mullen, Avishek Mukherjee, and Khandaker Rahman

A method for enhanced user authentication that relies on sensory data taken from a smartwatch while the user types the username and password has been explored. Eventually, these inherent gestures would work as an added layer of security to the current password-based authentication scheme in a hostile scenario assuming the username-password has been compromised. In our experiments, we recorded the 3D coordinate values given off by the accelerometer and gyroscope over a set of username-password typing combinations. For the sensor data collection, we developed an Android Wear OS smartwatch application, then proceeded to implement our method of sensor data processing and performed experiments to demonstrate the potential of this method. We experimented with 50 samples taken from five users, performed 1,800 genuine and impostor authentication attempts, and achieved an equal error rate (EER) as low as 0.07. With such low EER, the proposed method can be an effective solution to username-password breaches.

D. W. Silva et al. (Eds.): ICACGA 2022, LNCS 13771, p. 230, 2024.
https://doi.org/10.1007/978-3-031-34031-4

Computation of Multivector Inverses in Non-degenerated Clifford Algebras

Dimiter Prodanov

The development of Clifford algebras is based on the insights of Hamilton, Grasssmann, and Clifford from the 19th century. After a hiatus lasting many decades, the Clifford geometric algebra experienced a renaissance with the advent of contemporary Computer Algebra Systems (CAS). The poster demonstrates an algorithm for the computation of an arbitrary multivector inverse in a non-degenerate Clifford algebra of arbitrary dimension, which is in fact a proof certificate for the existence of an inverse. The algorithm is proven using an algorithmic, constructive representation of a Clifford number in the matrix algebra over the reals, but it by no means depends on such a representation. As a side product, the algorithm can compute the characteristic polynomial of the Clifford number and its determinant also without any resort to a matrix representation. The presented algorithm is based on the Faddeev–LeVerrier–Souriau algorithm for matrix inverse computation [1. Faddeev, D. K., Sominskij, I. S.: Sbornik Zadatch po Vyshej Algebre. Nauka, Moscow–Leningrad (1949), 4. Souriau, J.: Une methode pour la decomposition spectrale et l'inversion des matrices. Comptes Rend. 227, 1010–1011 (1948)]. The algorithm is implemented in the open-source CAS Maxima using the Clifford package [Prodanov, D., Toth, V. T.: Sparse representations of Clifford and tensor algebras in Maxima. Advances in Applied Clifford Algebras pp. 1–23 (2016). http://dx.doi.org/10.1007/s00006-016-0682-x]. The package can be downloaded from the Zenodo repository [Prodanov D. Clifford: v2.5.1 (2021). Zenodo. https://doi.org/10.5281/zenodo.5628350].

D. W. Silva et al. (Eds.): ICACGA 2022, LNCS 13771, p. 231, 2024.
https://doi.org/10.1007/978-3-031-34031-4

Quaternions Associated to Curves and Surfaces

Haohao Wang and J. William Hoffman

This paper investigates the use of quaternions in studying space curves and surfaces in affine 3-space. First, we generate a large variety of rational space curves and rational surfaces via quaternion multiplication by taking advantage of the fact that quaternions represent space rotations. Then, we prove that the curvature and the torsion of a space curve can be computed by a quaternion function that is associated to this space curve. Finally, we show that the Gaussian and the mean curvature of a surface can also be computed by a quaternion function that is associated to this surface.

D. W. Silva et al. (Eds.): ICACGA 2022, LNCS 13771, p. 232, 2024.
https://doi.org/10.1007/978-3-031-34031-4

A Geometric Algebra Framework for Event Driven Network Data Model: Taking Emergency Evacuation Under Gas Diffusion as an Example

Yuhao Teng and Zhao Yuan Yu

With the development of urbanization, there are numerous security problems that come along with it, exposing many problems in emergency management. The evacuation of the affected people, the rapid acquisition of the disaster situation and the real-time quickly dispatch of relief resources still face great challenges. Based on these, this study uses the mathematical structure of geometric algebra and computational operators to construct and express the road network in emergency evacuation, designs a network analysis model based on geometric algebra with multiple constraints, and constructs an event-driven data model to realize emergency evacuation under the scenario of hazardous gas diffusion.

In this paper, the characteristics and types of objects and events in emergency evacuation under the diffusion of harmful gases are studied, and the elements are abstracted and summarized. Then we study the characteristics and temporal-spatial relationship of each element, and constructed three mathematical models for destination selection, rescue material dispatch and affected people evacuation. Secondly, we study the extension method of path in the network based on the geometry algebra, including the extension rules between nodes, nodes and edges, edges and edges, and routes. Thirdly, we study the route generation and filtering method. In the route filtering, we learned four types of route filtering methods, including weight filtering, k-order filtering, node filtering and topology filtering. Finally, an event-driven data model is constructed. The dynamic update of the data is implemented during the network analysis and its new drivers and constraints are used to optimize and solve the model.

D. W. Silva et al. (Eds.): ICACGA 2022, LNCS 13771, pp. 233, 2024.
https://doi.org/10.1007/978-3-031-34031-4

Author Index

D. W. Silva et al. (Eds.): ICACGA 2022, LNCS 13771, pp. 235–236, 2024.
https://doi.org/10.1007/978-3-031-34031-4

Printed in the United States
by Baker & Taylor Publisher Services